*Gernot Krabbes, Günter Fuchs,
Wolf-Rüdiger Canders, Hardo May,
and Ryszard Palka*

**High Temperature
Superconductor Bulk Materials**

Related Titles

Clarke, J., Braginski, A. I. (Eds.)

The SQUID Handbook
SET

approx. 600 pages in 2 volumes with approx. 125 figures and approx. 10 tables
2004
Hardcover
ISBN 3-527-40411-2

Barone, A.

Physics and Applications of the Josephson Effect: Low and High-TC Superconductors, Second Edition

2006
Hardcover
ISBN 0-471-02666-2

Lee, P. J. (Ed.)

Engineering Superconductivity

672 pages
2001
Hardcover
ISBN 0-471-41116-7

O'Handley, R. C.

Modern Magnetic Materials
Principles and Applications

768 pages
2000
Hardcover
ISBN-0-471-15566-7

*Gernot Krabbes, Günter Fuchs,
Wolf-Rüdiger Canders, Hardo May,
and Ryszard Palka*

High Temperature Superconductor Bulk Materials

Fundamentals – Processing – Properties Control – Application Aspects

WILEY-VCH Verlag GmbH & Co. KGaA

The Authors

Gernot Krabbes,
Leibniz-Institut für Festkörper- und
Werkstoffforschung IWF Dresden
e-mail: g.krabbes@ifw-dresden.de

Günter Fuchs,
Leibniz-Institut für Festkörper- und
Werkstoffforschung IWF Dresden
e-mail: g.fuchs@ifw-dresden.de

Wolf-Rüdiger Canders,
Technische Universität Braunschweig
e-mail: w.canders@tu-bs.de

Hardo May,
Technische Universität Braunschweig
e-mail: h.may@tu-bs.de

Ryszard Palka,
Technische Universität Braunschweig
e-mail: r.palka@tu-bs.de

Cover Pictures
A cylindrical $YBa_2Cu_3O_7$ based sample in course of melt texturing in a tube furnace. The seed crystal is visible on top. The sketch of primary crystallization fields symbolizes the physico-chemical background of processing. (source: authors)

Rotor of a superconducting reluctance motor equipped with bulk HTS blocks. The diagram represents the field distribution in the stator Cu – winding and the rotor of the motor. Bulk HTS lamellas in the rotor concentrate the field into parallel steel segments. (source: Oswald Elektromotoren GmbH Miltenberg)

Block of two single domains joined by welding and the contour of trapped magnetic field measured on top. (source: authors)

■ All books published by Wiley-VCH are carefully produced. Nevertheless, authors, editors, and publisher do not warrant the information contained in these books, including this book, to be free of errors. Readers are advised to keep in mind that statements, data, illustrations, procedural details or other items may inadvertently be inaccurate.

Library of Congress Card No.: applied for

British Library Cataloguing-in-Publication Data
A catalogue record for this book is available from the British Library.

**Bibliographic information published by
Die Deutsche Bibliothek**
Die Deutsche Bibliothek lists this publication in the Deutsche Nationalbibliografie; detailed bibliographic data is available in the Internet at <http://dnb.ddb.de>.

© 2006 WILEY-VCH Verlag GmbH & Co. KGaA, Weinheim

All rights reserved (including those of translation into other languages).
No part of this book may be reproduced in any form – nor transmitted or translated into machine language without written permission from the publishers. Registered names, trademarks, etc. used in this book, even when not specifically marked as such, are not to be considered unprotected by law.

Printed in the Federal Republic of Germany.
Printed on acid-free paper.

Typesetting Kühn & Weyh, Satz und Medien, Freiburg
Printing betz-druck GmbH, Darmstadt
Bookbinding Litges & Dopf Buchbinderei GmbH, Heppenheim

ISBN-13: 978-3-527-40383-7
ISBN-10: 3-527-40383-3

Contents

Preface *XI*

1	**Fundamentals** *1*	
1.1	Introduction to Superconductivity in High-Temperature Superconductors (HTSCs) *1*	
1.1.1	Introductory Remarks *1*	
1.1.2	Internal Nomenclature *3*	
1.1.3	Critical Currents and Flux Motion in Superconductors *3*	
1.1.4	Magnetization Curve of a Type II Superconductor *7*	
1.1.5	Determination of Critical Currents from Magnetization Loops *9*	
1.1.6	Magnetic Relaxation *11*	
1.1.7	Electric Field–Current Relation *13*	
1.1.8	Peculiarities of HTSCs in Comparison to Low-Temperature Superconductors *15*	
1.1.9	Basic Relations for the Pinning Force and Models for its Calculation *16*	
1.2	Features of Bulk HTSCs *18*	
1.2.1	Bulk HTSCs of Large Dimensions *19*	
1.2.2	Potential of Bulk HTSC for Applications *20*	
1.3	Solid-State Chemistry and Crystal Structures of HTSCs *22*	
1.3.1	Crystal Structures and Functionality *22*	
1.3.2	Chemistry and Doping *24*	
1.3.3	Intrinsic Doping: Variations of Stoichiometry *25*	
1.3.4	Defect Chemistry *26*	
1.3.5	Extrinsic Doping *27*	
2	**Growth and Melt Processing of $YBa_2Cu_3O_7$** *31*	
2.1	Physico-Chemistry of *RE*-Ba-Cu-O Systems *31*	
2.1.1	Phase Diagrams and Fundamental Thermodynamics *31*	
2.1.2	Subsolidus Phase Relationships *33*	
2.1.3	The Influence of Oxygen on Phase Equilibria: the System Y-Ba-Cu-O *35*	
2.1.4	The Oxygen Nonstoichiometry in 123 phases: $YBa_2Cu_3O_{7-\delta}$ ($YBa_2Cu_3O_{6+x}$) *37*	

High Temperature Superconductor Bulk Materials.
Gernot Krabbes, Günter Fuchs, Wolf-Rüdiger Canders, Hardo May, and Ryszard Palka
Copyright © 2006 WILEY-VCH Verlag GmbH & Co. KGaA, Weinheim
ISBN: 3-527-40383-3

2.1.5   Phase Relationships in Y-Ba-Cu-O in the Solidus and Liquidus Range *39*
2.1.6   Phase Relationships and the Liquidus Surface in Systems Ln-Ba-Cu-O (Ln = Nd, Sm,..) *42*
2.1.7   Additional Factors *45*
2.2   Preparation of Polycrystalline *RE*123 Materials *45*
2.2.1   Synthesis of HTSC Compounds *45*
2.3   Growth of $YBa_2Cu_3O_7$ Single Crystals *46*
2.4   Processing of "Melt-Textured" YBaCuO Bulk Materials *47*
2.4.1   Experimental Procedure *47*
2.4.2   Mass Flow, Growth Rates, Kinetic and Constitutional Undercooling *50*
2.4.3   Developing Microstructures: Morphology, Inclusions, Defects *53*
2.5   Modified Melt Crystallization Processes For YBCO *60*
2.5.1   Variants of the $YBa_2Cu_3O_7$–Y_2BaCuO_5 Melt-Texturing Process *60*
2.5.2   Processing Mixtures of Y123 and Yttria *60*
2.5.3   Processing in Reduced Oxygen Partial Pressure *63*

3   Pinning-Relevant Defects in Bulk YBCO *67*

4   Properties of Bulk YBCO *75*
4.1   Vortex Matter Phase Diagram of Bulk YBCO *75*
4.1.1   Irreversibility Fields *75*
4.1.2   Upper Critical Fields *78*
4.1.3   Vortex Matter Phase Diagram *79*
4.2   Critical Currents and Pinning Force *82*
4.2.1   Transport Measurements *82*
4.2.2   Magnetization Measurements *83*
4.3   Flux Creep *86*
4.3.1   Flux Creep in Bulk YBCO *86*
4.3.2   Reduction of Flux Creep *87*
4.3.3   Pinning Properties from Relaxation Data *89*
4.4   Mechanical Properties *91*
4.4.1   Basic Relations *91*
4.4.2   Mechanical Data for Bulk YBCO *93*
4.5   Selected Thermodynamic and Thermal Properties *96*
4.5.1   Symmetry of the Order Parameter *96*
4.5.2   Specific Heat *98*
4.5.3   Thermal Expansion *99*
4.5.4   Thermal Conductivity *101*

5   Trapped Fields *105*
5.1   Low-Temperature Superconductors *105*
5.2   Bulk HTSC at 77 K *105*
5.3   Trapped Field Data at 77 K *109*
5.4   Limitation of Trapped Fields in Bulk YBCO at Lower Temperatures *110*

5.4.1	Magnetic Tensile Stress and Cracking	*111*
5.4.2	Thermomagnetic Instabilities	*113*
5.5	Magnetizing Superconducting Permanent Magnets by Pulsed Fields	*115*
5.6	Numerical Calculations of the Local Critical Current Density from Field Profiles	*119*
5.6.1	Inverse Field Problem: Two-Dimensional Estimation	*120*
5.6.2	Three-Dimensional Estimation	*122*
5.7	Visualization of Inhomogeneities in Bulk Superconductors	*125*

6 Improved $YBa_2Cu_3O_{7-\delta}$-Based Bulk Superconductors and Functional Elements *129*

6.1	Improved Pinning Properties	*129*
6.1.1	Chemical Modifications in $YBa_2Cu_3O_7$	*129*
6.1.2	Sub-Micro Particles Included in Bulk YBCO	*135*
6.1.3	Irradiation Techniques	*136*
6.2	Improved Mechanical Properties in $YBa_2Cu_3O_{7-\delta}$/Ag Composite Materials	*138*
6.2.1	Fundamentals of the Processing and Growth of YBCO/Ag Composite Materials	*138*
6.2.2	Processing and Results	*140*
6.2.3	Properties of Bulk YBaCuO/Ag Composite Materials	*141*
6.3	Near Net Shape Processing: Large Sized Bulk Superconductors and Functional Elements	*142*
6.3.1	Finishing and Shaping	*143*
6.3.2	The Multi-Seed Technique	*144*
6.3.3	Rings of 123 Bulk Materials	*144*
6.3.4	Joining of Separate Single Grains	*146*
6.4	Bulk Materials and Processing Designed for Special Applications	*148*
6.4.1	Infiltration Technique and Foams	*148*
6.4.2	Long-Length Conductors and Controlled-Resistance Materials	*149*
6.4.3	Bi2212 Bulk Materials and Rings	*151*
6.4.4	Batch Processing of 123 Bulk Materials	*151*

7 Alternative Systems *155*

7.1	Impact of Solid Solutions $Ln_{1+y}Ba_{2-y}Cu_3O_{7\pm\delta}$ on Phase Stability and Developing Microstructure	*155*
7.2	Advanced Processing of Ln123	*160*
7.2.1	Oxygen Potential Control	*160*
7.2.2	Oxygen-Controlled Melt Growth Process (OCMG)	*161*
7.2.3	Isothermal Growth Process at Variable Oxygen Partial Pressure (OCIG)	*161*
7.2.4	Composition Control in Oxidizing Atmosphere for Growing (CCOG)	*163*
7.3	Alternative Seeding Techniques	*165*

| 7.4 | Further $LnBa_2Cu_3O_7$-Based Materials 165
| 7.5 | Ag/LnBaCuO Composites with Large Lanthanide Ions 166
| 7.5.1 | Fundamentals of Processing 166
| 7.5.2 | Reactions Near the Seed–Melt Interface 167
| 7.5.3 | Growth and Properties of Ag/LnBaCuO Composites 168

8 Peak Effect 171
8.1 Peak Effect (due to Cluster of Oxygen Vacancies) in Single Crystals 172
8.2 Peak Effect in Bulk HTSC 175

9 Very High Trapped Fields in YBCO Permanent Magnets 179
9.1 Bulk YBCO in Steel Tubes 179
9.1.1 Magnetic Tensile Stress (in Reinforced YBCO Disks) 179
9.1.2 Trapped Field Measurements 181
9.2 Resin-Impregnated YBCO 184
9.3 Trapped Field Data of Steel-reinforced YBCO 185
9.4 Comparison of Trapped Field Data 188

10 Engineering Aspects: Field Distribution in Bulk HTSC 191
10.1 Field Distribution in the Meissner Phase 192
10.1.1 Field Cooling 192
10.1.2 Zero-Field Cooling 193
10.2 Field Distribution in the Mixed State 194
10.2.1 Field Cooling 194
10.2.2 Zero-Field Cooling 195

11 Inherently Stable Superconducting Magnetic Bearings 199
11.1 Principles of Superconducting Bearings 199
11.1.1 Introduction to Magnetic Levitation 199
11.1.2 Attributes of Superconducting Magnetic Bearings with Bulk HTSC 200
11.2 Forces in Superconducting Bearings 201
11.2.1 Forces in the Meissner and the Mixed State 201
11.2.2 Maximum Levitational Pressure in Superconducting Bearings 204
11.3 Force Activation Modes and Magnet Systems in Superconducting Bearings 207
11.3.1 Cooling Modes 207
11.3.2 Operational Field Cooling with an Offset 209
11.3.3 Maximum Field Cooling Mode 211
11.3.4 Magnet Systems for Field Excitation in Superconducting Bearings 211
11.3.5 Force Characteristics 216
11.4 Optimized Flux Concentration Systems for Operational-Field Cooling (OFCo) 220
11.4.1 Stray Field Compensation 221
11.4.2 Dimensional Optimization of System Components 221
11.5 Parameters Influencing the Forces of Superconducting Bearings 225

11.5.1	Critical Current Density 225
11.5.2	Temperature 226
11.5.3	Flux Creep 227
11.5.4	HTSC Bulk Elements Composed of Multiple Isolated Grains 229
11.5.5	Number of Poles of the Excitation System 232
11.6	Applications of Superconducting Bearings 233
11.6.1	Bearings for Stationary Levitation 234
11.6.2	Bearings for Rotary Motion 236
11.6.3	Bearings for Linear Motion 241
11.7	Specific Operation Conditions 245
11.7.1	Precise Positioning of Horizontal Rotating Axis 245
11.7.2	Bulk HTSCs Cooled Below 77 K 246
11.7.3	Cooling the Excitation System along with the Superconductor 247
11.7.4	Dynamics of Rotating Superconducting Bearings 247
11.8	Numerical Methods 249
11.8.1	Perfectly Trapped Flux Model (2D) 250
11.8.2	Perfectly Trapped Flux Model (3D) 252
11.8.3	Vector-Controlled Model (2D) 253

12 Applications of Bulk HTSCs in Electromagnetic Energy Converters 259

12.1	Design Principles 259
12.2	Basic Demonstrator for Application in Electrical Machines – Hysteresis or Induction Machines 261
12.3	Trapped-Field Machine Designs 263
12.4	Stator-Excited Machine Designs with Superconducting Shields – The Reluctance Motor with Bulk HTSC 269
12.5	Machines with Bulk HTSCs – Status and Perspectives 273

13 Applications in Magnet Technologies and Power Supplies 279

13.1	Superconducting Permanent Magnets with Extremely High Magnetic Fields 279
13.1.1	Laboratory Magnets 279
13.1.2	Magnetic Separators 280
13.1.3	Sputtering Device 283
13.1.4	Superconducting Wigglers and Undulators 284
13.2	High-Temperature Superconducting Current Leads 284
13.3	Superconducting Fault Current Limiters 285
13.3.1	Inductive Fault Current Limiters 286
13.3.2	Resistive Superconducting Fault Current Limiters 287
13.3.3	Status of High AC Power SFCL Concepts 288
13.4	High-Temperature Superconducting Magnetic Shields 290

List of Abbreviations 293

Index 295

Preface

Soon after *Müller* and *Bednorz* found superconductivity at temperatures above 30 K in perovskite related complex oxides, cuprate superconductors have been prepared with critical temperatures T_c well above 77 K – the boiling point of liquid nitrogen. During the nineties three fundamental structural families of oxide superconductors have been found, including Bi, Tl or Hg based materials with T_c up to about 130 K and rare earth barium cuprates with T_c typically between 90 and 97 K.

Although the physical nature of this "high temperature" superconductivity is not completely understood this has not been an obstacle for technical application, because of certain unique properties which make them highly interesting from the technical point of view and very soon engineers started to investigate their technical potential.

Typically they were developed as bulk materials in form of cast semi products or well crystallized single domain blocks of sizes up to 100 mm edge length or more with multiple seeding.

While scientific attempts are now concentrating to make the new class of superconductors available in superconducting wires or tapes, the bulk materials are meanwhile on the way to technical applications. Their applications are based on the fascinating properties – carrying loss free high currents and the capability to trap extremely high magnetic fields. The largest potential is expected from the use of the bulk materials in inherently stable contact-less bearings, fault current limiters, and electrical machines. Magnetic bearings may be configured as linear transport devices or as a rotating device in high speed machines. Electrical machines benefit from the high potentials for shielding (flux concentration and reducing armature reaction) and flux trapping (in superconducting permanent magnets). Superconducting current limiters use the quenching effect thus commutating short circuit currents from a superconductor to a normal conductor. They have a key function for operation of complex grids for high performance energy supply.

Meanwhile, a vast amount of papers exist dealing with details of physics, preparation, technical aspects and application of bulk superconductors. A comprehensive synopsis considering these aspects and the interrelationship is still pending. Thus, the engineer or physicist who wants to solve a technical problem by application of the new class of materials meets certain barriers:

- understanding the materials properties and specific features of processing and handling,
- understanding electromagnetic features of superconductivity and appropriate physical models,
- knowledge about methods of modelling which support optimal design of devices,
- knowledge and experience about chemical and physico-chemical fundamentals.

On the other hand, the chemist often does not recognize the chances for an optimal design of the materials properties, missing certain fundamentals of engineering.

Therefore, the book will present an introduction into the chemical and the physical nature of the material and facilitate the understanding of the behavior of the superconductor under electric and magnetic field.

It is the intention of the authors to provide scientists and engineers interested in working on this interdisciplinary field with the knowledge covering the disciplines as concerned. In this context, interrelationships between chemistry, physics and engineering should find a careful consideration.

1
Fundamentals

1.1
Introduction to Superconductivity in High-Temperature Superconductors (HTSCs)

1.1.1
Introductory Remarks

The two most important properties of type II superconductors are the disappearance of the electrical resistance below the superconducting transition temperature T_c and the special magnetic properties in an applied magnetic field H including perfect diamagnetism at low fields and the penetration of quantized magnetic flux (vortices or flux lines) at higher fields. The electrons in the superconducting state can be described by a macroscopic wave function which is undisturbed by scattering and, in the absence of applied fields and currents, uniform in space over macroscopic distances. This coherence of the superconducting state is preserved even in the presence of weak currents and magnetic fields (below the upper critical field H_{c2}), which is a necessary condition for the practical use of type II superconductors in strong magnetic fields. The further condition for applications of these superconductors is to avoid any motion of flux lines. This can be achieved through the introduction of pinning centers in the superconductor, interacting, in most cases, with the normal conducting core of the flux lines.

The microscopic origin of the macroscopic coherence has been explained within the BCS theory by the existence of Cooper pairs. In low-T_c superconductors, the average radius of a Cooper pair, the coherence length, is large compared to interatomic distances, and there are many Cooper pairs within the coherence length. Therefore, the coherence of the superconducting state is strong in low-T_c superconductors. High-T_c superconductors (HTSCs) have much smaller coherence lengths and thus a reduced coherence of the superconducting state. This manifests itself in a reduced critical current density under applied fields. The coherence length within the *ab*-plane is of the order of $\xi_{ab} \approx 1\text{--}2$ nm for all HTSCs, whereas the much smaller coherence length ξ_c depends on the anisotropy of these compounds. For the less anisotropic YBa$_2$Cu$_3$O$_{7-\delta}$ (Y-123), ξ_c is 0.3 nm, becoming even smaller than 0.1 nm for the very anisotropic Bi- and Tl-based HTSCs.

High Temperature Superconductor Bulk Materials.
Gernot Krabbes, Günter Fuchs, Wolf-Rüdiger Canders, Hardo May, and Ryszard Palka
Copyright © 2006 WILEY-VCH Verlag GmbH & Co. KGaA, Weinheim
ISBN: 3-527-40383-3

Shortly after the discovery of $La_{2-x}Ba_xCuO_4$, with T_c = 35 K [1.1], the high-T_c superconductor Y-123 with a transition temperature near 90 K was synthesized [1.2]. Meanwhile, $T_c \approx$ 130 K has been achieved in Hg-1223 [1.3], and the application of pressure to Hg-1223 pushed the current record up to $T_c \approx$ 155 K [1.4]. Nevertheless, Y-123 is one of the most promising HTSCs for applications at 77 K because its relatively small anisotropy allows high critical current densities also in the presence of applied magnetic fields. $YBa_2Cu_3O_{7-\delta}$ has its highest value of T_c = 93 K for x = 0.1. It has been found that T_c remains nearly unchanged if Y is partly or completely substituted by other rare-earth elements such as Nd, Sm, Eu, Gd, Dy, Ho, Er, or Tm. Improved pinning properties have been reported for compounds with two or three rare-earth elements on Y sites, which is a further advantage of this HTSC with regard to applications.

The mechanism of superconductivity in HTSCs is still controversially discussed. The general BCS concept of a macroscopic quantum state of Cooper pairs remains the common basis in almost all theories proposed for HTSCs. However, there is no consensus about the origin of the pairing. The lack of an isotope effect in HTSCs of highest T_c would favor a non-phonon pairing mechanism as, for instance, *d*-wave pairing, whereas its presence in several HTSCs with lower T_c suggests a phonon-mediated *s*-wave pairing. The most experimental features favor *d*-wave pairing. A direct test of the phonon mechanism for pairing in HTSCs is to check the sign change in the energy gap $\Delta(k)$ as function of momentum *k*. A *d*-state energy gap is composed of two pairs of positive and negative leaves forming a four-leaf clover. In contrast, the magnitude of an *s*-state may be anisotropic but always remains positive. Using a SQUID microscope, the magnetic flux has been investigated in YBCO tricrystals with three Josephson contacts near the tricrystal point [1.5]. The observation of a half-integer flux quantum can be considered as direct evidence for *d*-symmetry of the superconducting state of this hole-doped HTSC. The *d*-wave character of YBCO has been verified by low-temperature specific heat measurements on high-quality YBCO single crystals [1.6] and by studying the angular variation of the thermal conductivity in detwinned YBCO single crystals [1.7].

The flux line lattice in HTSCs is influenced by novel phenomena such as strong thermal and quantum fluctuations, resulting in several new features such as a vortex liquid phase below the upper critical field separated from a vortex glass phase by the irreversibility field. At low magnetic fields, another new transition, from a Bragg glass (taking, in HTSCs, the place of the usual Abrikosov lattice observed in conventional superconductors) into a vortex glass, has been found. Strong anisotropy of Bi-based HTSCs introduces a new feature, i.e. a layer decoupling transition, which changes flux lines into pancake vortices and thus modifies pinning and flux creep. All these new effects can be treated within a modern version of collective pinning theory including thermal and quantum fluctuations. Several of these novel features, such as the Bragg glass, can only be observed in clean single crystals, others, such as as the vortex-liquid phase, strongly influence the high-field performance of YBCO thin films and bulk textured materials. Bulk HTSCs with high critical currents, high trapped fields, and strong levitation forces, have

1.1.2
Internal Nomenclature

Naming of the complex HTSC compounds according to the IUPAC convention results in verbal monsters, and this has led workers in the field to create a compact jargon consisting of a sequence of characters or numbers or both. Typically, YBCO stands for the Y-Ba-Cu-O system, BSCCO for Bi-Sr-Ca-Cu-O, etc. *RE* denotes the rare earth elements (including Y) whereas *Ln* means lanthanide elements.

A sequence of figures indicates the idealized stoichiometry of a certain phase:

123 or *RE*123 for $RE_{1+y}Ba_{2-y}Cu_3O_{7-\delta}$, Y123 for $YBa_2Cu_3O_{7-\delta}$, Y211 for Y_2BaCuO_5, Nd123 for $Nd_{1+y}Ba_{2-y}Cu_3O_{7-\delta}$, Nd422 for $Nd_{4+y}Ba_{2-y}Cu_2O_{10-\delta}$, Bi2212 for $Bi_2Sr_2CaCu_2O_8$, 001 for $BaCuO_2$, Y200 for Y_2O_3, 001 for CuO, etc. Oxygen is not identified separately.

1.1.3
Critical Currents and Flux Motion in Superconductors

The highest intrinsic current that a superconductor can support is the so-called depairing critical current j_c^{dep}. This current, which destroys Cooper pairs and superconductivity, is of the order of H_c/λ, where H_c is the thermodynamic critical field and λ is the penetration depth. As $\mu_o H_c$ and λ of conventional type II superconductors are of the order 0.1 T and 100 nm, respectively, one gets

$$j_c^{dep} = H_c/\lambda \approx 10^8 \text{ A cm}^{-2} \tag{1.1}$$

for the maximum critical current in the superconductor. However, in real superconductors the critical current densities are several orders of magnitude lower than j_c^{dep}. The reason is that currents in type II superconductors are usually limited by flux motion in the presence of magnetic fields.

The behavior of type II superconductors is illustrated in Figure 1.1. Type II superconductors are ideal diamagnets for applied magnetic fields $H < H_{c1}$, where H_{c1} is the lower critical field. Above H_{c1}, magnetic flux penetrates the superconductor in quantized form, i.e. each flux line contains one quantum of magnetic flux $\Phi_o = 2.1 \times 10^{-15}$ T m². The number of flux lines which are extended parallel to the applied magnetic field increases with the field until the normal state is reached at the upper critical field H_{c2}.

A flux line (see Figure 1.2) is composed of a normal conducting core of radius ξ (Ginzburg-Landau coherence length) and a surrounding circulating supercurrent region where

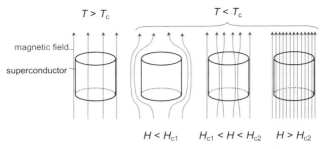

Figure 1.1 Type II superconductor in an applied magnetic field

the magnetic field and supercurrents fall off within a distance λ (penetration depth). The ratio

$$\kappa = \frac{\lambda}{\xi} \tag{1.2}$$

exceeding the value $1/\sqrt{2}$ for a type II superconductor, is the so-called Ginzburg-Landau parameter. In the absence of defects, the flux lines form a triangular lattice, as was predicted by Abrikosov in 1957 [1.8] and later experimentally confirmed by a magnetic decoration technique [1.9], by neutron diffraction [1.10] (see Figure 1.3) or by other experimental techniques. The flux line lattice in a YBCO thin film is shown in Figure 1.4 [1.11]. The vortices of this magnetic force microscope image

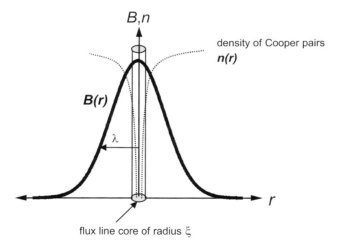

Figure 1.2 Isolated flux line (schematic). Shown are the distributions of magnetic field $B(r)$ (thick line) and of the density of Cooper pairs $n(r)$ (dotted line). $n(r)$ disappears within the flux line core (visualized as tube) of radius ξ (coherence length). $B(r)$ decays exponentially becoming very small for $r \gg \lambda$ (penetration depth)

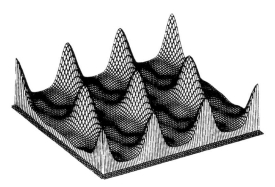

Figure 1.3 Local magnetic field distribution in Niobium at 4.2 K at low magnetic field obtained by neutron diffraction. Reprinted from [1.10] with permission from Elsevier

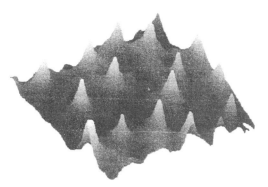

Figure 1.4 Local magnetic field distribution in a YBCO thin film at 60 K at low magnetic field obtained by magnetic force microscopy. Reprinted from [1.11] with permission from Elsevier

obtained at 60 K do not form a lattice because of the interaction with strong pinning centers present in the YBCO film.

The elastic properties of the flux line lattice have been reviewed by Brandt [1.12]. The flux line lattice of type II superconductors has three elastic moduli: the modulus of isotropic compression ($c_{11}-c_{66}$), the tilt modulus c_{44}, and the shear modulus c_{66}. In anisotropic superconductors, nonlocal elasticity has to be taken into account and the flux line lattice becomes much softer against compressional and tilt distortions than it is the case for uniform compression or tilt.

The number n of flux lines (per unit area) increases with increasing applied field. The magnetic flux density in the superconductor per unit area is Φ_o times n, corresponding to the average magnetic field inside the superconductor

$$B = n\, \Phi_o \tag{1.3}$$

The distance between two flux lines in the lattice becomes close to the diameter 2ξ of the flux line core when the applied magnetic field approaches the upper critical field

$$B_{c2} = \frac{\Phi_o}{2\pi \xi^2} \tag{1.4}$$

at which the superconductor goes into the normal state. In the anisotropic HTSCs, the coherence length ξ_{ab} within the ab plane is larger than the coherence length ξ_c in the c direction. The upper critical fields at $T = 0$ in both directions are given by

$$\begin{aligned} B_{c2}^c &= \frac{\Phi_o}{2\pi \xi_{ab}^2} \\ B_{c2}^{ab} &= \frac{\Phi_o}{2\pi \xi_c \xi_{ab}} \end{aligned} \tag{1.5}$$

From Eq. (1.5), one obtains

$$\frac{B_{c2}^{ab}}{B_{c2}^c} = \frac{\xi_{ab}}{\xi_c} \tag{1.6}$$

with $\Gamma = \xi_{ab}/\xi_c = \lambda_c/\lambda_{ab}$ as the anisotropy parameter.

If a current (current density j) is flowing perpendicular to the applied magnetic field, then a Lorentz force (per unit volume) acts on the vortex lattice, which is given by

$$F = j \times n\, \Phi_o = j \times B \tag{1.7}$$

This force tends to move the vortex lattice perpendicular to j and B as sketched in Figure 1.5. If the flux lines can freely move with constant speed v under the influence of this Lorentz force, then an electric field

$$E = v \times B \tag{1.8}$$

is generated in the superconductor which is perpendicular to both v and B. Hence, this electric field has the same direction as the transport current, giving rise to the ohmic loss (jE).

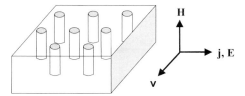

Figure 1.5 Lorentz force in a type II superconductor

and energy dissipation. Therefore, to achieve a finite loss-free current, the flux lines have to be pinned in order to prevent flux line motion. Flux lines become pinned to such energetically favorable sites as, for instance, inhomogeneities or normal conducting precipitates. The average pinning force density F_p balances a critical Lorentz force $F = j_c \times B$, i.e.

$$F_p = j_c \times B \tag{1.9}$$

where j_c is the critical current density. Significant dissipation due to flux motion starts for $j > j_c$.

1.1.4
Magnetization Curve of a Type II Superconductor

HTSCs are type II superconductors. Some of their basic properties can be described by the field-dependent magnetization, which shows, in addition, the requirements for large trapped fields of bulk superconductors.

The magnetization is defined as $M = \mu_0 H - \langle B \rangle$, with H_a as the external magnetic field and $\langle B \rangle$ as the average magnetic induction B in the bulk superconductor. The magnetization vs field dependence of a bulk YBCO sample is shown in Figure 1.6. In the Meissner state ($H < H_{c1}$ with H_{c1} as the lower critical field), a superconducting surface current screens off the external magnetic field so that the magnetic induction B in the bulk superconductor vanishes. In the mixed state between H_{c1} and the upper critical field H_{c2}, magnetic flux penetrates the superconductor in the form of flux lines as already mentioned. As the external field increases toward H_{c2}, the size of the superconducting region between the normal conducting cores of the flux lines shrinks to zero, and the sample makes a continuous transition to the normal state. The $M(H)$ dependence of a defect-free type II superconductor is reversible, and after switching off the external field no magnetic flux is trapped within the superconductor.

The $M(H)$ dependence becomes highly irreversible if the superconductor contains defects like dislocations, precipitates, etc., which interact with the penetrating flux lines. Flux lines cannot penetrate freely at $B_a = B_{c1}$; their density is large at the surface and decreases with increasing distance from the surface because of

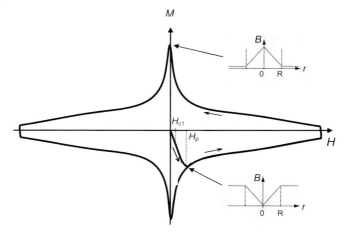

Figure 1.6 Field dependence of the magnetization in a type II superconductor (see text)

pinning. The pinning force between the defects in the material and the penetrating flux lines balances the driving force

$$F_L = j \times B = 1/\mu_0 \cdot (\text{rot } B \times B) \tag{1.10}$$

acting on the flux lines due to the density gradient of flux lines. That this force is related to the field gradient within the superconductor becomes apparent in the one-dimensional case, where the field gradient is given by

$$\frac{\partial B_z}{\partial x} = -j_y \tag{1.11}$$

resulting in a Lorentz force $F_l = dH/dx\, B$. The flux lines will continue their motion until $F_L = F_p$ at all points, with F_p as the volume pinning force:

$$F_p = j_c\, B = dH/dx\, B \tag{1.12}$$

According to this concept of the *critical state* proposed by Bean [1.15], the gradient of the magnetic field has its maximum value $dH/dx = j_c$ within regions of the superconductor penetrated by the field, whereas dH/dx is zero for those regions that never felt the magnetic field. In the original version of the critical state model, Bean assumed a field-independent critical current density. Later, other models with different field dependences of $j_c(B)$ were proposed. The field distribution within the superconductor can be calculated by substituting $j_c(B)$ into (1.11) and solving the resulting differential equation.

The insets of Figure 1.6 show the field distribution in the superconductor for several values of the applied magnetic field. Here, the field dependence of j_c is neglected according to Bean's model, which is an adequate approximation for this qualitative analysis. For an applied field $H_1 > H_{c1}$, the magnetic field starts to penetrate the superconductor. At the penetration field H_p, the internal magnetic field reaches the center of the superconductor. At this field, the magnetization has its maximum diamagnetic value. It is clearly seen that for $H_{c1} < H < H_p$, the irreversible magnetization curve shown in Figure 1.6 deviates only gradually from the straight line of perfect diamagnetism, demonstrating the strong shielding effect due to flux pinning. For fields $H \geq H_p$, Bean's model predicts a constant magnetization $M = H - \langle B \rangle = -1/2\,\mu_0 H_p$, whereas in Figure 1.6 the magnetization $|M|$ starts to decrease, which reflects the reduction of the critical current density with increasing magnetic field. For HTSCs, the irreversible magnetization becomes zero at $H = H_{irr}$ with H_{irr} as the irreversibility field, in contrast to the reversible magnetization which disappears at $H = H_{c2}$. This is shown in Figure 1.7. In Figure 1.6, the external field was reduced before H_{irr} was reached. As the external field is reduced, the gradient of the local field near to the sample edge changes its sign, but has the same absolute value as before. The magnetization now becomes positive because a magnetic field is trapped in the superconductor by the pinning effect. The field profile obtained after switching off the external field (see upper inset of Figure 1.6) is quite different from the constant field above a conventional permanent magnet. A direct consequence of this field distribution is that the maximum trapped field of a superconducting permanent magnet depends on its size. The gradient of the field profile is determined by the critical current density of the superconductor. Therefore, one needs large current loops in large bulk superconductors and high critical current densities in order to get large trapped fields in superconducting permanent magnets.

1.1.5
Determination of Critical Currents from Magnetization Loops

The characteristic features of an $M(H)$ loop of a type II superconductor are shown schematically in Figure 1.7. The reversible part of the magnetization is sketched as a thin line. The critical current density can be calculated from the irreversible magnetization loop provided that the sample is in the critical state. This is not the case either in the field range $H < H_p$ with H_p as the penetration field or in the field ranges around the field reversal where the magnetization changes its sign. On the other hand, the magnetization at zero field ($H = 0$) remains in the fully penetrated state if the applied field is cycled between $H = +H_o$ and $H = -H_o$. The irreversible magnetization

$$M_{irr} = 1/2\,(M^+ - M^-) = 1/2\,\Delta M \tag{1.13}$$

is derived from the vertical width ΔM of the hysteresis loop at a given field. M^+ and M^- are the magnetization branches in the fully penetrated state for decreasing and increasing applied fields, respectively. Because both M^+ and M^- contain the reversible magnetization, M_{rev} cancels out if the difference between M^+ and M^- is considered.

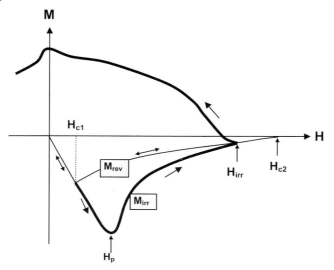

Figure 1.7 Irreversible and reversible magnetization of a HTSC (see text)

The critical current density of an infinitely long sample with rectangular cross-section ($2a \times 2b$) (with $a < b$) is obtained from the width ΔM of the $M(H)$ hysteresis loop using the expression

$$j_c(H) = \Delta M(H)/[a(1 - a/3b)], \tag{1.14}$$

which reduces to

$$j_c(H) = 3/2 \, \Delta M(H)/a \tag{1.15}$$

for a quadratic cross-section ($2a \times 2a$) and cylinders of radius a, and to

$$j_c(H) = \Delta M(H)/a \tag{1.16}$$

for slabs of thickness $2a$ [1.16]. It is convenient to use the expression

$$j_c(H) = 10 \, \Delta M(H)/[a(1 - a/3b)] \tag{1.17}$$

resulting from (1.14), if ΔM is measured in emu/cm^3, a and b in cm and j_c in A/cm^2. Note that j_c in (1.14)–(1.17) increases by a factor of two, if a sample with a cross-section ($a \times b$) is considered instead of ($2a \times 2b$).

These expressions were derived for infinitely long samples in the case of a field-independent critical current density. In the case of bulk YBCO, most measurements are performed on short samples, i.e., the dimensions of the superconductor parallel and perpendicular to the field are comparable. Furthermore, the critical current density of bulk YBCO exhibits a strong field dependence, especially at temperatures around 77 K. Because both conditions are not fulfilled, deviations from the true j_c are expected if expressions (1.14)–(1.17) are used for its determination. Fortunately, these deviations are mostly restricted to the low-field range $H < H_p$, where the $j_c(H)$ variation is particularly pronounced.

For a quantitative analysis, an exponential $j_c(H)$ function

$$j_c(H) = j_c(0)\exp(-H/H_o) = j_c(0)(1 + p)^{-H/H_p} \tag{1.18}$$

has been considered [1.17]. In this case, the field of complete flux penetration in a infinitely long cylinder of radius R is given by $H_p = H_o \ln(1 + p)$ with $p = j_c(0)R/H_o$. At $H=H_p$, $j_c(H_p)$ reduces to $j_c(0)/(1 + p)$ of its value $j_c(0)$ at $H=0$, which means that the parameter p is large for a strong $j_c(H)$ dependence and small for a weak $j_c(H)$ dependence. In the limit of $p = 0$, one gets $j_c(H) = j_c(0)$, which corresponds to the Bean model. In this case, the expression (1.15) for j_c of a long cylinder in axial field is correct in the whole field range. For $p = 3$, one obtains $j_c(H_p) = 0.25\, j_c(0)$. In this case, the applicability of Eq. (1.15) is restricted to fields $H > 0.5\, H_p$. For $p = 10$, the $j_c(H)$ dependence becomes very strong, with $j_c(H_p) = 0.09\, j_c(0)$, and expression (1.15) can be used only in the field range $H > H_p$ [1.17].

By using expressions (1.14) – (1.17) for samples of reduced sample length, the deviations from the true $j_c(H)$ at low magnetic fields become even smaller than for longer samples. The reason is that the field profiles within the superconductor in the case of short samples (with comparable dimensions parallel and perpendicular to the field) have a Bean-like shape already at low values of the applied magnetic field, whereas the field profiles within long samples remain strongly curved up to higher applied fields [1.17, 1.18].

1.1.6
Magnetic Relaxation

Flux motion for current densities exceeding the critical current density j_c has been explained in Section 1.1.3 by depinning of flux lines. However, thermally activated flux motion is observed also for current densities below j_c, i.e. for pinned flux lines, because at $T > 0$ there is a finite probability that the flux lines or flux bundles overcome the pinning barriers. This kind of flux motion which is caused by thermal activation is called flux creep. Magnetic relaxation in HTSCs has been reviewed by Yeshurun et al. [1.19]. Thermally activated flux motion is much stronger in HTSCs than in low-temperature superconductors. The reason for the strongly enhanced mobility of flux lines in HTSCs is the low value of the activation energy U for flux creep compared with the thermal energy kT. A typical value for the ratio U/k_BT is $U/k_BT \approx 2$. It should be noted that the activation energy becomes extremely low in anisotropic HTSCs, such as the Bi-based HTSC. In contrast, this ratio is $U/k_BT \approx 100$ in conventional superconductors. Therefore, flux creep is only a very small effect in conventional superconductors.

Studying the magnetic relaxation in conventional superconductors, a logarithmic time dependence of the magnetization has been observed [1.20] and interpreted as flux creep by Anderson and Kim [1.21, 1.22]. According to the Anderson-Kim model the field gradient within the superconductor decays with time because of thermally activated flux motion of flux lines resulting in relaxation of the persistent current and of the magnetization.

Considering a slab with flux lines along the z-axis moving along the x-axis (i.e. the macroscopically averaged current density j is directed along the y-axis and related to the gradient of the flux lines), the relation

$$U(j) = kT \ln(t/t_o) \tag{1.19}$$

between the activation energy for flux creep and the time dependence of the persistent current density j has been derived [1.23]. From this equation, $j(t)$ is determined by putting in the current dependence of the activation energy $U(j)$.

In the Anderson-Kim model, a linear $U(j)$ relation

$$U(j) = U_o(1 - j/j_{co}) \tag{1.20}$$

was considered, which is a reasonable approximation for low-temperature superconductors in which the persistent current is close to the critical current j_{co} at $T = 0$ in the absence of flux creep. From (1.19) and (1.20), one obtains the well-known logarithmic time dependence of the persistent current

$$j(t) = j_{co}\left[1 - \frac{kT}{U_o}\ln\left(\frac{t}{t_o}\right)\right] \tag{1.21}$$

For HTSCs, the non-linear $U(j)$ relation

$$U(j) = U_o[(j_{co}/j)^\mu - 1] \tag{1.22}$$

has been proposed within the model of collective creep [1.24] for persistent currents $j \ll j_{co}$. This model assumes weak random pinning and takes the elastic properties of the flux line lattice into account. The time dependence of the persistent current in the collective creep model is given by the so-called "interpolation formula"

$$j(t) = \frac{j_{co}}{[1 + (\mu kT/U_o)\ln(t/t_o)]^{1/\mu}} \tag{1.23}$$

resulting from (1.19) and (1.22). The additional factor μ in the denominator of Eq. (1.23) ensures that it is possible to interpolate between the Anderson-Kim model (for j near j_{co} at short times) and the long-time behavior. Indeed, relation (1.21) can be regarded as special case of Eq. (1.23) for $\mu = -1$.

In the collective creep model, the size of the flux bundles depends on the current density, in contrast to the Anderson-Kim model, where a constant volume of flux bundles is assumed. The parameter μ in Eqs. (1.22) and (1.23) is field and temperature dependent. Predicted values for the creep of flux lines are $\mu = 1/7$ at low fields and temperatures (creep of individual flux lines), $\mu = 3/2$ at higher temperatures and fields (collective creep of small flux bundles), and $\mu = 7/9$ at still higher temperatures and fields (collective creep of large flux bundles).

Magnetic relaxation experiments were performed for YBCO single crystals [1.25] and melt-textured YBCO samples [1.26]. In Figure 1.8, the time dependence of magnetization is shown for a bulk YBCO sample at $T = 77$ K and $\mu_o H = 1$ T. The relaxation rate $S = |dM/M|/d(\ln t)$ of about $S = 8$ % per time decade estimated from this dependence, is much stronger than that for low-temperature superconductors, where the relaxation rate at 4.2 K is in the range of $S \approx 0.1$% [1.20]. Deviations from the logarithmic time dependence become apparent from the curved shape of the $M(t)$ curve in the semi-logarithmic plot of Figure 1.8. Since $M \propto j$, the influence of collective creep can be analyzed by comparing the data with Eq. (1.23). The experimental curve in Figure 1.8 can be roughly described by $M^{-0.8} \propto \ln(t)$ suggesting collective creep of large flux bundles ($\mu = 7/9 \approx 0.8$). A modified analysis of relaxation data including data from different temperatures will be presented in Chapter 4.

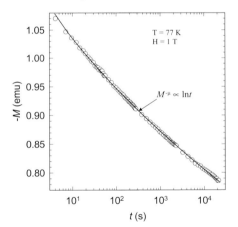

Figure 1.8 Time dependence of magnetization for a bulk YBCO sample

1.1.7
Electric Field–Current Relation

Thermally activated flux creep causes dissipation and is associated with a highly nonlinear electric field-current density relation $E(j)$ for $j < j_c$. The electric field E induced by flux creep can be expressed as

$$E = E_o \exp(-U(j)/kT) \tag{1.24}$$

where $U(j)$ is the activation energy for flux creep.

1 Fundamentals

For the linear $U(j)$ relation (1.20) of the Anderson-Kim model, one gets from (1.24)

$$E(j) \propto \exp\left(-\frac{U}{kT}\right) \sinh\left(\frac{jU}{j_{co}k_B T}\right), \tag{1.25}$$

if thermally activated jumps of flux bundles along the Lorentz force and in the opposite direction are taken into account. In (1.25) enters the critical current density $j_{co}(0,B)$ in the absence of flux creep. This relation predicts $E(j) \propto j$ for small current densities $j \ll j_c$ due to thermally assisted flux flow (TAFF) [1.13, 1.14] and an exponential $E(j)$ dependence

$$E(j) \propto \exp(jU/j_{co}kT) \tag{1.26}$$

for $j \leq j_c$. In conventional superconductors, usually this exponential flux creep relation is observed, whereas the electric fields within the TAFF region are too small to be measured.

For HTSCs, the power law $E(j) \propto (j/j_c)^p$ or power-like $E(j)$ relationships have been observed [1.27] which can be understood within the collective creep model. Using the non-linear $U(j)$ relation [Eq. (1.22)] of this model, one gets:

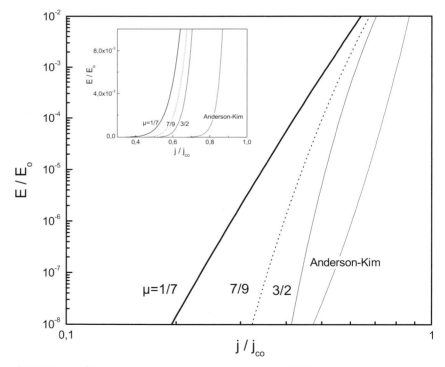

Figure 1.9 E–j relation in conventional superconductors and in HTSCs

$$E(j) \propto \exp\left[-U_o/\mu kT \left(\left[\frac{j_{co}}{j}\right]^\mu - 1\right)\right] \tag{1.27}$$

This expression corresponds for $\mu = 0$ to a power law, whereas the log E–log j curves for $\mu = 1/7$ (creep of individual flux lines), $\mu = 7/9$ (creep of large flux bundles), and $\mu = 3/2$ (creep of small flux bundles) exhibit negative curvatures. This is shown in Figure 1.9. For comparison, expression (1.25) of the Anderson-Kim model is shown in this plot for a typical activation energy of a low-T_c superconductor.

1.1.8
Peculiarities of HTSCs in Comparison to Low-Temperature Superconductors

As was discussed in the previous section, flux creep effects are much stronger in HTSCs than in low-temperature superconductors. Therefore, the mobility of the flux line lattice in HTSCs is very pronounced, and large portions of the H–T plane are dominated by thermal fluctuations. In particular, above the so-called irreversibility field $H_{irr} < H_{c2}$ the influence of thermal fluctuations on the vortex lattice becomes so strong that the vortex lattice changes in a vortex liquid. In this region currents cannot flow without losses, although the superconductor is not yet in the normal state. The field range between H_{irr} and H_{c2} in which the vortex liquid exists increases with the anisotropy of the HTSC.

In general, HTSCs exhibit a *large anisotropy* of the crystal structure and of the electronic properties, which has implications for both the physical and the mechanical properties. The electrical conductivity is highly anisotropic, with much higher conductivity within the CuO_2 planes than perpendicular to the planes. Therefore, supercurrents can flow only within the superconducting CuO_2 planes (see Sections 1.3.1 and 1.3.2), i.e. the trapped field generated by these supercurrents is directed along the *c*-axis. Also, the structure of the vortex lattice for fields applied perpendicular to the CuO_2 planes is influenced by the anisotropy. Whereas at moderate anisotropy (as for YBCO), pancake vortices in neighboring CuO_2 planes are strongly enough coupled to form flux lines, decoupling of flux lines into pancake vortices is observed in HTSCs with strong anisotropy, as, for instance, Bi-2212 or Bi-2223. In this case, the tendency to flux creep effects becomes even stronger and, thus, the irreversibility field strongly decreases. The brittleness of HTS is mainly caused by the anisotropy of the mechanical properties. For granular YBCO in particular, the anisotropic expansion coefficient results in large stresses causing microcracking, because by cooling from the crystal formation temperature to the superconducting transition temperature the lattice contracts far more along the *ab* planes than perpendicular to the planes. Microcracking is a central issue to be solved for any application of YBCO and especially for superconducting permanent magnets, in which large forces arise by the interaction between magnetic fields and supercurrents.

Another peculiarity of HTSCs is their small *coherence length* $\xi \propto v_F/T_c$ (with v_F as the Fermi velocity), which is comparable with the unit cell so that only a few

Cooper pairs overlap. In contrast, low-temperature superconductors have a much larger coherence length, and the space occupied by a Cooper pair is overlapped by many other Cooper pairs. The coherence length is a key parameter for the performance of superconductors for applications, because this parameter determines the size of the normally conducting core of the flux lines. In order to control the motion of flux lines one needs a microstructure with defects as small as the coherence length. The extremely small coherence length of HTS, which is for YBCO only 2.7 nm at 77 K within the ab-plane, is the reason that defects such as grain boundaries, which are very beneficial in low-temperature superconductors because they act as pinning defects, serve now as weak links and limit the critical current, especially in the presence of an external magnetic field.

1.1.9
Basic Relations for the Pinning Force and Models for its Calculation

Scaling of the volume pinning force

Mostly, experimental data for the critical current density are used in order to compare the pinning properties of superconductors. However, it is important to note that the fundamental measure of flux pinning, i.e. the interaction between the vortex lattice and the pinning defects within the superconductor, is the pinning force F_p per volume (the so-called volume pinning force) rather than the critical current density. The volume pinning force is given by $F_p = j_c \times \mu_o H$ with H as the applied magnetic field. This quantity turned out to be a very fruitful tool to investigate the pinning properties in conventional superconductors. For these superconductors, a scaling of the pinning force F_p with field and temperature was observed according to

$$F_p(H,T) = H_{c2}(T)^m \cdot f(h) \tag{1.28}$$

with $h = H/H_{c2}$ as the reduced magnetic field. The pinning force as a function of the reduced field h is 0 at $h = 0$ and $h = 1$ and goes through a maximum value F_{max} at the reduced field $h_m = H_m/H_{c2}$. The reduced field h_m was found to be independent of the temperature. Therefore, the normalized pinning force F/F_{max} measured at different temperatures is given by one curve

$$F_p/F_{max} = f(h) \tag{1.29}$$

As a first approximation, the function $f(h)$ of conventional superconductors can be described by

$$f(h) = h^p(1-h)^q \tag{1.30}$$

The parameters m, p and q depend on the size and type of the pinning mechanism, which can be classified according to the defect geometry (point-like, linear, planar), the type of interaction with the flux lines (core or magnetic), and the type of pinning center (normal-conducting or superconducting). Different cases resulting from combinations of these parameters have been described [1.28, 1.29].

A scaling of the volume pinning force was reported also for high-T_c superconductors [1.30]; however, the appropriate scaling field is here the irreversibility field, H_{irr}, instead of the upper critical field H_{c2}. The irreversibility field is the natural boundary for flux pinning in high-T_c superconductors. The resulting vortex lattice phase diagram will be considered in Section 4.2.

Models

In order to calculate the volume pinning force or the critical current density, one has to calculate the maximum elementary pinning force f_p of individual pinning centers, considering their interaction with the vortex lattice. Then one has to sum up these elementary pinning forces to get the volume pinning force, taking into account the elastic response of the vortex lattice to the pinning forces. This requires a knowledge of the elastic properties of the vortex lattice.

In most cases, elementary pinning forces act on the cores of flux lines. The optimum size of these pinning centers is in this case the coherence length. Pinning may be caused by spatial variation of T_c (δT_c pinning) or of the mean free path of the electrons (δl pinning) [1.32]. Normal conducting 211 particles in HTSCs provide pinning of the δl type. In $RBa_2Cu_3O_{7-x}$ compounds with the light rare-earth elements R= Nd,Sm,Eu,Gd, typically an R-rich solution between R and Ba with weak superconducting properties is formed. Pinning due to this solid solution is of the δT_c type.

In the following, the summation of elementary pinning forces will be briefly discussed. For strong pins, each pin can exert its maximum force f_p and the volume pinning force can be calculated by direct summation of the elementary pinning forces according to

$$F_p = n f_p \tag{1.31}$$

with n as the volume density of pins. For weak pins, the elastic distortions of the flux line lattice become important. A statistical summation of elementary pinning forces was proposed by Larkin and Ovchinnikov [1.31] for weak random pins in an elastic flux line lattice without dislocations. In the collective pinning theory, F_p is given by

$$F_p = \left(n \langle f_p^2 \rangle / V_c \right)^{1/2} \tag{1.32}$$

with V_c as a correlation volume. Whereas the pins are assumed to destroy the long-range order of the flux line lattice outside V_c, this order is preserved within flux bundles of size V_c. For 3D pinning, this size is given by $V_c \approx R_c^2 \cdot L_c$ with R_c and L_c as radius and length (along H) of the flux bundles, respectively. If pinning is strong, so that R_c becomes comparable to the flux line spacing a_o, then each flux bundle reduces to a single flux line, i.e. one arrives at collective pinning of individual flux lines. For very strong pinning, V_c reduces to $V_c \approx a_o^2 \xi = 1/n$, with

$L_c = \xi$ (coherence length). Then, Eq. (1.32) goes over into Eq. (1.31) and the collective pinning reduces to the direct summation of elementary pinning forces [1.12].

So far, it has been assumed that the volume pinning force is limited by vortex depinning and motion of the flux line lattice. However, at high magnetic fields the elastic properties of the flux line lattice and especially its shear strength becomes weak. Additionally, the flux line lattice can become weak by dislocations within the lattice. Therefore, plastic flux motion of weakly pinned flux lines around flux lines which are too strong to be broken is a possible scenario. A pinning model of plastic flux motion by synchronous shear of the flux line lattice around such strongly pinning flux lines was proposed by Kramer [1.33], later further developed [1.34], and applied to YBCO thin films and single crystals [1.35]. In these models, the pinning force at low fields $h < h_m$ is determined by the pinning properties of the superconductor, whereas the pinning force at high fields $h > h_m$ is limited by the elastic (and plastic) properties of the flux line lattice. The classical flux-line shear model [1.33] predicts

$$F_p \propto h (1-h)^2 (1-0.29h) \tag{1.33}$$

which reflects the field dependence of the shear modulus c_{66} of the flux line lattice for large Ginzburg–Landau parameters κ. This expression is similar to the function $f(h)$ in Eq. (1.30) for $p = 1$ and $q = 2$ and fits many experimental data found for conventional superconductors very well. In particular, a quadratic decrease of $F_p(H)$ near the upper critical field H_{c2}, which is a common feature of many experimental data, can be explained by this model.

1.2
Features of Bulk HTSCs

Great progress has been achieved in recent years in the development of HTSC thin films and tapes for different applications. Long Ag-sheathed $(Bi,Pb)_2Sr_2Ca_2Cu_3O_{10}$ (Bi-2223) tapes are now available which have been developed for energy cables and transformers working at 77 K or for windings in large motors intended for the temperature range between 20 and 30 K. Such tapes can carry high critical current densities at 77 K in self-field; however, j_c decreases strongly even at low applied magnetic fields. The reason is the decoupling of flux lines into pancake vortices observed in Bi-2223 (see Section 1.1.8) and the strongly enhanced mobility of the pancake vortices in applied magnetic fields. The development of YBCO-coated conductors for magnet field applications at 77 K is in progress. High critical current densities have already been achieved, but the lengths of the coated conductors are so far in the range 1–20 m. In this situation, the availability of bulk YBCO with high j_c in magnetic fields at 77 K is very useful. But even if long coated YBCO conductors are going to be available, several interesting applications, as superconducting magnetic bearings, can be better realized with bulk HTSC material.

1.2.1
Bulk HTSCs of Large Dimensions

The critical current density of bulk, polycrystalline HTSC material with randomly oriented grains is strongly limited by its pronounced weak link behavior. Only very small currents are able to flow across the grain boundaries (see Figure 1.10), especially if a magnetic field is applied. Typically, the $j_c(H)$ dependence is dominated by a $j_c(H) \propto 1/H$ law [1.36] reflecting the $j_c(H)$ dependence of a percolation network of Josephson junctions [1.37]. The rapid decrease of $j_c(H) \propto 1/H$ can be explained by assuming that the current through each junction corresponds to its Fraunhofer diffraction pattern whose oscillations are smeared out after averaging the junction orientations and lengths.

Weak links can be avoided by a *c*-axis texture (see Figure 1.11) using melt-texturing techniques. Several techniques have been found to be suitable for the growth of large single-domain HTSC materials without high-angle grain boundaries [1.38–1.40] (see Chapters 2 and 6). In most cases, seed crystals are used in order to orientate the *c*-axis of the HTSC sample perpendicular to the plane surface of the sample. Large superconducting domains with a *c*-axis texture can be produced in this way. In the present state of the art, the upper limit of the diameter of bulk, single-domain HTSC samples is about 10 cm.

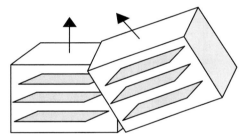

Figure 1.10 Grain boundary in granular superconductors (schematic)

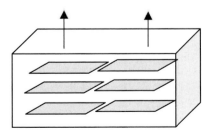

Figure 1.11 Grain boundary in a superconductor with *c*-axis texture (schematic)

The simplest way to control the grain structure is to investigate the distribution of the trapped field at 77 K. By applying a magnetic field in the *c*-direction flux lines penetrate the material and supercurrents are induced in the *a,b* plane of the YBCO sample, which is cooled by liquid nitrogen. After switching off the applied field the distribution of the trapped field in the superconductor is scanned by means of a Hall sensor with a small active area. Examples of field profiles obtained for single-domain YBCO samples are shown in Figures 5.3 and 5.12, the field profile of a multi-domain YBCO sample is plotted in Figure 5.13.

1.2.2
Potential of Bulk HTSC for Applications

The field profile of a superconducting permanent magnet is shown schematically in Figure 1.12 assuming a field-independent critical current density according to the Bean model [1.15]. The maximum trapped field B_o depends not only on the gradient of this field profile but also on the radius R of the current loops. The field gradient is determined by the critical current density j_c of the supercurrents or the pinning force exerted by the pinning centers acting on the flux lines.

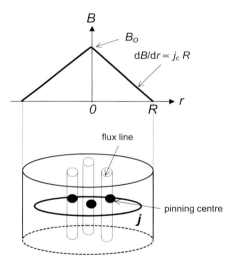

Figure 1.12 Field profile of a superconducting permanent magnet (schematic)

For a cylindrical sample of infinite length, B_o is given by

$$B_o/\mu_o = j_c R \tag{1.34}$$

Assuming $j_c = 10^4$ A/cm² and $R = 1$ cm one obtains from Eq. (1.34) a maximum trapped field $B_o = 12.6$ T within the superconductor or in the gap between two long superconducting samples. The trapped field on the surface of a long superconducting sample is half as large as this value, hence $B_o = 6.3$ T. It should be noted that the results obtained from Eq. (1.34) overestimate the trapped field, because in real superconductors the critical current density decreases with increasing magnetic field. Nevertheless, this approximation gives an impression about the relation between the critical current density and the trapped field in superconducting permanent magnets.

Trapped field magnets have to be magnetized whenever they have been heated above T_c. Permanent magnets, superconducting magnets, or pulsed field magnets can be used to magnetize trapped field magnets. Permanent magnets are the easiest to handle, but their field is too weak to saturate large high-quality bulk HTSC samples. Higher fields can be generated using superconducting or pulsed field magnets. Pulsed field magnets are very promising for applications at 77 K. The use of pulsed field magnets for magnetizing bulk YBCO at lower temperatures requires new ideas and solutions to overcome heating effects connected with the penetration of a large magnetic flux into the bulk superconductor during the short time at which the pulsed field is available. This will be discussed in more detail in Chapter 5.

Nevertheless, these new superconducting permanent magnets form a new class of ultrahigh-performance permanent magnets. The trapped flux density can be higher by an order of magnitude than the remanence of the best ferromagnetic permanent magnets. Since stored magnetic energies and forces scale quadratically with the trapped flux density, these new "cold magnets" offer magnetic energies and forces two orders of magnitude higher than conventional magnets. This means that completely new fields of application are accessible for permanent magnets.

In this context, there is a special interest in superconducting bearings. The stable levitation of a permanent magnet over a bulk HTSC is one of the most fascinating features associated with HTSC. This levitation is completely passive, because the magnetic flux surrounding the permanent magnet is pinned on defects within the superconductor. It should be borne in mind that no stable levitation is possible for two permanent magnets, and conventional magnetic bearings therefore require an active feedback control in order to stabilize the levitation. Passive superconducting bearings consisting of a superconductor and a permanent magnet are one of the most promising applications of bulk HTSC. They can be used as rotating bearings in flywheels for energy storage, in motors for liquid gas pumps or centrifuges, as well as for frictionless linear transportation systems. This should allow transportation of goods that are so extremely heavy that they cannot be moved by wheel systems, e.g., concrete ports or moving bridges. Alternatively, contactless transportation which does not create any wear debris can advantageously be used in clean rooms in the microelectronics industry.

1.3
Solid-State Chemistry and Crystal Structures of HTSCs

1.3.1
Crystal Structures and Functionality

All materials designated "high-T_c superconductors" (HTSCs) known at the present time are "cuprates", a group of ternary (or more complicated) metal oxides for which superconductivity had not been expected and whose superconductivity cannot be explained by theories applicable to conventional superconductors. In contrast, the low-T_c superconductors are metals, metallic alloys, or intermetallic compounds. High-T_c superconductivity was detected in $La_{2-y}Ba_yCuO_4$ only in 1986 by Bednorz and Müller, although solid solutions of relevant composition had been prepared considerably earlier [1.41, 1.42].

There is some overlap in T_c with some examples of doped fullerenes and MgB_2, which have not yet attained importance as bulk materials. Their physics and chemistry are rather different from those of superconducting oxides and they will not be considered in the present monograph.

The electronic structure in the normal state (i.e. above T_c) in all considered HTSC materials is of the metallic type, and is strongly correlated with their crystal structure, chemical bonding, and composition. Later it will be shown that secondary structural features (defects on the atomic, nanoscopic, or microscopic scale) also have a strong influence on the properties. The electronic configuration of a Cu atom is $3d^9s^1$, and if it is coordinated by O^{2-} anions with a coordination number (CN) 4, 5, or 6, the fully ionized approach leads to the configuration d^9 for Cu^{2+}, d^8 for Cu^{3+} or d^{10} for Cu^+, respectively. A large variety of structures were reported in individual cuprate compounds [1.43]. High T_c-superconductivity appeared in structures which contain anionic $[(CuO_2)^{(2-\Theta)-}]$ layers alternately stacked between compensating cationic metal or metal-oxide layers. These structures can be represented by a general formula

$$B^{b+}[(CuO_2)^{(2-\Theta)-}]_n C_{n-1}^{c+} \tag{1.35}$$

which distinguishes the building units according to their functionality [1.44].

The building units $[(CuO_2)^{(2-\Theta)-}]_\infty$ represent layers consisting of CuO_4 polyhedra, which share corners in an almost planar arrangement. These layers are essential for the superconducting behavior in these cuprate materials, forming the conducting block. In certain structures (e.g. Bi2212, Bi2223, Tl1212), this block consists of a number of consecutive $[(CuO_2)^{(2-\Theta)-}]_\infty$ layers (n > 1), which are separated by the intra-block cationic layers C^{c+} (typically electropositive fully ionized elements: Ca^{2+}, Y^{3+}, La^{3+} ...).

The cationic metal oxide spacing layer B is stacked between consecutive conducting blocks (inter-block spacing). The spacing layer B is the second indispensable structural component of HTS cuprates, whereas the intra-block layer C only appears for extended conducting blocks, with 2 or more layers in the conducting block. Three main families of cuprate superconductors can be distinguished:

Family (i) $Ln_{2-x}(Ba,Sr)_x CuO_{4-\delta}$, e.g., $(La_{2-x}Sr_xO_{2-\delta})^{(2-\Theta)}(CuO_2)^{(2-\Theta)-}$

Family (ii) $REBa_2Cu_3O_{7\pm\delta}$, e.g., $YBa_2Cu_3O_{7\pm\delta}$, $Nd_{1\pm y}Ba_{2\pm y}Cu_3O_{7\pm\delta}$

Family (iii) $B^{(2-n\Theta)+}[(CuO_2)^{(2-\Theta)-}]_n Ca^{2+}_{n-1}$, n = 1, 2, 3,
$B = (Bi,Pb)_2Sr_2O_{4-x}$ or $TlBa_2O_{2.5+x}$ or $Tl_2Ba_2O_{4+x}$
or $HgBa_2O_{2+x}$ and Θ increasing with x.

From the crystallographic point of view, the stacked sub-units represent building groups of the two prototype structures, the perovskite and rock-salt structure. Because of different electronic, magnetic and structural interactions, the stoichiometry, fine structure and properties of the resulting superconducting compounds will be strongly influenced by preparation and annealing conditions. Note that the stacking axis appears on the longest unit cell parameter, which is denoted by "c".

Generally, compounds of the *type (iii) family* tend to behave nearly 2-dimensionally, because of only weak interactions between the superconducting layer blocks. They are characterized by high T_c if the number of individual layers in the block is large, but they cannot carry high currents in magnetic fields because of the restriction of vortices to only one block ("pancake vortices"). This is not the case for the compounds of the structural family (ii), which are therefore preferably considered as a bulk material because of their fascinating behavior in high magnetic fields. Therefore, *this book will be concentrated on family (ii)*. The crystal structure of composition $YBa_2Cu_3O_7$, the most intensively investigated representative, is shown in Figure 1.13 in comparison to $(La,Sr)_2CuO_4$ and $Bi_2Sr_2CaCu_2O_8$ as representatives of the families (i) and (iii), respectively. Phases of $REBa_2Cu_3O_7$ structure are

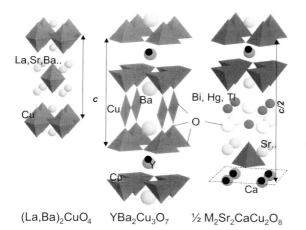

Figure 1.13 Crystal structure of $YBa_2Cu_3O_{7\pm\delta}$, a representative of the type (ii) family, which includes the most important bulk materials, compared with type (i) (La,Ba)CuO$_4$ and type (iii) M$_2$Sr$_2$CaCu$_2$O$_8$. Only half the unit cell is shown in the latter case

known for RE = Nd, Sm, Dy, Gd, Er, Tm, and Eu. Compounds of the nominal composition $YBa_2Cu_4O_8$ and $Y_2Ba_4Cu_7O_{15}$ also belong to family (ii). Their structures contain edge-sharing $[Cu_2O_5]_\infty$ double chains substituting every $[CuO_3]_\infty$ chain or every second chain, respectively [1.45, 1.46].

Superconductors belonging to family (iii) are important for application in conducting wires and cables by the "powder-in-tube" technique, whereas the variety of *family (i) superconductors* – the structures of which are less complicated – gives a deeper understanding of fundamental mechanisms.

1.3.2
Chemistry and Doping

Chemical bonding is highly covalent in the layers which carry superconductivity (i.e. in nearly planar $[(CuO_2)^{(2-\Theta)-}]_\infty$), and it is characterized by magnetic superexchange interactions along Cu-O-Cu bonds which suppress the appearance of a magnetic moment in the $[(CuO_2)^{(2-\Theta)-}]_\infty$ layer. On the other hand, Figure 1.14 indicates a superconducting phase appearing only in a certain region of carrier concentration in this layer, and therefore appropriate carrier doping plays an important role in controlling superconducting properties. The possibilities of doping superconductors of the 123 type are closely connected with the features of its crystal structure and may be divided into intrinsic and extrinsic "impurity" doping, the latter with additives or substituents.

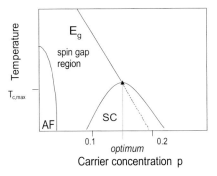

Figure 1.14 Schematic representation of the low-temperature phase diagram of p-type HTSC indicating the stability range of the superconducting phase dependent on carrier concentration

It should be emphasized here that the use of the terms "doping" and "dopant" should be restricted to the immediate influence on and incorporation into the crystal lattice of the considered (superconducting) phase, either considering the electronic aspect ("carrier doping") or the chemical aspect by "chemical dopants" (nonstoichiometry and "impurities"). Additives resulting in the formation of secondary phases cannot be considered as "dopants" from either physical or chemical aspects.

1.3.3
Intrinsic Doping: Variations of Stoichiometry

The simplest approach to illustrate "pseudo-intrinsic" doping is a "two-band model" permitting charge transfer between structural units. This is schematically illustrated in Figure 1.15.

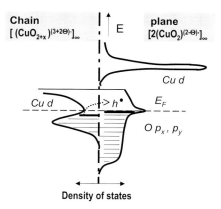

Figure 1.15 Sketch of the two-band model to explain doping by charge transfer between the structure elements "chain" and "plane" in Figure 1.13

The left-hand band is localized in the $[Ba^{2+}(CuO_{2+x})^{(3+2\Theta)-}Ba^{2+}]$ units, which contain the characteristic $(CuO_3)^{(3+2\Theta)-}$ chains in the case of full oxygenation, whereas the right hand band represents the superconducting layer. The parameter Θ is characteristic for the degree of hole doping in the conducting (nearly) planar $[(CuO_2)^{(2-\Theta)-}]_\infty$ layers due to a charge transfer between the two sub-systems (i.e. transfer of holes from "chains" to "planes"). It is obvious from this consideration that the total positive charge (which is distributed between the two kinds of Cu sites) depends on the degree of oxygenation, i.e. on the occupation x on the bridging oxygen sites in the $(CuO_{2+x})^{(3+2\Theta)-}$ chains (i.e. in the $[Ba^{2+}(CuO_{2+x})^{(3+2\Theta)-}Ba^{2+}]$ blocks. In $YBa_2Cu_3O_{6+x}$ (synonymously $YBa_2Cu_3O_{7-\delta}$ with $1 - x = \delta$) and related superconductors, the charge doping concentration was found to be within the limits $\Theta \approx 0$ for $(1-\delta) = x \leq 0.3$ and $\Theta \approx 0.2$ for $(1-\delta) = x = 1$.

From these facts one can conclude that carrier concentration ("hole doping") and hence the superconducting transition temperature T_c can be controlled by the oxygen stoichiometry. Vice versa, *any influence on the superconducting temperature T_c observed after chemical or physical treatment cannot be unambiguously explained unless the oxygen stoichiometry has been fixed.*

Another case of "intrinsic" doping is realized with $RE_{1+y}Ba_{2-y}Cu_3O_{7-\delta}$ ($RE_{1+y}Ba_{2-y}Cu_3O_{6+x}$). Solid solutions of this type can be formed for RE = Nd, Sm, Eu, (Gd), i.e. for RE elements with larger ionic radii which become comparable with that of Ba^{2+}. Considering compositions with the same oxygen content (i.e. x and δ kept constant), the increasing positive charge on Ba sites due to substitution of Ba^{2+} by RE^{3+} has to be compensated by decreasing positive charge on Cu sites, in other words by localization of holes in the block layer $B = [Ba_2 CuO_{2+x}^{(3+2\theta)-}]$.

1.3.4
Defect Chemistry

Defect chemistry models are a useful tool to understand mechanisms to control carrier type and concentration in ionic compounds under conditions of sample preparation.

In the simplest case the deviation from the stoichiometric composition appears at elevated temperature as point defects. They are generated by intrinsic reactions, preserving the oxidation state, or by extrinsic or "pseudo-intrinsic" reaction with the environment, resulting in a change of the oxidation state ("doping"). Point defects are generated by chemical reactions at preparation temperatures typically above 700 K, which is sufficient to permit *charge* and *mass* transfer.

Within the formalism, it is useful to define the oxygen-poor state by x = 0 in the formula $YBa_2Cu_3O_{6+x}$ as the reference state. Then, the oxygen sites in the chain segments can be treated either as *interstitial* O_i (if occupied) or as unoccupied *vacancies* V_i. They can carry (localized) negative or positive charges indicated by primes or dots, respectively (compared with the unoccupied lattice).

Environmental oxygen – filling the unoccupied sites in the cuprate – binds electrons, thus leaving a hole ("oxidation"), which can be either *itinerant* or *localized* in the neighboring Cu – O bonds. This is formally expressed by Eq. (1.36), using the Kröger–Vink notation:

$$\tfrac{1}{2} O_2 + V_i \rightleftarrows O_i'' + 2 h^\bullet \tag{1.36}$$

Because of strong Coulomb interactions, the holes are expected to remain associated with the corresponding defects, represented by the equilibrium

$$O_i'' + 2 h^\bullet \rightleftarrows O_i' + h^\bullet \rightleftarrows O_i^\circ \tag{1.37}$$

Considering the site balance, the equilibrium concentration of V_i is $(1 - x)$, if x represents the concentration of interstitials, O_i°, thus

$$p(O_2) = K^2(T)\left(\left(\frac{x}{1-x}\right)\right)^2. \tag{1.38}$$

Providing h the concentration of holes remains low (i.e. $[O_i^\circ]$ nearly unchanged), the product $[O_i'] h = K^*$ represents an equilibrium constant. The hole concentration is then mainly determined by the right hand equilibrium in Eq. (1.37) [1.47]. Taking into account the balance information, $h = [O_i']$, one gets

$$p_{O_2} = K^2(T)\left(\frac{h^2/K^*}{1 - h^2/K^*}\right)^2 \approx const \cdot h^2 \tag{1.39}$$

which is in accordance with Eq. (2.7).

It should be emphasized that doping by "impurities" (i.e. strange elements) can also be considered in the frame of defect chemistry [1.47]. Substitution of Y^{3+} by Ca^{2+}, e.g., can be expressed by

$$2\ CaO + 2\ (Y_Y^{3+}) + 1/2 O_2 \rightleftarrows Y_2O_3 + 2\ (Ca_Y^{2+})' + 2\ h^\bullet \tag{1.40}$$

The combination of this equation with Eqs. (1.38) and (2.7) clearly indicates the interrelationship between thermodynamic activities (chemical potentials) of the dopant, $p(O_2)$, oxygen stoichiometry, and hole concentration: if $p(O_2)$ is fixed (depending on the environment), *holes as generated by impurities will be partially consumed to form oxygen vacancies on interstitial sites* (to reduce x) until an equilibrium state is achieved! Carrier concentrations, as prepared by treatment at elevated temperature, will be preserved after cooling to room temperature.

Experiments indicate significant deviations due to the non-ideal behavior resulting from the high defect density, so that activities instead of concentrations should be introduced with activity coefficients far from the value 1. Nevertheless, defect chemistry predicts a relationship to control oxygen content and initial carrier concentration in preparation procedures. Strong interactions between defects in high concentration result in clustering and secondary defect structures. As an example, in the forthcoming chapters it will be shown that oxygen content and cation composition cannot be changed independently.

1.3.5
Extrinsic Doping

Extrinsic doping by cation substitution can be realized in *RE*123 at the different cation sites appearing in the general formula, namely
1. on Cu(1) sites of $(CuO_{2+x})^{(3+2\Theta)-}$ chains in the *spacing layer* B,
2. on the Cu(2) site in $[(CuO_2)^{(2-\Theta)-}]_\infty$, the *conducting layer*,
3. on the Ba^{2+} sites as *another part of the spacing layer B*, and
4. on the cationic site RE^{3+} in the *intra-block layer*.

Obviously, substitution by heterovalent cations will affect the actual doping state and hence T_c. Furthermore, changing size of substituents will influence orbital overlap and hence magnetic superexchange. Magnetic superexchange will also be drastically affected by isovalent ions having a spin differing from $1/2$ as for Cu^{2+} on Cu sites. In this case a resulting magnetic moment leads to "deterioration of superconductivity", i.e. decrease of T_c, the rate of which is 6 – 10 K for in-plane substitution or 3 – 4 K for in-chain substitution per at% (i.e. x × 100%) in $YBa_2Cu_{3-3x}M_{3x}O_7$ [1.48]. Examples with relevance to critical currents in bulk materials will be discussed in detail within Chapters 6 and 7 [1.49, 1.50].

Whether or not substitution is possible at a certain lattice is determined to a large extent by the compatibility of ionic radii, a compilation of which is presented in Table 1.1.

General trends will be introduced in the following paragraphs considering the influence of chemical alteration on charge doping in $REBa_2Cu_3O_{7-\delta}$ phases.

Substitution on Ba^{2+} sites

Ba^{2+} can be substituted by large cations, which prefer coordination numbers (CN) larger than 8. In the case of *substitution by M^{n+}* with ionic charge n > 2, the substitution of Ba^{2+} will have an analogous influence on carrier doping and charge transfer as discussed above for "intrinsic" doping by large RE^{3+} ions.

Substitution on RE^{3+} sites

Substitution by Ca^{2+} for Y^{3+} gave a fundamental insight into the influence of extrinsic ("impurity") doping in $YBa_2Cu_3O_{7-\delta}$ bulk materials. Tallon et al. [1.51] confirmed the existence of an "optimum" doping state, representing a maximum value of T_c. Figure 1.14 represents the results obtained for compositions $Y_{1-y}Ca_yBa_2Cu_3O_6$. Since these experiments kept x = 0, the changing carrier concentration can be attributed to the Ca^{2+} dopant concentration. The empirical equation

$$T_c/T_{c,max} = 1 - 82.6\,(p - 0.16)^2 \tag{1.41}$$

derived from the experiments is accepted to have a general meaning if applied in a properly normalized form. The appearing maximum is considered as a "quantum-critical point", and the same behavior has meanwhile been found also for p-type superconductors of the (i), (ii), and (iii) families [1.52, 1.53].

Substitution of Y^{3+} by M^{3+} will influence charge doping in the same manner, like "intrinsic" substitution by RE^{3+}.

Substitution of Y^{3+} by RE^{3+}

Substitution of Y^{3+} or RE^{3+} in the RE^{3+} layers is possible, sometimes at high rates resulting in solid solutions. If the valence remains 3+, the influence on charge doping is expected to be of second order and to be exceeded by effects on processing, magnetic properties or microstructure.

In concluding this section, it has to be emphasized that the considered correlations between structure, composition, and doping represent more or less a static and non-superconducting state which is representative of temperatures up to slightly above room temperature. They give information neither about preparation conditions nor coexistence with the environment etc. More information about the latter will be found in the following chapter.

References

1.1 J. G. Bednorz, K. A. Müller, Rev. Mod. Phys. **60**, 585 (1988).

1.2 M. K. Wu, J. R. Ashburn, C. J. Torng, P. H. Hor, R. L. Meng, L. Gao, Z. J. Huang, Y. Q. Wang, C. W. Chu, Phys. Rev. Lett. **58**, 908 (1987).

1.3 A. Schilling, M. Cantori, J. D. Guo, H. R. Ott, Nature **363**, 56 (1993).

1.4 C. W. Chu, L. Gao, F. Chen, Z. J. Huang, R. Meng, Y. Y. Xue, Nature **366**, 323 (1993).

1.5 J. R. Kirtley, C. C. Tsuei, Martin Rupp, J. Z. Sun, Lock See Yu-Jahnes, A. Gupta, M. B. Ketchen, K. A. Moler, M. Bhushan, Phys. Rev. Lett. **76**, 1336 (1996).

1.6 Y. Wang, B. Revaz, A. Erb, A. Junod, Phys. Rev. B **63**, 094508 (2001).

1.7 H. Aubin, K. Behnia, M. Ribault, L. Taillefer, R. Gagnon, Z. Phys. B **104**, 175 (1997).

1.8 A. A. Abrikosov, Zh. Eksp. Teor. Fiz. **32**, 1442 (1957).

1.9 U. Essmann, H. Träuble, Phys. Lett. **24A**, 526 (1967).

1.10 H. W. Weber, Flux Pinning, in Handbook on the Chemistry and Physics of Rare Earths, Vol. 31: High Temperature Superconductors – II, ed. by K. A. Gschneider, Jr., L. Eyring, M. B. Maple (Elsevier, Amsterdam, 2001) p. 187.

1.11 A. Moser, H. J. Hug, B. Stiefel, H.-J. Güntherodt, J. Magnetism and Magn. Mater. **190**, 114 (1998).

1.12 E. H. Brandt, Rep. Progr. Phys. **58**, 1465 (1995).

1.13 T. T. Palstra, B. Batlogg, R. B. van Dover, L. F. Schneemeyer, J. V. Waszczak, Phys. Rev. B **41**, 6621 (1990).

1.14 T. K. Worthington, E. Olsson, C. S. Nichols, T. M. Shaw, D. R. Clarke, Phys. Rev. B **43**, 10538 (1991).

1.15 C. P. Bean, Rev. Mod. Phys. **36**, 31 (1964).

1.16 D. Chen, R. B. Goldfarb, J. Appl. Phys. **66**, 2489 (1989).

1.17 A. Sanchez, C. Navau, Supercond. Sci. Technol. **14**, 444 (2001).

1.18 E. H. Brandt, Phys. Rev. B **58**, 6506 (1998).

1.19 Y. Yeshurun, A. P. Malozemoff, A. Saulov, Rev. Mod. Phys. **68**, 911 (1996).

1.20 Y. B. Kim, C. F. Hempstead, A. R. Strnad, Phys. Rev. **131**, 2486 (1963).

1.21 P. W. Anderson, Phys. Rev. Lett. **9**, 309 (1962).

1.22 P. W. Anderson, Y. B. Kim, Rev. Mod. Phys. **36**, 39 (1964).

1.23 V. B. Geshkenbein, A. I. Larkin, Sov. Phys. JETP **68**, 639 (1989).

1.24 M. V. Feigel'man, V. B. Geshkenbein, A. I. Larkin, V. M. Vinokur, Phys. Rev. Lett. **63**, 2303 (1989).

1.25 J. R. Thompson, Yang Ren Sun, L. Civale, A. P. Malozemoff, M. W. McElfresh, A. D. Marwick, F. Holtzberg, Phys. Rev. B **47**, 14440 (1993).

1.26 S. Gruss, G. Fuchs, G. Krabbes, P. Schätzle, J. Fink, K. H. Müller and L. Schultz, IEEE Trans. on Magnetics **34**, 2099 (1998).

1.27 Küpfer, S. N. Gordeev, W. Jahn, R. Kresse, R. Meier-Hirmer, T. Wolf, A. A. Zhukov, K. Salama, D. Lee, Phys. Rev. B **50**, 7016 (1994).

1.28 D. Dew-Hughes, Phil. Mag. **30**, 293 (1974).

1.29 A. M. Campbell, J. E. Evetts, Adv. Phys. **72**, 199 (1972).

1.30 M. R. Koblischka, M. Murakami, Supercond. Sci. Technol. **13**, 738 (2000).

1.31 A. I. Larkin, Yu. N. Ovchinnikov, J. Low Temp. Phys. **73**, 109 (1979).

1.32 G. Blatter, M. V. Feigel'man, V. B. Geshkenbein, A. I. Larkin, V. M. Vinokur, Rev. Mod. Phys. **66**, 1125 (1994).

1.33 E. J. Kramer, J. Appl. Phys. **44**, 1360 (1973).

1.34 A. Pruymboom, P. H. Kes, E. van der Drift, S. Radelaar, Appl. Phys. Lett. **52**, 662 (1988).

1.35 R. Wördenweber, Phys. Rev. B **46**, 3076 (1992).

1.36 G. Fuchs, A. Gladun, K. Fischer, C. Rodig, Cryogenics **32**, 591 (1992).

1.37 R. I. Peterson, J. W. Ekin, Phys. Rev. B **42**, 8014 (1990).

1.38 M. Murakami, in Processing and Properties of High Tc Superconductors, edi-

ted by Sungho Jin (World Scientific, Singapore, 1993), Vol.1, pp. 215-268.
1.39 P. McGinn, in High Temperature Superconducting Materials Science and Engineering, edited by Donglu Shi (Elsevier, Amsterdam, 1995), pp. 345–382.
1.40 G. Krabbes, P. Schätzle, W. Bieger, G. Fuchs, U. Wiesner, G. Stöver, IEEE Trans. Appl. Supercond. **7**, 1735 (1997).
1.41 I. S. Shaplygin, B. G. Kakhan, V. B. Lazarev, Zh. Neorg. Khim (in Russian) **42**, 1478 (1979).
1.42 N. Ngyen, J. Choisnet, M. Hervieu, B. Raveau, J. Solid State Chem. **39**, 120 (1981).
1.43 H. Müller-Buschbaum, Angew. Chem. **103**, 741 (1991).
1.44 H. Eschrig, High T_c Superconductors: Electronic Structure, in K. H. J. Buschow, Encyclopedia of Materials, Elsevier, Oxford 2003, 7
1.45 J. Karpinski, E. Kaldis, E. Jilek, S. Rusiecki, B. Bucher, Nature **336**, 660 (1988).
1.46 D. E. Morris, N. N. Asmar, J. H. Nickel, R. L. Sid, J. Y. Wie, J. E. Post, Physica C **159**, 287 (1989).
1.47 J. Meier, H. L. Tuller, Phys. Rev. B **47**, 8105 (1993).
1.48 S. Zagulaev, P. Monod, J. Jegoudez, Phys. Rev. B **52**, 1074 (1990).
1.49 G. Krabbes, G. Fuchs, P. Schätzle, S. Gruß, J. W. Park, F. Hardinghaus, G. Stöver, R. Hayn, S.-L. Drechsler, T. Fahr, Physica C **330**, 181 (2000).
1.50 L. Shlyk, G. Krabbes, G. Fuchs, K. Nenkov, P. Verges, Appl. Phys. Lett. **81**, 5000 (2002).
1.51 J. L. Tallon, J. W. Loram, G. V.M. Williams, J. R. Cooper, I. R. Fisher, J. D. Johnson, M. P. Staines, C. Bernhardt, Phys. Stat. Sol. B **215**, 531 (1999).
1.52 M. R. Presland, J. L. Tallon, R. G. Buckley, R. S. Liu, N. E. Flower, Physica C **176**, 95 (1991).
1.53 B. C. Sales, B. C. Chakoumakos, Phys. Rev. B **43**, 12994 (1991).

2
Growth and Melt Processing of YBa$_2$Cu$_3$O$_7$

The term "melt-textured superconductor" was chosen to describe the highly textured structure observed in pelletized YBCO material after partial melting [2.1]. The procedure, however, is based on the "science and art of growing crystals" rather than on texturing mechanisms known from the science of metallurgy or ceramics technology. Consequently, this chapter will first consider the fundamentals of processing, control of the properties, and growth of YBa$_2$Cu$_3$O$_7$ crystals before turning to the peculiarities of melt-grown bulk superconductors.

2.1
Physico-Chemistry of *RE*-Ba-Cu-O Systems

2.1.1
Phase Diagrams and Fundamental Thermodynamics

A set of stable phases coexisting in chemical equilibrium is determined by the minimum of the *Gibbs* free energy G which exists for each mixture of elementary components E_i (representing Y, *RE*, Ba, Cu, O...). The chemical potential μ of an element i is defined as the partial *Gibbs* free energy

$$\mu_i = (\partial G/\partial n_i)_{P,T,n_{(l \neq i)}} \tag{2.1}$$

where n_i and n_l are the molar amounts of elements E_i and E_l. The minimum is defined by $dG = 0$, where

$$G = \Sigma \, (n_i \mu_i) \tag{2.2}$$

The chemical potential of element E_i in any phase ϕ is a function of p, T, and $\kappa_1(\phi) .. \kappa_{k-1}(\phi)$, the concentrations in the phase ϕ. Furthermore, *Gibbs* free energy G is a first-order homogeneous function. The following fundamental relationships are important to characterize the equilibrium state:

1. Chemical potentials in any phase cannot change independently (*Gibbs–Duhem* relation):

$$\Sigma (n_i \partial \mu_i) = 0 \qquad (2.3)$$

2. Chemical potentials of any component are equal in all phases in the equilibrium state.

$$\mu_1(\phi 1) = \mu_1(\phi 2) = \mu_1(\phi 3)$$
$$\mu_2(\phi 1) = \mu_2(\phi 2) = \mu_2(\phi 3)$$
$$.....$$
$$\mu_k(\phi 1) = \mu_k(\phi 2) = \mu_k(\phi 3). \qquad (2.4)$$

3. The phase relations of a k-component system need $k + 1$ variables to be fully represented.

The equilibrium state, which will usually be approached during any transformation or reaction, is depicted in the equilibrium phase diagram. Graphs of phase relationships (phase diagrams) have to follow topological rules. Phase relationships of any system can be described by a set of invariant ($\Phi = 0$) and univariant ($\Phi = 1$) reactions, i.e. reactions which have no degrees of freedom, or only one, for the thermodynamic variables according to *Gibbs's* phase law

$$\Pi + \Phi = k + 2. \qquad (2.5)$$

Π is the number of phases in equilibrium, Φ the number of parameters free to the experimenter's choice ("degrees of freedom"), and k the number of "components" in the system, i.e. units which will not be disconnected in the experiment (in the most general case, chemical elements).

The coexistence between 2 condensed phases and the gas in a 2-component system leaves one parameter variable (resulting in a line in a 2-axis diagram) or 2 parameters in a ternary system (resulting in a plane), which needs a 3-axis representation. The number of independent variables which are necessary to identify the composition is already 3 for undoped *RE*-Ba-Cu-O. Therefore, an intelligent choice of sections through the phase space has to be used appropriate to the considered problem. The number of dimensions of the representation can be reduced by one for a section in which one physical quantity is set to a fixed value. By this procedure, planes change to lines and lines change to dots. An invariant reaction appears as a point in the phase diagram, which, however, is the case only for a unique set of crossing sections.

If the exact equilibrium conditions are left (e.g. by changing T or $p(O_2)$ or the concentration of a certain component), the reaction proceeds into the adjacent phase field.

2.1.2
Subsolidus Phase Relationships

HTS cuprates with relevance to bulk materials application are ternary or quaternary compounds appearing in the $REO_{1.5}$-MO-CuO_x quasi-ternary phase diagram. Which phase can appear depends to a large extent on the relation between the ionic radii of RE^{3+} and M^{2+}. The tolerance parameter t defined by Eq. (2.6) is appropriate as a criterion for the formation of a stable phase.

$$r_{RE^{3+}} - t\sqrt{2}\, r_{Cu^{n+}} = r_{O^{2-}}(t\sqrt{2} - 1) \tag{2.6}$$

From the geometry of the perovskite lattice, perovskite can be expected to be stable for $0.89 < t \leq 1$ and a distorted perovskite lattice appears for $0.8 < t \leq 1$. This consideration can be applied also for La_2CuO_4 type phases, and the more complicated

Table 2.1 Ionic radii r^{n+}(nm) of Y^{3+} and Ln^{n+} ions

Shannon's parametric system [2.55]. $T(m_1)$ is the invariant peritectic temperature at $p(O_2) = 0.21\,bar$. CN is the coordination number. Values marked by askerisk are taken for CN = 9 in case that data for CN10 have not been available.
Data $T(m_1)$ in °C from [2.53] as far as not cited otherwise.

	La^{3+}	Ce^{3+}	Pr^{3+}	Nd^{3+}	Sm^{3+}	Eu^{3+}	Gd^{3+}	Tb^{3+}	Dy^{3+}	Y^{3+}
CN 8	0.116	0.114	0.113	0.111	0.108	0.107	0.105	0.104	0.103	**0.102**
CN 10	0.127	0.125	0.118*	0.116*	0.11*	0.112*	0.120	0.110*	0.107*	
							(0.111*)			
$T(m_1)$	1090			1090	1070	1050	1047		1030	1020
					[2.28]		[2.29], [2.34]		[2.54]	[2.26], [2.27]

	Y^{3+}	Er^{3+}	Ho^{3+}	Tm^{3+}	Yb^{3+}	Lu^{3+}	Ce^{4+}	Pr^{4+}
CN 8	0.102	0.100	0.101	0.100	0.099	0.099	0.097	0.096
CN 10			0.112	0.106*	0.105*	0.104*	0.107	
$T(m_1)$	1020	1005 [2.54]	980	960	900	880		

Ionic radii for Cu and alkaline earth elements (*Shannon's* parameters)

	Ca^{2+}	Sr^{2+}	Ba^{2+}		Cu^{2+}		Cu^+
CN 8	0.112	0.126	0.142	CN4	0.057	CN2:	0.056
CN 10	0.123	0.136	0.152	CN6	0.073		

structures (like 123 etc.) appear for slightly larger deviations. The relation also gives roughly an estimate for structure compatibility in the case of site substitution and formation of solid solutions.

Typical subsolidus constitutions are represented by the quasiternary sections in Figure 2.1. The thermodynamic variables $p(O_2)$ and T are fixed in each Figure.

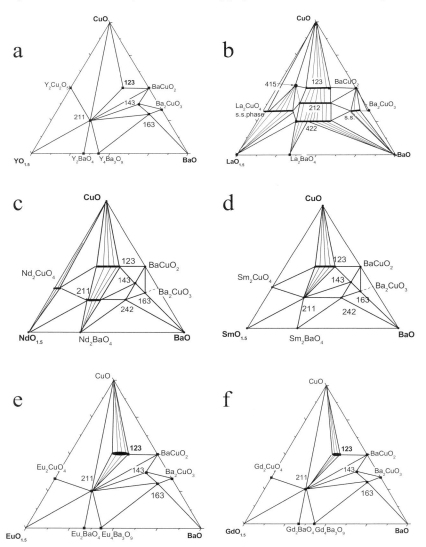

Figure 2.1 Constitution of the quasiternary isothermal subsolidus sections in RE-Ba-Cu-O systems at $p(O_2) = 0.21$ bar and temperatures around 880 °C [2.2–2.4]. Note that at 1 bar O_2 the phases 247 and 124 become stable phases with small ions RE (for Y below 879 and 859 °C, respectively) [2.57]

Superconductivity below 45 K appears in the La-rich phase of the restricted solid solutions $La_{2-y}M_y(=Sr$ or $Ba)CuO_4$. In contrast, the superconducting 123 phase is one of the phases which newly appear only with the larger Ba^{2+} ion. Significant homogeneity regions are typical for large RE^{3+} (La^{3+} and the light lanthanide ions) because the solid solution, e.g., substitution of RE^{3+} by Ba^{2+} or Sr^{2+}. This appears not only in the superconducting phases but also, e.g., in a phase of the nominal ratio 2:1:1, which plays an important role in bulk superconductors and their technology. However, the crystal structure of the solid solution phases with larger RE^{3+} is quite different [2.5], and according to the larger unit cell, the notation $RE_{4-2x}Ba_{2+2x}Cu_2O_{10}$ or RE 422 is applied.

Under elevated $p(O_2)$, two more superconducting phases appear at equilibrium conditions, $Y_2Ba_4Cu_7O_{14}$ (Y247, $T_c \cong 40$ K) and $YBa_2Cu_4O_8$ (Y124, $T_c \cong 80$ K). Although Y124 can be prepared by solid-state reaction from metal oxides at 1 bar O_2 [2.6], a crystallization field exists only under high pressure [2.7].

2.1.3
The Influence of Oxygen on Phase Equilibria: the System Y-Ba-Cu-O

There are a number of reasons for the cuprate systems in Figure 2.1 to be considered as 4-component systems, with oxygen as the fourth component [2.8]:
1. Several phases like Cu_2O and $RECuO_2$ contain Cu with a valency +1, whereas its valency is between 2 and 3 in the superconductor phases.
2. Phase stability depends sensitively on $p(O_2)$.
3. The oxygen content is variable in a number of phases and the phase compatibility alters depending on $p(O_2)$.
4. The oxygen concentration in the appearing melt depends on melt composition *and* $p(O_2)$.

The phase compatibility of $YBa_2Cu_3O_{6+x}$ within the 4-component coordinate system is schematically illustrated in Figure 2.2 for the subsolidus state at preset values $p(O_2) = 0.21$ bar and $T = 850\,°C$.

The change of phase relationships dependent on the oxygen content as a coordinate can also be visualized in a T–x section of the fictive "pseudo-binary" system AO_x [2.9].

2 Growth and Melt Processing of YBa$_2$Cu$_3$O$_7$

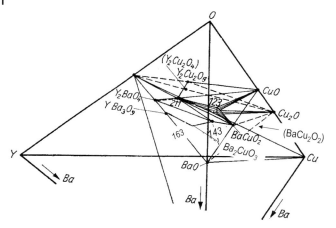

Figure 2.2 The isobar–isothermal subsolidus section of the *quaternary* system Y-Ba-Cu-O 0.21 bar, 850 °C [2.8]

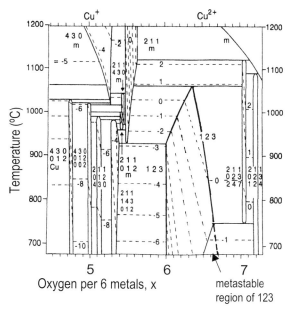

Figure 2.3 Phase diagram of Y-Ba-Cu-O in a (YBa$_2$Cu$_3$) – xO *pseudo*binary representation. The T–x diagram indicates phase stability and stoichiometry dependent on log $p(O_2)$ as a parameter [2.9]

Figure 2.3 represents the section A – xO in the range 5 < x < 7.5 with A = (YBa$_2$Cu$_3$). $p(O_2)$ isobars are used to identify the chemical potential of oxygen at any point according to Eq. (2.7),

$$\mu(O) - \mu°(O) = 1/2\ RT \ln p(O_2)/p°(O_2) \qquad (2.7)$$

with $p°(O_2) = 1$ bar. Note that in this representation isobars appear horizontally in the invariant state, whereas all other fields may represent either 1, 2, or 3 phase fields (in contrast to real 2-component systems) in accordance with Gibbs's phase rule.

Furthermore, Figure 2.3 indicates that $YBa_2Cu_3O_{6+x}$ is not a thermodynamically stable phase in the range $0.5 \leq x < 1$. Whether or not compositions of the phase in this region are formed depends on the treatment of samples. Below 750 °C and at $p(O_2) \leq 1$ bar, bulk $YBa_2Cu_3O_{6+x}$ appears in a metastable state in which it can be oxidized up to $x \cong 0.98$ without deterioration of the cationic sublattice (broken line in Figure 2.3).

2.1.4
The Oxygen Nonstoichiometry in 123 phases: $YBa_2Cu_3O_{7-\delta}$ ($YBa_2Cu_3O_{6+x}$)

As early as 1987/88 a number of parallel investigations were published indicating the strong variation of the oxygen content $x = 1 - \delta$ in the $YBa_2Cu_3O_{7-\delta}$ phase, the results of which are compiled in [2.4].

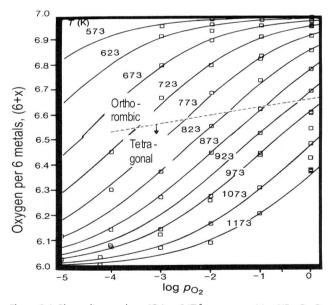

Figure 2.4 Phase diagram $\log p(O_2)$ vs $1/T$ for a composition $YBa_2Cu_3O_{7-\delta}$ [2.10]

Because of the high diffusivity of oxygen at elevated temperatures, the results represent a reproducible state (equilibrium or metastable, depending on T). They have been obtained by different methods for oxygenation or reduction of samples.

Experimental results in Figure 2.10 indicate significant deviations from an ideal point defect model and values of activity coefficients are far from unity.

Oxygen ordering and phase separation

The actual defect concentration in $YBa_2Cu_3O_{6+x}$ and related materials is at the rather high level of 10^{21} cm^{-3}, which gives rise to clustering and formation of ordered secondary structures. Such phenomena – not detected by global analytical and microscopic methods – have been visualized by TEM and neutron diffraction studies (e.g. [2.11, 2.12]). The structure is characterized by an alternating arrangement of chains in which the interstitial oxygen sites (O_i) are occupied or empty (V_i) resulting in a gross stoichiometry $YBa_2Cu_3O_{6.5}$. This phase was attributed to a sharp superconducting transformation step observed at 60 K ("60 K phase" [2.13]). Whether stepwise or broad transmissions were observed – even if prepared under careful stoichiometry control – depends on the thermal treatment. Slow cooling tends to give broadening [2.14, 2.15]. A *phase separation* was indicated by electrochemical determination of the oxygen chemical potential [2.16] and absorption/desorption measurements [2.17]. However, the compositions $YBa_2Cu_3O_{<6.3}$ and $YBa_2Cu_3O_{\geq 6.7}$ were found to appear stable at $T < 500\,°C$, whereas in between (near $x = 6.5$) the isotherms show inflexion points instead of the isopotential behavior which is expected for coexisting equilibrium phases.

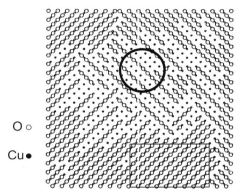

Figure 2.5 Snapshot of partial clustering, showing the nonequilibrium deviations from the ideal oxygen occupation in Cu-O-Cu-O- chains. For numerical simulation see Ref. [2.19]. The core of "ortho II" phase is surrounded by "fully oxidized" ortho I after 5000 Monte Carlo steps per site modeling "polycrystalline samples"

Thermodynamic calculations based on simple interaction models, e.g., in [2.18], clearly show the *dynamic development* of separated phases appearing on nanoscopic scale. Cu^{2+} is arranged in "filled" chains either neighbored by another filled chains, or between two empty chains, or between one empty and one filled chain, or at the head of a chain or at a branching position (Figure 2.5). The different positions could be distinguished by ^{63}Cu NMR, and the qualitative evaluation indicates a mean size of the appearing cumulated defects [2.20] which is on the nanometer scale.

Furthermore, calculations performed for the $T - x$ range covered by Figure 2.3 lead to the conclusion that indeed, a homogeneous composition $YBa_2Cu_3O_{6+x}$ with $0.2 < x < 0.8$ is thermodynamically in a metastable state below about 500 °C. This was also proven by annealing experiments (see, e.g., [2.21]). It should be borne in mind here that compositions with $x > 0.5$ are metastable also with respect to the separation of the Cu-rich double chain phase $Y_2Ba_4Cu_7O_{15}$ [2.22].

The sluggish reactions to achieve thermodynamic equilibrium result from the restricted mobility in the lattice below ~700 °C for cations and below 450 °C for oxygen.

Another type of irregular site occupation in the oxygen sub-lattice appears in combination with certain substitutions on the Cu1 (chain) sites. An important example is the partial substitution by Al^{3+} ions. They arrange in a 4-fold coordination sphere of oxygen in the form of strongly irregular tetrahedra. Breaking and bifurcation of Cu-O-Me-O-Cu chains, clustering of defects, or chain fragments are the consequences, as well as trapping of holes [2.23–2.25].

2.1.5
Phase Relationships in Y-Ba-Cu-O in the Solidus and Liquidus Range

Phase relationships at the liquidus surface have a crucial influence on both melt processing and crystal growth. Primary crystallization fields, shown for systems $1/2(RE)_2O_3$-BaO-CuO (RE = Y, Sm) in Figures 2.6 and 2.7, represent composition and temperature fields on the liquidus surface and indicate the appropriate solid phase which appears in equilibrium with a melt. The composition of the melt corresponds to the coordinates on the axes in the pseudo-ternary coordinate system. Note that the temperature is not constant at the curved liquidus surface although it is exactly determined at any point. In the Figures the liquidus surfaces are projected onto the triangulation of the systems at a fixed subsolidus temperature. Note also that these figures do not provide any information about oxygen content.

The liquidus surface in Y-Ba-Cu-O

The primary crystallization field of $YBa_2Cu_3O_{7-\delta}$ in Figure 2.6 is limited by the equilibrium conditions of four univariant reactions if the chemical potential of oxygen is pre-determined by $p(O_2)$, the partial pressure of oxygen in the environment. The fixed reaction temperature T is given in parentheses for 0.21 bar oxygen partial pressure:

$$m_1 \quad YBa_2Cu_3O_{7-\delta} = k_1\, Y_2BaCuO_5 + k_2\, L(m_1) + k_3\, O_2 \quad (1020\,°C)$$
$$L(m_1) = Y_{0.04}Ba_{0.365}Cu_{0.595}O_{\sim 0.821} \tag{2.8}$$

$$e_1 \quad YBa_2Cu_3O_{7-\delta} + l_1\, BaCuO_2 + l_2\, CuO = l_3\, L(e_1) + l_4\, O_2 \quad (899\,°C)$$
$$L(e_1) = Y_{0.012}Ba_{0.288}Cu_{0.70}O_{\sim 0.787} \tag{2.9}$$

$$p_1 \quad YBa_2Cu_3O_{7-\delta} + n_1\, CuO = n_1\, Y_2BaCuO_5 + n_3\, L(p_1) + e\, n_4O_2 \quad (940\,°C)$$
$$L(p_1) = Y_{0.037}Ba_{0.223}Cu_{0.740}O_{\sim 0.776} \tag{2.10}$$

2 Growth and Melt Processing of $YBa_2Cu_3O_7$

$$p_3 \quad YBa_2Cu_3O_{7-\delta} + n'_1 \, BaCuO_2 = n'_2 \, YBa_4Cu_3O_9 + n'_3 \, L(p_3) + n'_4 \, O_2 \, (991\,°C)$$
$$L(p_3) = Y_{0.004}Ba_{.492}Cu_{.504}O_{-0.851} \tag{2.11}$$

A more detailed investigation following the section Y123 – 011 indicated one more transformation which must be attributed to a peritectic reaction $p_3^\times$ [2.26]:

$$p_3^\times \quad YBa_2Cu_3O_{7-\delta} + n''_1 \, YBa_4Cu_3O_{9'} = n''_2 \, Y_2BaCuO_5 + n''_3 \, L + n''_4 \, O_2 \, (1010\,°C) \tag{2.12}$$

This reaction restricts the temperature range for the primary crystallization of Y123 from Ba-rich melts rather than the peritectic reaction p_3! On the other hand, the equilibrium $p_3^\times$ [Eq.(2.12)], separates a primary crystallization field of Y143, which, however, covers only a very small range of concentration. Therefore, Figure 2.6 does not distinguish between p_3 and $p_3^\times$.

The fundamental "peritectic" reaction (m_1) which corresponds to a cusp on the liquidus surface [2.56] proceeds in a *univariant equilibrium* for all gross compositions corresponding to Y123 or mixtures of Y123 + Y211 because of the fixed composition of the 123 phase in this system. The well-known section in Figure 2.8 represents the reversible decomposition of Y123 at $T(m_1) = 1020\,°C$ under the partial pressure 0.21 bar O_2. The oxygen index in Y123 under this condition is $\delta = 0.74$. Literature data for the peritectic liquidus composition $L(m_1)$ stretch between 0.6 and 3.7 mol% $YO_{1.5}$ in 0.21 bar O_2. Generally, further reduction is observed by about 50% at the reduced oxygen partial pressure $p(O_2) = 0.02$ bar. Liquidus compositions in Eqs. (2.8) – (2.11) refer to [2.27].

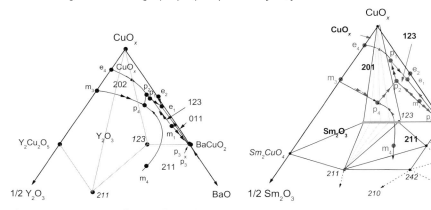

Figure 2.6 Primary crystallization fields of YBCO (according to [2.27] and [2.61])

Figure 2.7 Primary crystallization fields of SmBCO (according to [2.28])

Footnote: Symbols m denote melting (here: incongruent melting) at a local maximum on the liquidus surface, e stands for eutectic and p for peritectic type reactions. They are subsequently numbered in a given system, but not all appear in the considered part of the liquidus. The stoichiometric coefficients $k_1... n''_4$ in each equation refer to the amount of a phase appearing in the reaction normalized to 1 mole $YBa_2Cu_3O_{7-\delta}$. L stands for the melt the composition of which appears in the corresponding equilibrium state.

2.1 Physico-Chemistry of RE-Ba-Cu-O Systems | 41

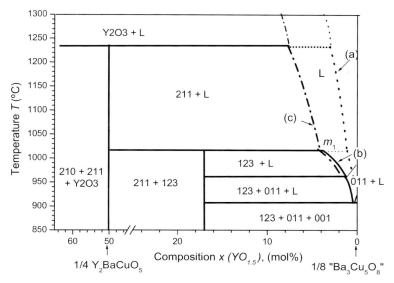

Figure 2.8 The vertical section Y123 – Y211 at 0.21 bar O_2. Experimental data for the liquidus composition stretch between lines a and b obtained by dip-coating (Krauns et al. [2.29]) and soaking (Bieger in [2.27]), respectively. Line c is calculated by Lee [2.30]. The concentration of $YO_{1.5}$ at m_1 must be larger than at e_1 where the value near 1 mol% has been determined by dynamic calorimetry [2.26]. Experience from growth experiments indicates that the solubility of $YO_{1.5}$ at m_1 is close to the limit at e_1

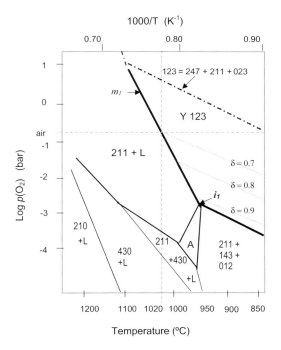

Figure 2.9 Part of the first-order log $p(O_2)$ vs T^{-1} phase diagram of the system Y-Ba-Cu-O, compiling results from [2.9], [2.10], [2.57], [2.58], [2.59], [2.60]. A indicates the 3 – phase region 211 + 012 + L

The influence of $p(O_2)$ on phase equilibria and the phase diagram

The first-order phase diagram log $p(O_2)$ vs. T^{-1} in Figure 2.9 reminds us of the fact that the temperatures of the equilibria m_1, e_1, p_1, and p_3 are univariant and appear *invariant* only at a fixed value of the oxygen partial pressure (i.e. 0.21 bar in Figure 2.8); otherwise their temperatures are influenced by the oxygen chemical potential. It is worth noting that both the fraction of the appearing phases and the composition of the equilibrium melt also vary.

The reaction scheme as represented by Eqs. (2.8) – (2.12), but is limited by the invariant point i_1 (Figure 2.9) which results from the invariant equilibrium i_1 [Eq. (2.13)] at 950 °C and 0.0015 bar oxygen [2.31]:

$$i_1 \quad YBa_2Cu_3O_{7-\delta} = k^*_1 Y_2BaCuO_5 + k^*_2 BaCu_2O_2 + k^*_3 YBa_4Cu_3O_{11} + k^*L(i_1) + k^*_5 O_2 \quad (2.13)$$

The stability of Y123 at $p(O_2) < 0.0015$ bar is determined by the univariant *solid state reaction* (2.14).

$$sr \quad YBa_2Cu_3O_{6+x} = 4/9\, Y_2BaCuO_5 + 10/9\, BaCu_2O_2 + 1/9 YBa_4Cu_3O_{8.5} + (1/36)(11+18x)\, O_2 \quad (2.14)$$

An important conclusion with respect to melt processing is that the melt – from which Y-123 can be grown – is always depleted in both yttrium and oxygen. In air, the amount of available oxygen from the environment is sufficient and the diffusion of yttrium through the melt becomes process-limiting.

On the other hand, as a general tendency for CuO and Cu(II) cuprates, the oxygen content in the liquid is remarkably less that in the solid phase in equilibrium. Thus, the composition of solid Y123 *in air* at the peritectic point m_1 is $YBa_2Cu_3O_{6.3}$ (i.e. $YO_{1.5} \cdot 2BaO \cdot 3CuO_{0.933}$), whereas the oxygenation state of Cu in the peritectic melt corresponds to a hypothetical oxide $CuO_{0.65}$. The equilibrium oxygen concentrations and oxygenation state of Cu in both the liquid and 123 phase in the univariant peritectic equilibrium m_1 are *unique functions* of the chemical potential $RT\ln p(O_2)$, (Figure 2.9). Therefore, the stoichiometry of the crystal at the peritectic point [also $T(m_1)$] depends on $p(O_2)$ in the environmental gas (under extreme conditions: on altitude [2.56] and weather!).

2.1.6
Phase Relationships and the Liquidus Surface in Systems Ln-Ba-Cu-O (Ln = Nd, Sm,..)

The formation of solid solutions $Ln_{1+y}Ba_{2-y}Cu_3O_{7\pm\delta}$ is characteristic for the large ions of the light lanthanides (Figure 2.1). A typical temperature maximum appears in the $T - x$ representation of the homogeneity range, which corresponds to an apex in the primary crystallization field on the liquidus surface. The topological situation is equivalent to the thermodynamic condition $(\partial G/\partial X) = 0$ with the concentration vector X. Consequently, the number of free parameters (*Gibbs phase rule*) is reduced by 1, and *only here*, the "peritectic" reaction m_1' proceeds as a quasibinary and congruent reaction i.e. without change of molar ratios $LnO_{1.5}$: $BaO:CuO_x$ on both the left and the right hand sides in Eq. (2.15):

$$m_1' \quad Ln_{1+y}Ba_{2-y}Cu_3O_{7-\delta} \rightleftarrows k^{\#}_1 \, Ln_2BaCuO_5{}^a + k^{\#}_2 \, (Ln_\alpha \, Ba_\beta \, Cu_{1-\alpha-\beta})_{liquid} \quad (2.15)$$

The invariant temperature $T(m_1')$ decreases with decreasing ionic radius from La^{3+} to Lu^{3+} (Table 2.1).

Careful investigations of the phase realationship in correlation to structural and superconducting parameters [2.32] have proven that for $Ln = Nd$ the maximum appears with a certain deviation from the 1:2:3 stoichiometry (y ≅ 0.12 at 0.21 bar O_2). The stoichiometric composition y = 0 appears within a limit ±0.02 near the phase field boundary in a certain temperature range (between 980 and 1030 °C for $Nd_{1+y}Ba_{2-y}Cu_3O_{7\pm\delta}$). Recent results on $Sm_{1+y}Ba_{2-y}Cu_3O_{7\pm\delta}$ agree with the situation in Nd-Ba-Cu-O [2.28]. Below this temperature range, $Ln_{1.0}Ba_{2.0}Cu_3O_{7-\delta}$ decomposes into $BaCuO_2$ and Ln-rich Ln123ss, and a peritectic decomposition results in Ln211 (or Nd 422) + melt, leaving also Ln-rich Ln123ss.

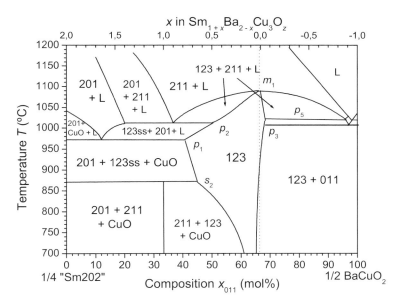

Figure 2.10 The vertical section "$SmCuO_2$" – $BaCuO_2$, involving the solid solution $Sm_{1+y}Ba_{2-y}Cu_3O_{7-\delta}$, $p(O_2) = 0.21$ bar. Reprinted from [2.28] with permission from Elsevier

a   or $Ln_4Ba_2Cu_2O_{10}$ in the case of Nd

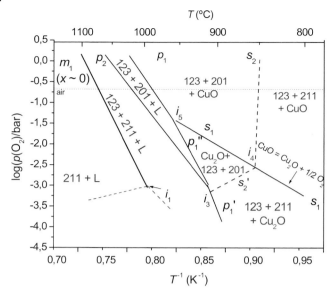

Figure 2.11 The 1st order representation log $p(O_2)$ vs T^{-1} of the $Sm_{1+y}Ba_{2-y}Cu_3O_{7-\delta}$ phase stability field. Reprinted from [2.28] with permission from Elsevier

The extension of this homogeneity range will be *remarkably reduced* if the chemical potential of oxygen decreases. Thus – under reduced $p(O_2)$ – the composition within the $RE123$ phase, which is in equilibrium with a melt, must be close to the stoichiometry 1:2:3.

On the other hand, the exceptional properties observed with $RE_{1+y}Ba_{2-y}Cu_3O_7$ solid solutions prompted speculation about a phase separation into a "336" $RE_3Ba_3Cu_6O_{13}$ phase as well as 123 ("246"). Although short-range ordered structures were detected by EXAFS in the intermediate composition range in polycrystalline samples prepared under special cooling conditions, they never were obtained in bulk crystallized materials. This fact indicates that preparation conditions for several phases – if not appearing under equilibrium conditions – should be described by Time-Temperature-Transition diagrams [2.33] rather than by the equilibrium state. Alternatively, the phase stability in the system Sm-Ba-Cu-O can also be represented in a log $p(O_2)$ vs T^{-1} first-order phase diagram to visualize the behavior under variable $p(O_2)$. Note that the composition is not fixed in Figure 2.11, which represents the equilibria at the Sm-rich boundary of Sm123. Reaction m_1 is univariant (a line in the representation) for y = 0, while the composition y in Sm123 increases from m_1 to p_2.

(The situation on the Ba-rich boundary and its impact for processing will be discussed in Chapter 7).

The liquidus surface of Ln-Ba-Cu-O: Ln = Nd, Sm,..

The topology of the liquidus surface near the primary crystallization field of Sm123ss (Figure 2.7) appears equivalent to that of Y-Ba-Cu-O despite the appearance of the solid solution of Ln123. In the Ba-poor part, however, crystallization of Ln_2CuO_4 appears for Ln = Nd, Sm instead of the primary crystallization of $Y_2Cu_2O_5$. Furthermore, the solid solution $Nd_{4-x}Ba_{2+x}Cu_2O_{10}$ (Nd422) appears in Nd-Ba-Cu-O instead of the 211phase. This, however, is of minor influence on the topology and on the appearing primary crystallization field for $Nd_{1+y}Ba_{2-y}Cu_3O_{7\pm\delta}$, obviously due to the fact that the stoichiometric ratio in the solid solution which appears in the quasibinary reaction m_1 (i.e. on the apex at the liquidus surface) is exactly 2:1:1, i.e. the same as for Y or Sm. On the other hand, a phase Nd143 does not appear in the Ba-rich region, and it is still unclear which phase appears instead, either Nd163 or a solid solution $(Ba,Nd)_2CuO_3$ [2.62].

2.1.7
Additional Factors

Any further admixed component will influence the activities of the main constituents Y(Ln)BaCuO, which may even result in changes of the topology of the corresponding phase diagram and may consequently affect the processing. This is especially the case if Ag is admixed. Therefore, phase relationships in the presence of Ag will be discussed in relation to processing in a separate chapter.

Numerous admixtures have also been reported to promote the "melt texturing" process, like Pt, CeO_2, etc. which – if homogeneously distributed and below an individual threshold composition – have only a marginal influence on the phase relationships. Therefore, they can often be neglected when considering thermodynamic aspects of processing, despite the fact that they have in certain cases a considerable influence on surface/interface energies and/or the viscosity of the appearing melt.

2.2
Preparation of Polycrystalline *RE*123 Materials

2.2.1
Synthesis of HTSC Compounds

With the exception of thin films, the preparation of high-T_c superconducting (HTSC) materials – powdered or bulk – needs processing with at least one high-temperature step. The most important information about the parameters which limit the conditions for the existence of the superconductor, its coexistence with secondary phases, and its chemical composition within a homogeneity region, is represented by the phase diagram.

The conventional preparation route for $YBa_2Cu_3O_{7-\delta}$ starts from homogenous mixtures of powdered Y_2O_3, a Ba compound (frequently $BaCO_3$) and a Cu compound (typically CuO). The blend is carefully ground and mixed and fired at temperatures below those leading to the appearance of a (partial) melt (< 900 °C in air). Grinding, homogenization, and firing have to be repeated until the pure phase is formed. Although this technique is also transferable to very large batches, alternative precursors or alternative methods have been applied such as co-precipitation, spray drying, freeze drying, sol-gel techniques, ignition techniques, and spray pyrolysis. Some of them are scalable to large scale synthesis. The last two processes proceed at high temperatures, whereas the others need at least one additional high temperature calcination and reaction step.

The powders from different providers or laboratories differ, sometimes remarkably, with respect to grain size, phase purity, and impurity content. Furthermore, T_c depends on the final oxygenation of the as-prepared powders.

The preparation of $RE123$ phases proceeds analogously. The appropriate temperatures have to be changed slightly according to the individual phase diagrams (next chapter). Special care has to be taken of the sample stoichiometry in compounds which form solid solutions $RE_{1+y}Ba_{2-y}Cu_3O_{7-\delta}$. Here, the composition which can be achieved in a single phase powder is controlled by thermodynamic restrictions of the phase diagrams. Superconductivity in the precursor powder is not relevant to its suitability for the melt process; it can, however, give indications about parameters like homogeneity, stoichiometry, and purity.

2.3
Growth of $YBa_2Cu_3O_7$ Single Crystals

Two representative examples of the liquidus surface were shown in Figures 2.6 and 2.7 for Y-Ba-Cu-O and Sm-Ba-Cu-O, respectively. The fields of primary crystallization designated as 123 indicate the melt compositions from which single crystals of $YBa_2Cu_3O_{7-\delta}$ or $Sm_{1+y}Ba_{2-y}Cu_3O_{7\pm\delta}$ can be grown. It should be emphasized that solidification of these cuprates proceeds incongruently with respect to the oxygen content also, since the melt is poor in oxygen and consequently the gas atmosphere has to serve as an oxygen reservoir.

The crystallization path starting from a composition marked by an arrow runs along the extended tie line between the arrowhead and the composition 123 towards the RE(Y, Sm,..)-rich boundary line of the primary crystallization field. The latter is then followed until the eutectic point e_1 is achieved. The corresponding liquidus composition contains $YO_{1.5}$ only in amounts of less than 3.8% (0.6%), and the $YO_{1.5}$ concentration in the eutectic composition is in the order of 1% (0.1%). Scattering values (indicated by the values in brackets) result from different authors applying various methods of investigation (see Section 2.1). Therefore, even in the optimal case when the ingot composition is on the tie line between 123 and e_1, the theoretical yield of Y123 is only 16% (2.5%). Continuously feeding the melt of rather limited volume is used in the crucible-free "travelling solvent" methods, which unfortunately resulted in crystals of only restricted quality and size – at least for $YBa_2Cu_3O_7$ – because of the strong wetting and creeping of the Cu-rich melt. Alternatively, the top seeding solution growth (*Czochralski* technique) with a reservoir of RE_2O_3 is a powerful method, which was developed by *Shiohara* and co-workers. Preferably, the appropriate RE_2O_3 ceramics are used as the crucible material itself [2.34]. A low concentration of Y or *Ln* and the incongruent behavior make the crystallization process strongly dependent on diffusion, and therefore the process needs an extremely low rate.

Crystallization rates increase in a peritectic solidification

$$L + a \leftrightarrows \gamma \qquad (2.16)$$

when the crystallization of the γ phase $REBa_2Cu_3O_{6+x}$ proceeds from the peritectic liquid L in which RE_2BaCuO_5 – the peritectic a phase – is suspended. Single crystals have been grown for RE = Y, Nd,..., applying a traveling solvent technique. Although the use of rods of diameter 6 mm and length 27 mm have been reported [2.35], the size of single crystalline grains is only a few mm.

The morphology of the growing 123 – liquid interface is strongly dependent on temperature gradient G_T and growth rate R. It was shown that a ratio $G_T/R > 3300$ K h cm^{-2} is required to obtain planar interfaces, corresponding to growth rates between 0.1 and 0.5 mm/h and temperature gradients from 10 to 50 K/mm, respectively [2.36]. Exceeding the critical rate results in the formation of defects, dislocations, and grain boundaries, which – if they appear in an uncontrolled fashion – are detrimental to the investigation of fundamental properties with single crystals.

The progress in growing high-quality single crystals to be used, e.g., for substrate materials or fundamental research has been reviewed in Ref. [2.34]. Different techniques for crystals growing from the peritectic system have also been applied, such as zone melting, directional cooling, or pulling. Although the process of peritectic solidification is less suitable for growing high-quality single crystals, it is the approved basis for bulk materials technology because of its advantages with respect to technology and its capability for scaling up, which will be treated in the following paragraphs.

2.4
Processing of "Melt-Textured" YBaCuO Bulk Materials

2.4.1
Experimental Procedure

The *melt-textured growth* (MTG) process introduced by Jin et al. [2.1] is based on the reversible reaction m_1 proceeding on the *Y (or Ln)-rich boundary* of the primary crystallization field of 123, as is also the case for peritectic growth of single crystals. Indeed, the boundary between "semi-solid crystal growth" and the "melt-texturing processes" is not sharp. The MTG process originally started from a body of pre-sintered Y123 powder, which was heated to temperatures above the peritectic temperature $T(m_1)$ and then slowly cooled through the peritectic temperature. The temperature vs time schedule is shown in Figure 2.1. The crystallized material typically contains several large grains of the Y123 phase with microscopic inclusions of Y_2BaCuO_5 (211 phase) and some polycrystalline material solidified from a segregated melt. Salama et al. [2.37] observed the particles of Y211 refined by reducing dwell temperature and time when the mixture was kept above $T(m_1)$, (MMTG, Figure 2.12). The large grains have a preferential orientation originating from the applied temperature field.

Further modifications were directed to finer powders used in the precursor state, to control size and distribution of 211 inclusions, such as "Quench-Melt Growth" (QMG) and "Melt-Powder-Melt" Growth (MPMG), characterized by a quenching of the 211+melt mixture (QMG) and intermediate grinding of the quenched product (MPMG) before the powder is pressed to a cylinder or rod, which then undergoes the final growth step [2.39]. These modifications – like the standard process – are characterized by applying a starting material of Y123 powder or any precursor mixture the gross composition of which corresponds to n parts Y123 + m parts Y211. Therefore, the conventional melt texturing proceeds in a first approach on line with the projection through the points Y211 and Y123 in the Gibbs triangle (compare Figures 2.6 and 2.8). A breakthrough was achieved by setting a seed on top of the pressed body, controlling nucleation and orientation of the grown material.

A standard process can be summarized as follows. A precursor mixture 123 + 211 is mixed and usually blended with certain additives like Pt or CeO_2. The

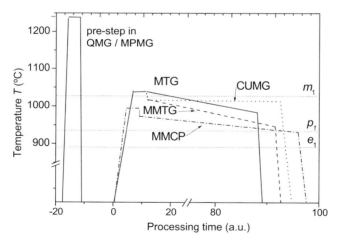

Figure 2.12 Temperature vs time schedule for variants of melt processing: MTG Melt Texturing Growth (original variant) [2.1], MMTG Modified Melt Texturing Growth [2.37], CUMG: Constant Undercooling Growth [2.38], MMCP: Modified Melt Crystallization Process with an alternative solidification path [2.50]. Partial melt-assisted precursor preparation is typical for Quench Melt Growth (QMG) and MPMG. The latter is the modification with intermediate grinding [2.39]

"green body" is pressed to a cylindrical or parallelepipedic shape and heated to a temperature about 10 to 30 K above the peritectic equilibrium temperature $T(m_1)$. Thereby, the appearing melt is soaked between the network of solid Y211 particles. After a short dwell time, a body which follows the MMTG line in Fig. 2.12 will be rapidly cooled to a temperature $T(m_1) - \Theta$ which is below the equilibrium temperature (Figure 2.8). Θ represents the undercooling, which causes a supersaturation in the melt with respect to the equilibrium state. It is now a standard technique to apply seed crystals (e.g. Sm123, MgO) to control the orientation. Seeding initializes the growth with less supersaturation, thus avoiding parasitic nucleation.

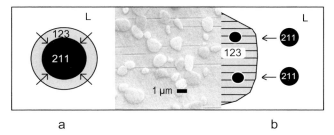

Figure 2.13 Illustration of the two mechanisms proceeding according to Eq. (2.16): **a** peritectic reaction and **b** primary crystallization. The micrograph of a polished section parallel to (100) in a properly melt grown material corresponds to schedule **b**

The solidified bodies are then cooled down to about 400 or 500 °C and treated with oxygen gas to adjust the oxygen stoichiometry to give δ between 0.02 and 0.07. Figure 2.13 illustrates that a large single grain bulk superconductor can grow only by primary crystallization of the γ phase from the melt. This process is determined by the diffusivity in the liquid and interface kinetics, in contrast to a reaction at the solid–solid interface.

The outlined process conditions have been optimized to assist this primary crystallization mechanism.

Figure 2.13 shows a cylindrical YBCO sample as grown. The seed is still on top (see also the cover). Typical (110) facets appear on the circumference, and four sectors independently in <100> directions are visible (*a*-growth sectors) on the photo.

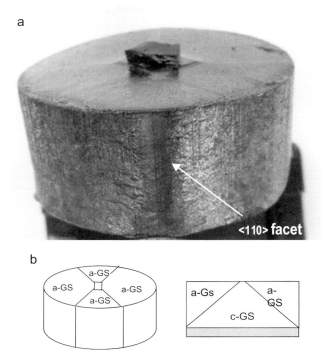

Figure 2.14 Photograph of a cylindrical YBCO sample as grown, and sketch of the cross section indicating *a*-growth and *c*-growth sectors

2.4.2
Mass Flow, Growth Rates, Kinetic and Constitutional Undercooling

The crystallization process can be considered as a sequence of virtual process units such as mass transport from the 211/422 source, nucleation, interface reaction, and growth, the driving force of which is the excess ∂G of Gibbs free energy related to the equilibrium state. This excess ∂G corresponds to processing parameters, which are supersaturation Δc and undercooling Θ on the concentration or temperature scales, respectively. Continuous primary crystallization from the incongruent peritectic melt can proceed only under steady-state conditions, which result from the balance between mass flow through the growing interface ("Stefan flow" j_γ) and the flow originating from the dissolving 211 particles of the peritectic a-phase. As a prototype, the situation is schematically shown for the simple peritectic system in Figure 2.15a, applied for melt texturing of $YBa_2Cu_3O_7$ (or Y123-Y211 mixtures) or Ln123 phases which do not form a solid solution. In this case, the corresponding composition ratio in the equilibrium melt is fixed at the peritectic temperature $T(m_1)$ (i.e. 2 BaO : 3 CuO inYBCO at 1020 °C in air), and the situation is illustrated by Figure 2.15. The concentration of the melt in equilibrium with the dissolving 211 particle (the "source") is $c(L_{m1})$. On the other hand, the influence of interface kinetics has to be taken into account, as may be concluded from the appearance of faceted faces (Figure 2.14).

In such a case, mass flow across a diffusion boundary layer, thickness b, is commonly accepted to be representative for the net flow:

$$j_b = Db^{-1}\left[c(L'_{211}) - c(L^0_{123})\right] \tag{2.17}$$

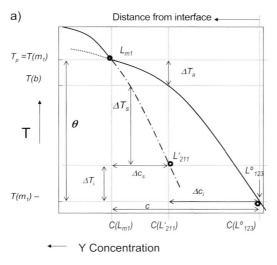

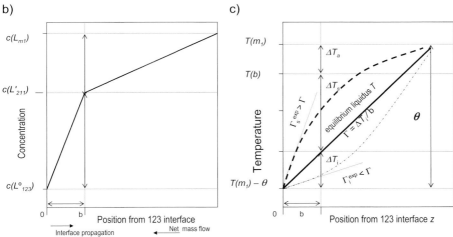

Figure 2.15 Schematic representation of the interrelationship between phase diagram and concentration gradients near the interface

Here, $c(L_{123}^0)$ is the equilibrium liquidus concentration at $T_0 = T(m_1) - \Theta$ and $c(L'_{211})$ is the metastable liquidus concentration of the supercooled 211 phase at distance b (i.e. on the metastable branch of the 211 liquidus line). The corresponding composition profile is shown in Figure 2.15b. D is the diffusion coefficient.

The moving growth front corresponds to the net mass flow j_γ according to (2.18)

$$j_\gamma = R\left[c_{123}^s - c(L_{123}^0)\right] \qquad (2.18)$$

which – in the steady state – has to be balanced by j_b:

$$R\left[c_{123}^s - c(L_{123}^0)\right] = \frac{D}{b}\left[c(L'_{211}) - c(L_{123}^0)\right] \qquad (2.19)$$

Here, R is the linear growth rate and c_{123}^s the volume concentration in the solid 123 phase.

Concentrations in Eqs. (2.17) – (2.19) have to be considered as k-tuples, i.e. the concentration has to be balanced *for each component*. Therefore, the stoichiometry of the growing 123 (γ-) phase is determined by the actual steady state mass flow concentration and becomes a function of the progress in growth. This is essential for those Ln123 which have a homogeneity range. In the case of Y123, however, the steady state growth rate is unambiguously represented by the flux of Y, the minority component in the melt. Figure 2.15c represents a sketch of the virtual temperature profile with the gradient $\Gamma = (dT/dz)$ near the interface, which, *under equilibrium conditions,* is represented by Eq. (2.20).

$$\Gamma = \Delta T_i\, b^{-1} \qquad (2.20)$$

ΔT_i is the difference between the temperature at the diffusion layer boundary and the interface temperature T_i in accordance with Figure 2.15b.

The virtual temperature profile is in accordance with the schematic phase diagram of Figure 2.15a and has to be compared with an experimental temperature profile (gradient Γ_{exp}). Whereas Θ is the total undercooling, representing the overall driving force δG, the quantity ΔT_i and the corresponding $\Delta c_i = \left[c(L'_{211}) - c(L_{123}^0)\right]$ are attributed to the "kinetic undercooling" or "interfacial supersaturation", respectively. They represent the driving force for interface reactions $\delta G_i - RT\Delta c_i$. The thickness of the diffusion boundary in melt-texturing experiments is assumed to be in the order of the mean spacing between 211 particles (Cima et al. defined half the particles spacing [2.40]).

The appearing ΔT_a and $\Delta c_a = \left[c(L_{m1}) - c(L'_{211})\right]$ are representative of the force driving the diffusion through the bulk melt and can be neglected for a high density of 211 particles. It should be noted that oxygen is not balanced in Eqs. (2.17) – (2.19) and will be treated separately.

A contribution ΔT_s remains from the total undercooling Θ if the experimental gradient Γ near the interface is less than the gradient representing the equilibrium curve:

$$\Delta T_s = \Theta - \Delta T_a - \Delta T_i = \Theta - \Delta T_a - b\Gamma \tag{2.21}$$

The quantity ΔT_s represents a *constitutional undercooling*, i.e. the suppression of the actual temperature at the boundary of the diffusion layer *below the equilibrium temperature* on the liquidus line (Figure 2.15c).

2.4.3
Developing Microstructures: Morphology, Inclusions, Defects

Crystal Morphology

A causal correlation exists between anisotropy of growth rates and crystal morphology, resulting in the appearance of large faceted faces for slowly growing directions and vice versa. The linear growth rate is correlated with the kinetic undercooling according to Eq. (2.19). Experimental investigations revealed higher growth rates in the a direction than in the c over a wide range of rates R and undercooling Θ (the experimental parameter instead of Δc_i) and an overlap at about $\Theta = 18$ K and $R \cong 6$ mm/h with the potential laws for R_a and R_c [2.36] $R_a = 0.45 \times 10^{-6} \Delta T^{1.9}$ (mm/s) and $R_c = 2.8 \times 10^{-6} \Delta T^{1.3}$ (mm/s). The nearly quadratic dependency of ΔT on R_a on the one hand (as for a spiral growth mechanism), and on the other hand the typical appearance of small facets of (110) type on the circumference of cylindrical melt textured samples indicate a more complex "mix" of mechanisms. Furthermore, the experimental ratio R_a/R_c – even at its maximum – remains below the value 1.3, thus conflicting with the values between 5.4 and 7.3 predicted from the "periodic bond chain" (PBC) theory [2.41]. The discrepancy should be attributed to the strong influence of the chemistry and kinetics in the melt. A wide range of aspect ratio (from 0.5 to 15) found in self-flux grown *single crystals* was attributed to different growth mechanisms [2.43].

Constitutional undercooling

Constitutional undercooling (ΔT_s) is accepted to affect essentially the morphologic stability of growing single crystals. Morphological instabilities such as cellular or blocky growth appear if the actual temperature in the liquid ahead of the growth front T_L remains below the equilibrium temperature T_E, which corresponds to the actual concentration in the liquid at the same distance. Thus, the relation $\Gamma_{\exp} < \Delta T_i\, b^{-1}$ is the criterion for the appearance of morphologic instability. In this case, any bump appearing occasionally on the surface will grow faster, thus giving rise to an irregular pattern. Accordingly, growth conditions to obtain a planar interface (i.e. a single crystal) are restricted to low growth rates and large temperature gradients.

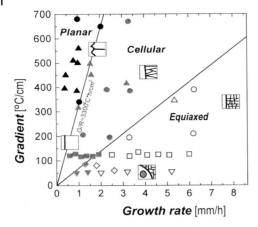

Figure 2.16 Morphology types and stability limits for peritectic crystallization of $YBa_2Cu_3O_{7-\delta}$ [2.34]

Results on growth stability in Figure 2.16 were originally obtained for peritectic crystal growth with little or no excess of 211 phase ([2.37, 2.40]). They have been reproduced at least qualitatively also for the melt-texturing technique. Peculiarities of the latter – which also starts from the "semi-solid" liquid – will be emphasized here:

1. Containerless processing of a semi-solid body with an excess of the peritectic a-phase
2. No pulling applied (in most cases); therefore the internal temperature gradient and actual undercooling are "self-organizing" ("Stefan problem")
3. Indirect control of growth rate and undercooling by external temperature field, cooling rate, or constant external undercooling.

Particle inclusions during melt texturing

Bulk melt-textured materials are composites consisting of the superconducting phase and particles of the peritectic a-phase (211 or 422). Although the peritectic phase is reactive during crystallization, an excess (even if only local) of 211 particles near the growth front behaves like an inert particle undergoing the "push"–"trap" criterion as investigated as a fundamental problem in crystal growth [2.42]. Figure 2.17a schematically shows the resulting force acting on a small particle (radius r) in front of the interface (growing at a rate R) and the viscous flow of the melt, which causes the drag force F_D (depending on flow rate and gap between particle and interface). The force F_i is the interface energy resulting from the sum $\sigma_{s-p} - \sigma_{l-p} - \sigma_{s-l} = \Delta\sigma_0$, the interfacial energies matrix–particle, liquid–particle and matrix–liquid, respectively.

Particles will be pushed from the growing crystal if $F_i > F_D$, whereas otherwise particles will be trapped (i.e. in all systems with $F_i < 0$ particles). Theoretical analysis results in the criterion to distinguish trap–push regions given in the relation

$$R^* \propto \frac{\Delta\sigma_0}{\eta r^*} \tag{2.22}$$

which expresses inverse proportionality between a critical growth rate R^* and a critical particle size r^* (assumed to be spherical; η is the melt viscosity), dividing the parameter field into "pushing" (preferably for small particles) and "trapping" (high rates) (Figure 2.17b). Trapping appears if a critical size is exceeded. Figure 2.17 illustrates schematically experimental results showing that the push–trap boundary is shifted to larger critical particle size for c growth direction compared with a-grown materials. The critical particle size is drastically decreased because of increased undercooling and growth rate, leading to trapping of small particles at $\Delta T = 30$ K which would be pushed at $\Delta T = 10$ K. In the adverse case (e.g. large excess of 211 small particles – as would be desired for physical properties), pushing may result in complete blocking of the interface diffusion due to a thick layer of 211 particles.

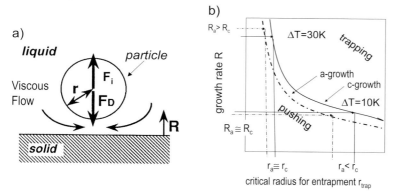

Figure 2.17 Push–trap discrimination of small particles. **a** Forces discriminating the push–trap behavior of a particle near the crystal–liquid interface. **b** Influence of growth direction on push–trap boundary [2.36]

Even though thermophysical data of the melt are pending, the qualitative correlation to the expected behavior for interaction between liquid and solid is surprising, especially remembering the fact that the semisolid blocks are very shape-resistant during all processing steps, i.e. far from the behavior of a fluid medium. On the other hand, the critical radius still allowing trapping of particles at a reasonable growth rate increases with increasing $\Delta\sigma_0$, (Figure 2.17b). This is the case when interfacial energies between solids and liquid become weaker. The surface energy can be significantly influenced by adding even small amounts of "surfactants". This appears with Pt additions, as found occasionally at an early stage by using Pt crucibles, when Pt is dissolved in the melt in oxidized form; it can also usefully be added as Pt-containing compounds. Although numerous additives have been tested, a quantitative investigation of interface properties is still pending. The same status still exists for viscosity data, which also have a crucial influence on growth properties. Impurities from other crucible materials etc. are also known to affect shape and morphology. On the other hand, additives – even in small concentrations – have also been investigated as "dopants" (Chapter 6). Unfortunately, many papers do not verify whether additives really influence the chemistry of the solid or the properties of the melt or the interface between the two.

Size of 211 particles

An additional contribution δG_σ to the driving force of peritectic solidification resulting from the excess interfacial energy of small particles was analyzed in Ref. [2.44]. Gibbs free energy of small particles steeply rises compared with the standard state. Therefore, the chemical potential of 211 will increase if particles with a small radius of curvature are involved. The consequences are:

1. Small particles of the peritectic a phase 211 will be preferably dissolved in the peritectic and high temperature melt;
2. The rising supersaturation at $T(m_1)$ increases recrystallization of 211 particles by *Ostwald* ripening, resulting in the coarsening of 211 particles.
3. The low end of the particle size distribution in the solidified body will be cut (additionally to the effect of pushing).

Microstructure and the interface 123–211

Inclusions of 211 (or 422) phase particles are the most striking defects to be visible by optical or scanning electron microscopy (SEM). Though the majority of 211 inclusions belong to grain size fractions > 1μm, TEM investigations also reveal – depending on the applied growth method – a number of 211 grains of size between 10 and 100 nm.

Further, optical microscopy in polarized light clearly reveals the formation of micro-twins in the orthorhombic 123 phase after oxygenation [2.45]. The angle between twin stripes easily permits us to determine the orientation of the crystallite. Furthermore, it indicates the progress of oxygenation in large bulk samples.

Typical defects in melt-textured materials are dislocations and low-angle grain boundaries. Figures 2.18 and 2.20a show the typical behavior of a tilted grain boundary which results from an agglomeration of dislocations (from an HREM study) [2.46]. Screw dislocations in melt-grown YBCO have also been illustrated (Figure 2.19). These are typical features of a two-dimensional nucleation step, but can also result from strain [2.47]. A further type of two-dimensional defects appears as stacking faults (Figure 2.20b, [2.63]), which may originate from fluctuations of concentration or temperature during the growth or from decomposition of metastable phases during post-annealing treatments. Their appearance is supported by the small change in Gibbs free energy between certain polytype-like phases, as was the case for the intergrowth of $YBa_2Cu_3O_7$ with double-chain fragments which regularly appear in the structures of $YBa_2Cu_4O_8$ or $Y_2Ba_4Cu_7O_{14}$ [2.48].

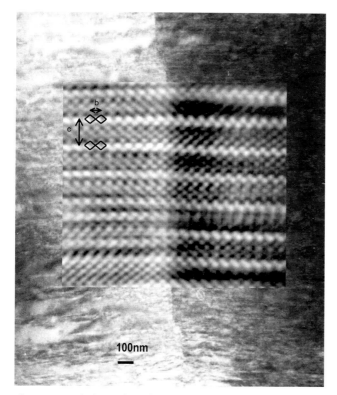

Figure 2.18 Tilted grain boundary in melt-grown $YBa_2Cu_3O_{7-\delta}$: HRTEM image [2.46]

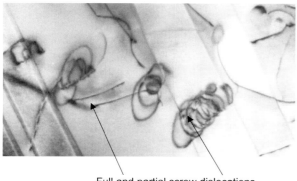

Full and partial screw dislocations
(generated under strain in 45° to (001))

Figure 2.19 TEM image indicating the presence of screw dislocations [2.47]

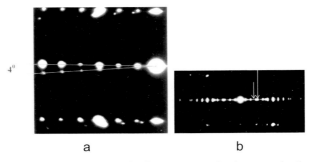

Figure 2.20 SAED pattern of bulk YBCO material indicating **a** the tilted low-angle grain boundary [2.46] and **b** "$Y_2Ba_4Cu_7O_{14}$" stacking fault [2.63]

In spite of incoherent crystal structures of RE123 with the corresponding 211 (or 422) phases, a number of mutual orientations with a commensurate grain boundary exist along the interface between the Y123 matrix and the strongly curved particles of 211 phase. The Fourier filtered pattern from selected area electron diffraction is displayed for a pair from (100) matrix and (021) inclusion planes in Figure 2.21 [2.49]. The Figure enables us to define a minimum transition range from the undisturbed regular matrix, limited by g_1 to the undisturbed 211 inclusion with the (tangential) boundary g_2. The angle between g_1 and g_2 is 11° and the width of the transition area in the considered case is 1 nm. Typical transition areas do not exceed a few nm, and the transition is realized by a sequence of dislocations.

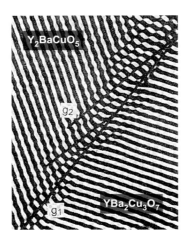

Figure 2.21 Sequence of planar defects at the interface between 123 matrix and 211 grain (Fourier filtered TEM) [2.49]

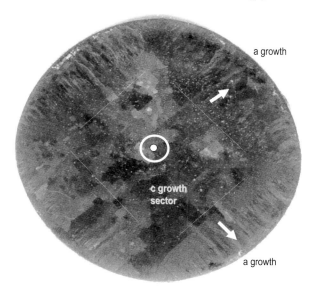

Figure 2.22 Cross section through a cylindrical YBCO bulk sample indicating the mosaic structures appearing in different growth sectors

Growth sectors and mosaic structure

Sections through the bulk material clearly exhibit different growth sectors. Typically one "*c*-growth sector" and four "*a*-growth sectors" appear in *c*-axis-oriented YBCO cylinders or plates (seeded with RE 123) according to the growth front propagating along <001> or <100>, respectively. The opening angle in the *c*-growth sector represents the growth rate ratio between *c*- and *a*-growth direction:

$$R_c/R_a = \tan(a/2) \tag{2.23}$$

Therefore, the extension of the *c*-growth sector (angle *a*) can be controlled by undercooling. Furthermore, the relationship in Eq. (2.8) implies a different distribution of 211 particles in *a*- and *c*-growth sectors since trapping is favored by faster growth rates.

Typical mosaic-like subgrain structures appear on both *a*- and *c*-growth sectors (Figure 2.22). In most cases the boundaries between subgrains behave as low-angle grain boundaries [2.45, 2.48]. The subgrain boundaries result from twisting around the *c*-axis (*c*-growth sector) or tilting the *c*-axis (*a*-growth sectors). The size of subgrains increases with distance from the seed, and secondary low-angle grain boundaries appear therein, indicated by increasing half-widths of local rocking curves with increasing distance from the seed. On the other hand, a kink-like pattern near the surface and narrow X-ray rocking curves along (110) indicate a terrace–kink growth following leading edges.

The mosaic structure can easily be observed on cleaved (not polished) surfaces of the as-grown tetragonal bulk material, thus proving their origin from the growth process. Obviously, the coincidence of a high density of dislocations near the 123–211 interface and the existing constitutional undercooling support the formation of the mosaic structure in bulk materials grown by melt-texturing methods.

2.5
Modified Melt Crystallization Processes For YBCO

2.5.1
Variants of the $YBa_2Cu_3O_7$–Y_2BaCuO_5 Melt-Texturing Process

Modifications of the MTG were directed to finer powders used in the precursor state to control size and distribution of 211 inclusions. These included "Quench-Melt-Growth" (QMG) and Melt-Powder-Melt-Growth (MPMG), characterized by a quenching of the 211 + melt mixture (QMG) and intermediate grinding of the quenched product (MPMG) before the final growth step [2.39]. These modifications – like the standard process – are characterized by using a starting material of Y123 powder or any precursor mixture the gross composition of which corresponds to n parts Y123 + m parts Y211.

Modifications may also differ in the applied precursor materials (the proper ratio is sometimes realized by mixing appropriate amounts of CuO, $BaCO_3$, and Y_2O_3). Furthermore, modified techniques were used, such as directional solidification and the Bridgeman and zone-melting techniques [2.34].

2.5.2
Processing Mixtures of Y123 and Yttria

In this modified melt crystallization process (MMCP), the green body is prepared from a mixture of Y123 and nY_2O_3. The objective of MMCP is to define conditions which permit improved stability for growth, avoiding the influences of constitutional undercooling and fluctuations of concentration and local temperature [2.50]. Figure 2.23 shows the vertical (polythermal) section 1/6 Y123 – 1/2 Y_2O_3 at $p(O_2)$ = 0.21 bar. One important feature of this figure is that Y_2O_3 and Y123 cannot coexist in equilibrium. Y_2O_3 will react with a part of the Y123 precursor powder pressed into the green body in an irreversible solid-state reaction in the pre-heating ramp before the solidus line will be achieved. Thereby, CuO and Y211 are formed the latter as small grains and homogeneously distributed in the pressed pellet. The CuO excess leads-after melting-to a Cu rich melt.

The second feature concerns the different situation during the solidification from the 211 + liquid state. It can easily be seen that the "peritectic" reaction which is responsible for the crystallization no longer proceeds with invariant temperature; the peritectic crystallization of an initial composition (marked by the arrow in Figure 2.23) will proceed passing the borderline from the equilibrium field Y211 + L toward the field Y211 + Y123. This fact opens a significantly enlarged process window for stable growth conditions. On the other hand, the composition of the equilibrium melt depends on the part of Y123 in the ingot. The optimized formula in MMCP contains $n = 0.2$–0.4 mole yttria per 1 mole Y123 phase. Admixtures of, e.g., Pt or CeO_2 are applied, which influence the properties of the melt (viscosity), and act even at low concentrations as a "surfactant", influencing the interface energies between liquid and 123 or 211. The latter is of importance with respect to trapping and coarsening of 211. Sm123 seeds are applied as usual. Another advantage of seeding is that the supersaturation in the melt can be kept small and nearly constant during the ramping down because of the slowly decreasing equilibrium temperature. The crystallization proceeds permanently close to thermodynamical equilibrium, which results in a controlled solidification process.

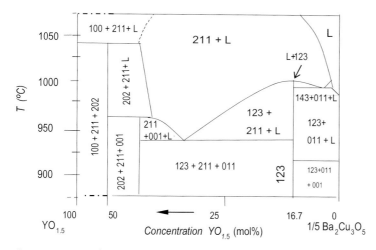

Figure 2.23 Vertical section $1/6$ Y123 – $1/2$ Y_2O_3

Melt crystallization experiments according to this MMCP have been performed within the zones of moderate temperature gradients either in tube or box furnaces. Typical conditions are: seeding with Sm123 before the heating, ramping up to 1050 °C followed by slow cooling (about 1 K/h) down to 940 °C, and final oxidation at 380 °C for between 200 and 250 h.

Crystallization of Y123 can start below the equilibrium temperature T_e (940 °C < T_e < $T(m_1)$) which corresponds to the ingot composition and proceeds along the phase field boundary until the peritectic temperature at 940 °C is achieved (T_e should be replaced by $T_e - \Theta$ if undercooling is considered). In contrast to conventional Y123 – Y211 mixtures, the composition of the remaining liquid phase will change if Y211 is trapped by the crystal (due to the fact that the composition Y211 does not appear at the "pseudo-binary" section in Figure 2.23). Therefore, a more realistic approach has to consider the entrapment of excess Y-211 particles in the

Y-123 matrix. Experimentally, the volume fraction of Y-211 is found to increase in the direction of propagating solidification. This variation can be described by means of an effective distribution coefficient K_{eff} [2.61], [2.64],

$$K_{eff} = \frac{c_{211}(s)}{c_{211}^{ex}(l)} \qquad (2.24)$$

which represents the ratio between volume concentrations of the Y-211 insertions in the solid, $c_{211}(s)$, and the excess of 211 particles in the quasi-peritectic suspension, $c^{ex}_{211}(l)$. The volume concentration of excess 211 results from the difference between the total amount of 211 suspended in the melt and the (virtual) part, which will react with melt near the interface, the reverse of Eq. (2.8), during the crystallization. A homogeneous distribution is represented by $K_{eff} = 1$, whereas $K_{eff} > 1$ indicates forced trapping. The progress is demonstrated in Figure 2.24 after a melt suspension of composition **A** was cooled down below the equilibrium line. (**A** in Figure 2.24). Crystallization proceeds from the undercooled melt (temperature T_A-Θ) the composition of which is determined by the liquidus boundary of the field for $YBa_2Cu_3O_{7-\delta}$ primary crystallization. The crystallization follows the path from A to E, its projection on the basal plane indicates the actual gross composition of the suspension. The composition of the solidified bulk (consisting of 123 plus 211 entrapped) is indicated by diamonds along the Y211 – Y123 tie line. It results from the mass balance analysis considering the composition of neighboring volume elements in a subsequent iterative step, etc. Solidification is completed with the peritectic reaction p_1 after the equilibrium temperature $T(p_1)$ was achieved. Parameters of the model are the undercooling Θ and the effective distribution coefficient K_{eff}, which can be estimated by a fit procedure [2.61], [2.64].

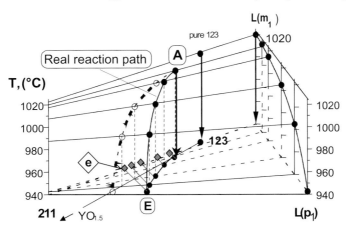

Figure 2.24 The reaction path in the modified melt crystallization process Y123 + 0.24 Y_2O_3 from the initial gross composition **A** to terminating solidification at **E**. For comparison, the dashed line represents the hypothetical equilibrium path assuming complete pushing (i.e. $K_{eff} = 0$ [2.61], [2.64]. The coordinate system used in Fig. 2.6 is mirrored and rotated for better presentation.)

Figure 2.25 represents the distribution of Y211 in a cylindrical sample as determined from Y and Cu EDX measurements following traces on cross sections along <100> and <001> in a– and in c-growth sectors, respectively. The sample was prepared from an ingot $1YBa_2Cu_3O_7 + 0.24\ Y_2O_3 + 0.02\ Pt$, seeded by Sm-123. K_{eff} closely approaches the value 1 in the c-growth sector, whereas the 211 distribution in the a-growth sector is fitted by $K_{eff} = 0.45$ and the undercooling $\Theta = 5$ K. The reaction path for this situation is shown in Figure 2.24 from the initial gross composition A toward the final composition E, which is solidified at the p_1-peritectic temperature 940 °C from the CuO-rich melt.

This inhomogeneity is an inherent feature of the modified process originated by the thermodynamical situation represented in the phase diagram. The apparent K_{eff} correlates to supersaturation or undercooling during the growth.

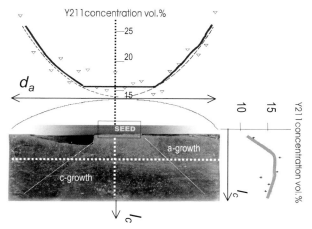

Figure 2.25 Volume concentration of 211 particles along the crystallization path in **a** and **c** growth sectors, respectively after [2.61]

2.5.3
Processing in Reduced Oxygen Partial Pressure

First attempts to take advantage of the reduced peritectic temperature at decreasing oxygen partial pressure were directed to the preparation of $YBa_2Cu_3O_7$ layers on a silver support (tape) [2.51]. The invariant reaction i_1 [Eq. (2.13)] determines the lowest limit for thermodynamical parameters at which crystallization of Y123 can proceed, namely $p(O_2) = 0.0015$ bar and $T = 960$ °C, respectively, with $\delta \approx 0.85$. In practice, however, the oxygen supply is not sufficient to support melt texturing even at $p(O_2) \approx 0.01$ bar, in contrast to an air environment, where a sufficient amount of oxygen is available to drive a continuous diffusion of oxygen to the crystallization zone [2.52]. On the other hand, processing at low oxygen pressure is more essential for RE123 with large RE ions and will be discussed in Chapter 7.

References

2.1 S. Jin, T. H. Tiefel, R. C. Sherwood, M. E. Davis, R. B. van Dover, G. W. Kammlott, R. A. Fastnacht, H. D. Keith, Appl. Phys. Lett. **52**, 2074 (1988).

2.2 P. Karen, O. Braaten, A. Kjekshus, Acta Chem. Scand., **46**, 805 (1992).

2.3 K. Osamura, W. Zhang, Z. Metallkd. **84**, 522 (1993).

2.4 W. Wong-Ng, L. P. Cook, B. Paretzkin, M. D. Hill, J. K. Stalick, J. Am. Ceram. Soc. **77**, 2354 (1994).

2.5 H. Müller-Buschbaum, Angew. Chem. **103**, 741 (1991).

2.6 M. Ritschel, H. Stephan, K. Gloe, P. Verges, N. Mattern, G. Krabbes, W. Bieger, J. Alloys Compd. **195**, 65 (1993).

2.7 J. Karpinski, E. Kaldis, S. Rusiecki, J. Less Common Metals,**150**, 207 (1989).

2.8 G. Krabbes, G. Auerswald, A. Bartl, H. Eschrig, B. Lippold, M. Ritschel, H. Vinzelberg, E. Wolf, U. Wiesner, Cryst. Res. Technol. **23**, 1161 (1988).

2.9 J. Hauck, Supercond. Sci. Technol. **9**, 1033 (1996).

2.10 T. B. Lindemer, J. F. Hunley, J. E. Gates, A. L. Sutton, J. Brynestad, C. R. Hubbard, J. Am. Ceram. Soc. **72**, 1775 (1989).

2.11 G. van Tendeloo, H. W. Zandbergen, S. Amelinkx, Solid- State Commun. **63**, 603 (1987).

2.12 J. D. Jorgensen, M. A. Beno, D. G. Hinks, C. U. Segre, K. Zhang, M. S. Kleefish, Phys. Rev. B **36**, 3608 (1987).

2.13 R. J. Cava, B. Battlog, C. H. Chen, E. A. Rietman, S. M. Zahurak, D. Werder, Phys. Rev. B **36**, 5719 (1987).

2.14 H. Verweij, W. H. M. Brungk, J. Phys. Chem. Solids **50**, 75 (1989).

2.15 W. Bieger, G. Krabbes, P. Verges, M. Ritschel, J. Thomas, J. Alloys Compds. **195**, 463 (1993).

2.16 M. Tetenbaum, B. Tani, B. Czech, M. Blander, Physica C **158**, 377 (1989).

2.17 K. Kanematsu, Japan. J. Appl. Phys. **29**, L906 (1990).

2.18 L. T. Wille, A. Berera, D. de Fontaine, Phys. Rev. Lett. **60**, 1065 (1988).

2.19 M. Ohkubo, T. Hioki, Physica C **185–189**, 921 (1991).

2.20 I. Heinmaa, H. Lütgemeier, S. Pekker, G. Krabbes, M. Buchgeister, Appl. Magn. Reson. **3**, 689 (1992).

2.21 S. Yang, H. Claus, B. V. Veal, R. Wheeler, A. P. Paulikas, J. W. Downey, Physica C **193**, 243 (1992).

2.22 S. A. Degtyarev, G. F. Voronin, Russ. J. Phys. Chem. **67**, 1217, 2155 (1993).

2.23 T. Siegrist, L. F. Schneemeyer, J. V. Waszczak, N. P. Singh, R. L. Opila, B. Batlogg, L. W. Rupp, D. W. Murphy, Phys. Rev. B **36**, 8365 (1987).

2.24 M. Guillaume, P. Allenspach, W. Henggeler, J. Mesot, B. Roessli, U. Staub, P. Fischer, A. Furrer, V. Trounov, J. Phys.: Condens. Matter **6**, 7963 (1994).

2.25 M. Scavini, R. Bianchi, J. Solid-State Chem. **161**, 396 (2001).

2.26 U. Wiesner, W. Bieger, G. Krabbes, Thermochim. Acta **290**, 115 (1996).

2.27 G. Krabbes, W. Bieger, U. Wiesner, M. Ritschel, A. Teresiak, J. Solid-State Chem. **103**, 420 (1993).

2.28 C. Wende, B. Schüpp, G. Krabbes, J. Alloys Compds. **381**, 320 (2004).

2.29 Ch. Krauns, M. Sumida, M. Tagami, Y. Yamada, Y. Shiohara, Z. Phys. B **96**, 207 (1994).

2.30 B. J. Lee, D. N. Lee, J. Am. Ceram. Soc. **72**, 78 (1991).

2.31 G. Krabbes, W. Bieger, U. Wiesner, K. Fischer, P. Schätzle, J. Electron. Mater. **23** 1135 (1994).

2.32 E. A. Goodilin, E. A. Trofimenko, E. A. Pomerantseva, A. P. Soloshenko, A. V. Kravchenko, I. S. Bezverkhy, J. Hester, V. V. Petrykin, N. N. Oleynikov, Yu. D. Tretyakov, Physica C **341 – 348**, 619 (2000).

2.33 Y. D. Tretyakov, E. A. Goodilin, Russ. J. Inorg. Chem. **46**, S 203 (2001).

2.34 Y. Shiohara, E. A. Goodilin in Handbook on the Physics and Chemistry of Rare Earths, edited by K. A. Gschneidner et al. Elsevier Science B. V. 2000, Vol. 30, p. 67.

2.35 A. Oka, T. Ito, Physica C **227**, 77 (1994).

2.36 Y. Shiohara, A. Endo, Mater. Sci. Eng. **R19**, 1 (1997).

2.37 K. Salama, V. Selvamanickam, L. Gao, K. Sun, Appl. Phys. Lett. **54**, 2352 (1989).

2.38 S. Marinel, J. Wang, I. Monot, M. P. Delmare, J. Provost, G. Desgardin, Supercond. Sci. Technol. **10**, 147 (1997).

2.39 M. Murakami, Appl. Supercond. **1**, 1157 (1993).

2.40 M. J. Cima, M. C. Flemings, A. M. Figueredo, M. Nakade, H. Ishii, H. D. Brody, J. S. Haggerty, J. Appl. Phys. **72**, 179 (1992).

2.41 B. N. Sun, P. Hartmann, C. F. Woensdregt, H. Schmid, J. Cryst. Growth **108**, 473 (1991).

2.42 A. A. Chernov, D. E. Temkin in Curr. Top. Mater. Sci., edited by E. Kaldis, H. J. Scheel, North Holland Comp. 1977, vol. 2, p. 3.

2.43 Th. Wolf, J. Cryst. Growth **166**, 810 (1996).

2.44 T. Izumi, Y. Nakamura, Y. Shiohara, J. Cryst. Growth **128**, 757 (1993).

2.45 P. Diko, N. Pelerin, P. Odier, Physica C **247**, 169 (1995).

2.46 W. Mader, B. Freitag, Research Report, 1997 not published.

2.47 J. Plain, F. Sandiumenge, J. Rabier, A. Proult, I. Stretton, T. Puig, X. Obradors, PASREG 2003, International Workshop on Processing and Application of Superconducting RE Large Grains, Jena 2003, to be published in Supercond. Sci. Technol. (2004).

2.48 F. Sandiumenge, S. Pinol, X. Obradors, E. Snoeck, C. Roucau, Phys. Rev. B **50**, 7032 (1994).

2.49 J. Thomas, P. Verges, P. Schätzle, A. Wetzig, U. Krämer, Physica C **251**, 315 (1995).

2.50 G. Krabbes, P. Schätzle, W. Bieger, U. Wiesner, G. Stöver, M. Wu, T. Strasser, A. Köhler, D. Litzkendorf, K. Fischer, P. Görnert, Physica C **244**, 145 (1995).

2.51 K. Fischer, G. Leitner, G. Fuchs, M. Schubert, B. Schlobach, A. Gladun, C. Rodig, Cryogenics **33**, 97 (1993).

2.52 G. Krabbes, P. Schätzle, U. Wiesner, W. Bieger, Physica C **235-240**, 299 (1994).

2.53 M. Murakami, Supercond. Sci. Technol. **9**, 1015 (1996).

2.54 T. Iwata, M. Hikita, S. Tsumari, *Advances in Superconductivity*, Proc. Int. Seminar on Superconductivity 1988, ed. by K. Kitazawa, T. Ishiguro, Springer Tokyo, p. 197.

2.55 R. D. Shannon, C. T. Prewitt, Acta Cryst. B **25**, 925 (1969) and **26**, 1046 (1970).

2.56 T. Aselage, K. Keefer, J. Mater. Res. **3**, 1279 (1988)

2.57 B. Lindemer, F. A. Washburn, C. S. MacDougall, R. Feenstra, O. B. Cavin, Physica C **178**, (1991) 93

2.58 R. Beyers, B. T. Ahn, Annu. Rev. Mater. Sci. **21**, 335 (1991)

2.59 N. Marushkin, G. D. Nipan, V. B. Lazarev, J. Chem. Thermodynamics **27**, 465 (1995)

2.60 Pitschke, W. Bieger, G. Krabbes, U. Wiesner, Powder Diffraction **10**, 165 (1995)

2.61 G. Krabbes, Processing of High Performance (RE)BaCuO Superconductor Bulk Materials: A Thermodynamic Approach in Advances in Condensed Matter and Materials Research Vol. 4, edited by F. Gerard (Nova Science Publishers Hauppauge N.Y. 2003) pp. 179–219

2.62 V. I. Kosyakov, L. N. Zelenina, C. Wende, G. Krabbes, Research Report INCH Novosibirsk – IFW Dresden, 2001, not published

2.63 J. Thomas, Research report 1994 IFW Dresden, not published

2.64 W. Bieger, Control of Y-211 entrapment in the MMCG process in Texturierte HTSL-Massivmaterialien: Synthese und Charakterisierung, Research Report, Part II, IFW Dresden, Project code 13N6934/4 Bundesminister Bildung, Wissenschaft, Forschung und Technologie, 1998

3
Pinning-Relevant Defects in Bulk YBCO

It has already been mentioned that high critical current densities require a defined microstructure with defects as small as the diameter of the flux lines, which is given by $2\xi_{ab}$ where ξ_{ab} is the coherence length within the ab-plane. In YBCO the coherence length increases from ξ_{ab} = 1.8 nm (at T = 0 K) to ξ_{ab} = 2.7 nm (at T = 77 K). Contributions to flux pinning come from most defects present in bulk YBCO, such as normal conducting Y_2BaCuO_5 (Y-211) precipitates, arrays of stacking faults, dislocations, and twin boundaries. These defects, which are described in Section 2.4.3, are inherent features of the actual material, appear at different structural levels, and are generated at different stages of processing. Additional defects such as point and extended defects can be generated and influenced by chemical or physical alteration during the process or by appropriate post-treatment, e.g., by introducing small particles, by chemical doping (pinning centers in the CuO_2 planes by Zn, Ni and Li substitution on Cu sites), or by irradiation. Examples for such artificial pinning sites are discussed in Chapter 6. This chapter focuses on inherent pinning sites relevant to the field orientation $H||c$, which is the important configuration for applications.

It is not easy to relate the microstructure in a type II superconductor to the volume pinning force or the critical current density. If one considers the interaction between individual pins and the vortex lattice and the summation of elementary pinning forces, one has to identify the most important pins which contribute to the volume pinning force. The role played by the different defects in the pinning effect in melt-textured YBCO material has been studied very extensively and is described in several review articles [3.1–3.3].

Stacking faults basically consist of the intercalation of a perovskite layer in the basal plane or completing CuO chains to double chains, and have been studied by TEM (Figure 2.20). Their inhomogeneous distribution around the 211 particles is an indication of the strain inhomogeneity in the surroundings of 211 particles. Stacking faults in a melt-textured YBCO sample have been found to be disk shaped, with diameters ranging from a few nm to about 30 nm. Flux pinning due to stacking faults has been considered for magnetic fields both $H||c$ and $H||ab$ [3.7]. Here, only the case $H||c$ will be briefly summarized. For a density of stacking faults of ca. 10^{15} cm^{-3} obtained from TEM images, a critical current density in the

High Temperature Superconductor Bulk Materials.
Gernot Krabbes, Günter Fuchs, Wolf-Rüdiger Canders, Hardo May, and Ryszard Palka
Copyright © 2006 WILEY-VCH Verlag GmbH & Co. KGaA, Weinheim
ISBN: 3-527-40383-3

range between $j_c = 10^4$ A cm^{-2} and 10^5 A cm^{-2} has been estimated at $T = 77$ K and $\mu_o H = 1$ T assuming direct summation of pinning forces. For the low estimate, the stress field around the 211 particles has been neglected, whereas for the higher estimate it has been assumed that the maximum strain around a stacking fault in the c direction is 3 unit cells above and below the plane of the fault [3.9]. This gives a pinned length of a flux line in the c direction of $L \approx 7$ nm, whereas for the stress-free case L is given by $L = 2\xi_c$.

The high critical current densities in small crystallites separated from melt-textured YBCO have been explained by *dislocations* around 211 precipitates [3.8]. A dislocation density of $N \approx (1-2) \times 10^{10}$ cm^{-2} and small 211 particles with an average diameter of about 35 nm have been identified in these samples. From magnetization measurements, a pronounced peak effect has been found in the $j_c(H)$ dependence at 70 K with a broad maximum of $j_c \approx 5 \times 10^4$ A cm^{-2} at applied fields between 1.5 and 2 T. At $T = 4$ K, $j_c \approx 2 \times 10^6$ A cm^{-2} has been measured for applied magnetic fields of 5 T. Estimating the pinning effect both of the dislocations around the 211 particles and of the 211 particles themselves, j_c values of $j_c = (4 - 7) \times 10^4$ A cm^{-2} have been obtained at $T = 70$ K and $\mu_o H = 2$ T in both cases for $H \parallel c$. At $T = 4$ K and $\mu_o H = 5$ T, pinning due to the 211 particles has been found to be too weak to explain the experimental j_c values. Therefore, the pinning effect due to dislocations has been favored to explain the experimental data [3.8]. However, it should also be noted that estimations of j_c using the direct summation of pinning forces become very uncertain at high magnetic fields and low temperatures because of the predominance of collective pinning of flux line bundles.

The *Y-211 precipitates*, which can be observed by optical microscopy, have extensions between about 0.5 and 20 μm. Shortly after the first melt-textured YBCO samples became available it was reported [3.4] that the critical current density scales with the surface area of Y-211 precipitates within the bulk YBCO, indicating that Y-221 particles act as pinning sites. However, this interpretation has been questioned [3.5–3.7], because the Y-211 precipitates are 2–3 orders of magnitude larger than the flux line cores in YBCO, which have a diameter of $2\xi = 5.4$ nm at 77 K. Instead of this it has been proposed that stacking faults [3.7] or dislocations [3.8] which are observed in the strained regions around the Y-221 precipitates may be responsible for flux pinning in bulk YBCO [3.7], or at least contribute to the pinning of flux lines [3.8].

The pinning effect due to 211 particles has been carefully analyzed [3.10]. It has been pointed out that pinning at the interface between the superconducting matrix and the large Y-211 precipitates due to the abrupt change of the order parameter over a short distance can explain many j_c results found for melt-textured YBCO at low magnetic fields. In the single-vortex pinning regime at low applied fields, the volume pinning force F_p can be estimated by direct summation of the elementary pinning forces f_p according to

$$F_p = j_c B = N f_p = N_p (d/a_o) f_p \tag{3.1}$$

with $a_o = 1.08 (\Phi_o/B)^{0.5}$ the flux line lattice spacing, N_p the number of Y-211 inclusions per unit volume and d their mean diameter. Because of $d > a_o$, each Y-211 inclusion can pin several flux lines. Their number is given by (d/a_o). When the core of a flux line (with radius ξ) intersects with a normal-conducting Y-211 inclusion, then the energy gain U corresponds to the condensation energy, which is $(1/2\mu_o H_c^2)$ per unit volume times the penetrated volume given by $(\pi\xi^2 d)$. Hence, the elementary pinning force can be expressed as

$$f_p = U/2\xi = \pi/4 \, \mu_o H_c^2 \, \xi d, \qquad (3.2)$$

where H_c is the thermodynamical critical field. Combining Eqs. (3.1) and (3.2), one gets for the critical current density

$$j_c = \pi/4 \, \mu_o \, H_c^2 \, \xi d^2 \, N_p / (a_o B) = \pi/4 \, \mu_o H_c^2 \, \xi d^2 \, N_p / (\Phi_o B)^{0.5} \qquad (3.3)$$

This relation predicts (i) $j_c \propto N_p d^2 = V/d$, with V the volume fraction of Y-211 inclusions, and (ii) a field dependence $j_c \propto B^{-0.5}$. Experimental results confirmed these predictions over a wide range of the volume fraction V and of the average size of Y-211 precipitates, in particular in the range of low fields and high temperatures, where single vortex pinning is expected to be predominant. The field dependence $j_c \propto B^{-0.5}$ has been found in the range of fields below 1.5 T and at temperatures above 45 K [3.10].

At higher fields and lower temperatures, other pins (including point pins, stacking faults, and dislocations) interacting with bundles of flux lines become important [3.11]. In principle, the interaction of these pins with flux bundles can be described within the collective pinning model [3.12], however, a detailed analysis is difficult because different types of pins interact simultaneously with flux bundles of different sizes. Several regions have been identified in the H-T phase diagram. At low fields and high temperatures, single vortex pinning at Y-211 interfaces with a field dependence $j_c \propto B^{-0.5}$ was found, as discussed above. Different $j_c(H)$ dependencies were observed at higher magnetic field. They have been attributed to collective pinning of small and large flux bundles [3.11]. Analyzing the temperature dependence of the self-field critical current density j_{co} two contributions have been identified: collective pinning of weak pins dominating the low temperature range and strong, single-vortex pinning dominating at high temperatures [3.13]. An exponential law $j_{co}(T) \propto \exp(-T/T_o)$ observed at low temperatures has been attributed to collective pinning of dislocations in the surroundings of the 211 particles, whereas a typical bump in the $j_{co}(T)$ dependence at high temperatures has been explained by the smooth temperature dependence $j_{co}(T) \propto \exp(-3[T/T^*]^2)$ of a pinning term due to strong pinning at 123/211 interfaces. It has been found that by a post-processing treatment at 600 °C under high oxygen pressure, the density of dislocations and stacking faults increases within the entire 123 matrix. A dramatic decrease of the irreversibility line observed in this case has been attributed to vortex cutting [3.13].

A promising way to enhance the critical current density in bulk YBCO is to refine the Y-211 particles. The relation $j_c \propto V/d$ derived from Eq. (3.3) has been tested so far in a limited range of the size d of the 211 particles. For most preparation routes, the smallest 211 particles in melt-textured YBCO samples have extensions of about 500 nm. A refinement of the 211 particles with sizes of a few hundred nanometers has been achieved through the addition of fine 211 particles to the starting material [3.14–3.17]. As an example, a refined Gd-211 phase in the matrix of (Nd,Eu,Gd)-123 [3.18] is shown in Figure 3.1.

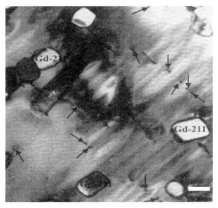

Figure 3.1 TEM image of a bulk (Nd,Eu,Gd)-123 sample with fine Gd-211 particles taken from [3.18]. Arrows indicate additional nanosized particles of a phase containing Zr which was precipitated in the presence of ZrO_2 impurities arising from the material of the ball mill. Reprinted from [3.18] with permission from Elsevier

Very fine Y-211 starting powders with grains 110 nm in diameter have been obtained by ball milling, leading to a very high critical current density of $j_c = 1.1 \times 10^5$ A cm^{-2} at 77 K in self-field [3.17].

Furthermore, the size of the 211 particles has been found to be reduced by additions of $BaSnO_3$, Pt, or Ce. $BaSnO_3$ additions promote the 211 dissolution, resulting in a refinement of the 211 particles [3.19]. Additions of Pt are assumed to refine 211 particles by altering the 211/liquid interfacial energy and/or the Y diffusivity [3.19]. High critical current densities have been reported in bulk YBCO with Ce additions [3.20, 3.21], which are suggested to increase the viscosity of the melt and to inhibit the growth of 211 particles [3.22, 3.23]. The effects of Ce and Pt addition on the critical current density of bulk YBCO at 77 K are compared in Figure 3.2. Addition of Ce was found to enhance j_c in the range from zero field to the highest applied magnetic field and leads, in addition, to a distinct shift of the irreversibility field to a higher value. In this approach, up to 2 wt% CeO_2 was admixed to the bulk YBCO resulting in a large proportion of 211 inclusions with a size between 200 and 600 nm [3.21]. Furthermore, data for Li-doped YBCO obtained with Pt admixtures are included in Figure 3.2. This approach, which is presented in more detail in Chapter 6, was observed to strongly enhance j_c at low applied magnetic fields, but also to cause a drastic reduction of the irreversibility field down to 4 T at 77K. In contrast, additions of nanoscaled precipitates of an Ir-doped $YBa_6Cu_3O_{10}$ phase forming by admixing 0.5 wt% IrO_2 [3.24] preferably increase j_c at high applied magnetic fields. This nanocomposite material with extended precipitates of diameters between 10 and 150 nm and lengths in the micrometer range exhibit similarly high irreversibility fields to those of the bulk YBCO with Ce additions as shown in Figure 3.2.

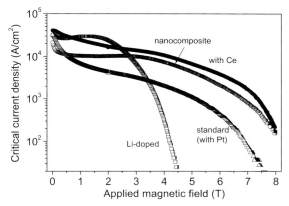

Figure 3.2 Field dependence of the critical current density of bulk YBCO at 77 K, showing standard material with Pt additions (open triangles), YBCO with Ce additions (filled circles), Li-doped YBCO (open squares) and nanocomposite material containing extended precipitates (open circles).

Extremely fine 211 particles with extensions below 100 nm have been reported for small crystallites of melt-textured YBCO [3.8] and in YBCO prepared by the QMG process (Section 2.4.1) or from a Cu-rich melt [3.25, 3.26]. However, in the latter case only a small density of these extremely fine 211 particles has been found. Therefore, it has been concluded [3.26] that at higher magnetic fields the critical current density is probably limited by shearing of weakly pinned parts of the flux line lattice against the strongly pinned flux lines interacting with these fine 211 particles. This pinning mechanism has been proposed by Kramer [3.27].

Twin boundaries between grains directed in the crystallographic [100] and [010] direction are formed during the phase transformation from the tetragonal to the orthorhombic crystal structure. The resulting twin planes are extended mainly along the c direction. Pinning by twin boundaries has been reported for YBCO thin films [3.28], YBCO single crystals [3.29, 3.30], and melt-textured YBCO [3.31]. For melt-textured YBCO, the out-of-plane anisotropy of the irreversibility field has been determined from magnetoresistance measurements, applying the transport current parallel to one of the twin boundary families. The irreversibility field H_{irr} decreasing with increasing angle between ab plane and applied field was found to enhance in a region around $H\|c$ where H_{irr} is expected to be dominated by twin boundary pinning [3.31]. These data suggest that the angular dependence of the irreversibility line is divided into two contributions, a background pinning due to randomly oriented defects which scales with the intrinsic anisotropy of the superconductor and correlated pinning due to the effect of twin boundaries.

It should be mentioned that stacking faults and in-plane dislocations surrounding the stacking faults are considered as candidates for pinning centers for applied fields $H\|ab$ [3.32]. However, the dominant pinning effect for this field orientation is due to *intrinsic pinning* [3.33]. In this case, the flux lines are aligned parallel to the superconducting CuO_2 planes. The Lorentz force acts in the c direction and the pinning effect arises from the modulation of the order parameter along the c axis. Because the distance between the CuO_2 planes is comparable to the coherence length ξ_c, the maximum pinning force is obtained if the flux lines are located in the weakly superconducting layers between the CuO_2 planes along their whole length. Studying the angular dependence of j_c of melt-textured YBCO by transport measurements, flux pinning due to intrinsic pinning has been observed [3.15, 3.34]. In Figure 3.3, the critical current density j_c at 77 K and $\mu_o H$ = 1.5 T of two YBCO samples is shown in dependence on the angle between applied magnetic field and c axis. The narrow and large j_c peak for $H\|ab$ is due to intrinsic pinning, whereas the height of the broad j_c peak for $H\|c$ depends on the extrinsic pinning sites. This peak remains very shallow for the sample without 211 particles and was found to be strongly increased by the addition of 15 wt% 211 precipitates [3.15].

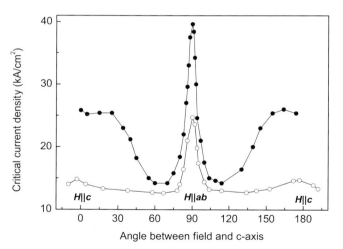

Figure 3.3 Angular dependence of the critical current density of melt-textured YBCO at 77 K and $\mu_o H$ = 1.5 T, showing a comparison of two samples containing no 211 precipitates (open circles) and 15 wt% 211 precipitates (filled circles). Data from Ref. [3.15].

Summarizing flux pinning in YBCO for $H\|c$, it has been established that submicron sized Y-211 precipitates play an important role for flux pinning. The pinning effect arising at the interface between the 211 particles and the superconducting matrix can be considerably improved by refining the 211 particles. In principle, stacking faults and dislocations appearing in the strained region around 211 particles have the potential to explain j_c values in the range of several 10^4 A cm^{-2} at 77 K. However, more systematic studies are necessary to clarify the role of these pinning sites in bulk YBCO.

References

3.1 M. Murakami, in *Melt Processed High-Temperature Superconductors*, edited by M. Murakami (World Scientific, Singapore, 1992) chapter 3, p. 21.

3.2 P. McGuinn, in *HighTemperature Superconducting Materials Science and Engineering*, edited by Donglu Shi (Elsevier, Amsterdam, 1995) chapter 8, p. 345.

3.3 H. W. Weber, *Flux Pinning*, in *Handbook on the Physics and Chemistry of Rare Earths*, edited by K. A. Gscheidner, Jr., L. Eyring, M. B. Maple (Elsevier, Amsterdam, 1999).

3.4 M. Murakami, S. Gotoh, N. Koshizuka, S. Tanaka, T. Matsuhita, S. Kanube, K. Kitazawa, Cryogenics **30**, 390 (1990).

3.5 S. Jin, T. H. Tiefel, G. W. Kamlott, Appl. Phys. Lett. **59**, 540 (1991).

3.6 P. McGinn, N. Zhu, W. Chen, S. Sengupta, T. Li, Physica C **176**, 203 (1991).

3.7 Z. L. Wang, A. Goyal, D. M. Kroeger, Phys. Rev. B **47**, 5373 (1993).

3.8 M. Ullrich, D. Müller, W. Mexner, M. Steins, K. Heinemann, H. C. Freyhardt, *Phys. Rev. B* **48**, 7513 (1993).

3.9 A. F. Marshall, K. Char, R. W. Barton, A. Kapitulnik, S. S. Laderman, J. Mater. Res. **5**, 2049 (1990).

3.10 B. Martinez, X. Obradors, A. Gou, V. Gomis, S. Pinol, J. Fontcuberta, H. Van Tol, Phys. Rev. B **53**, 2797 (1996).

3.11 F. Sandiumenge, B. Martinez, X. Obradors, Supercond. Sci. Technol. **10**, A93 (1997).

3.12 M. V. Feigel'man and V. M. Vinokur, Phys. Rev. B **43**, 6263 (1990).

3.13 J. Plain, T. Pluig, F. Sandiumenge, X. Obradors, J. Rabier, Phys. Rev. B **65**, 104526 (2002).

3.14 D.F. Lee, X. Chaud, K. Salama, Japan. J. Appl. Phys. **31**, 2411 (1992).

3.15 K. Salama, D.F. Lee, Supercond. Sci. Technol. **7**, 177 (1994).

3.16 C.-J. Kim, H.-W. Park, K. B. Kim, K.-W. Lee, G.-W. Hong, Jpn. J. Appl. Phys. **34**, L671 (1995).

3.17 S. Nariki, N. Sakai, M. Muratami, I. Hirabayashi, Supercond. Sci. Technol. **17**, S30 (2004).

3.18 M. Muralidhar, N. Sakai, M. Jirsa, M. Murakami, N. Koshizuka, Physica C **412-414**, 739 (2002).

3.19 C. Varanasi, M. A. Black, P. J. McGinn, Supercond. Sci. Technol. **7**, 10 (1994).

3.20 D. Litzkendorf, T. Habisreuther, R. Muller, S. Kracunovska, O. Surzhenko, M. Zeisberger, J. Riches, W. Gawalek, Physica C **372**, 1163 (2002).

3.21 L. Shlyk, K. Nenkov, G. Krabbes, G. Fuchs, Physica C **423**, 22 (2005).

3.22 I. Monot, K. Verbist, M. Hervieu, P. Lafez, M. P. Delemare, J. Wang, G. Desgardin, G. van Tendeloo, Physica C **274**, 253 (1997).

3.23 C. Leblond, I. Monot, J. Provost, G. Desgardin, Physica C **311**, 211 (1999).

3.24 L. Shlyk, G. Krabbes, G. Fuchs, K. Nenkov, Appl. Phys. Lett. **86**, 092503 (2005).

3.25 D. Shi, S. Sengupta, J. S. Luo, C. Varanasi, P. J. McGinn, Physica C **213**, 179 (1993).

3.26 S. Sengupta, Donglu Shi, J. S. Luo, A. Buzdin, V. Gorin, V. R. Todt, C. Varanasi, P. J. McGinn, J. Appl. Phys. **81**, 7396 (1997).

3.27 E.J. Kramer, J. Appl. Phys. **44**, 1360 (1973).

3.28 B. Roas, L. Schultz, G. Saemann-Ischenko, Phys. Rev. Lett. **64**, 479 (1990).

3.29 W. K. Kwok, U. Welp, G. W. Crabtree, K. G. Vandervoort, R. Hulscher, J. Z. Liu, Phys. Rev. Lett. **64**, 966 (1990).

3.30 H. Küpfer, A. A. Zhukov, A. Will, W. Jahn, R. Meier-Hirmer, T. Wolf, V. I. Veronkova, M. Kläser, K. Saito, Phys. Rev. B **54**, 644 (1996).

3.31 T. Puig, F. Galante, E. M. Gonzalez, J. L. Vicent, B. Martinez, X. Obradors, Phys. Rev. B **60**, 13099 (1999).

3.32 B. Martinez, T. Puig. A. Gou, V. Gomis, S. Pinol, J. Fontcuberta, X. Obradors, Phys. Rev. B **58**, 15198 (1998).

3.33 M. Tachiki, S. Takayashi, Solid State Commun. **72**, 1083 (1989).

3.34 V. Selvamanickam, K. Forster, K. Salama, Physica C **178**, 147 (1991).

4
Properties of Bulk YBCO

4.1
Vortex Matter Phase Diagram of Bulk YBCO

4.1.1
Irreversibility Fields

Applications of bulk YBCO superconductors are limited to magnetic fields below the irreversibility field. In the field range $B_{irr} < \mu_o H < B_{c2}$, the effect of thermal fluctuations on the vortex lattice becomes so strong that currents cannot flow without losses although the superconductor is not yet in the normal state. The origin of this resistive behavior may be a first-order melting transition of the vortex lattice, the transition from a vortex glass to a vortex liquid, or a thermally driven depinning transition. In the latter case, the activation energy for flux creep becomes so small that the flux lines start to move under the influence of any driving force. The strongly enhanced mobility of the magnetic flux within the superconductor makes applications above the irreversibility line impossible. It is important to determine the irreversibility line of bulk YBCO in a large temperature range, not only in order to know the field range for applications, but also for a better understanding of the origin of the irreversibility line. From a practical point of view, the question arises to what extent the irreversibility line can be shifted to higher temperatures by improved flux pinning.

In Figure 4.1, resistive transition curves of a small bulk YBCO sample are shown for magnetic fields up to 50 T applied parallel to the *c* axis of the melt-textured sample. These measurements were performed in a pulsed-field facility in which the rise time of the field pulse was 10 ms. The irreversibility fields were determined from these curves at the onset of resistance using an electric field criterion of $1\mu V/cm$. This definition was found to be consistent with magnetization vs temperature data obtained for the same sample in dc magnetic fields. This is illustrated in Figure 4.2, where $M(T)$ data are shown in an external field of 7 T. The comparison of field cooled and zero-field cooled data in Figure 4.2 shows a transition from irreversible to reversible behavior at 77 K, which means that $B_{irr}(77\ K) = 7\ T$.

Irreversibility field data of this YBCO sample obtained by different techniques including ac susceptibility and resistance measurements in static field and mag-

High Temperature Superconductor Bulk Materials.
Gernot Krabbes, Günter Fuchs, Wolf-Rüdiger Canders, Hardo May, and Ryszard Palka
Copyright © 2006 WILEY-VCH Verlag GmbH & Co. KGaA, Weinheim
ISBN: 3-527-40383-3

netization and resistance measurements in pulsed field are shown in Figure 4.3. A rather good agreement was found between the B_{irr} data obtained in pulsed and static field. The dotted line in Figure 4.3 corresponds to the expression

$$B_{irr}(T) = B_{irr}(0)\,(1-T/T_c)^n \tag{4.1}$$

where $B_{irr}(0) = 95$ T, $T_c = 90$ K and $n = 1.4$ which describes the experimental data in the field range up to 25 T. A similar temperature dependence of the irreversibility field is expected both in the case of thermally activated flux creep ($n = 1.5$) [4.1] and for the transition from a vortex glass to a vortex liquid ($n = 1.3$) [4.2]. Not understood is the almost linear $B_{irr}(T)$ dependence observed in the large field range between about 10 T and 45 T.

Surprisingly, a very similar linear $B_{irr}(T)$ dependence was also reported for YBCO thin films [4.3]. For the thin films, experimental data were determined in the whole temperature range, and $B_{irr}(0) = 64$ T was found at $T = 0$. By extrapolating the linear part of $B_{irr}(T)$ in Figure 4.3 to $T = 0$, a value of $B_{irr}(0) = 70$ T can be estimated.

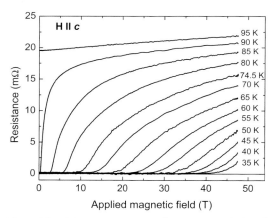

Figure 4.1 $R(H)$ transition curves of a YBCO sample for $H\|c$ at different temperatures. The measurements were performed in pulsed magnetic fields up to 50 T

4.1 Vortex Matter Phase Diagram of Bulk YBCO | 77

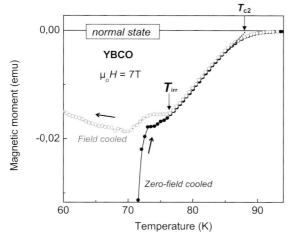

Figure 4.2 Magnetization of a YBCO sample plotted against the temperature at a dc field $\mu_o H = 7$ T applied along the c axis. The field was applied after cooling the sample (zero-field cooling). Then the temperature was increased to 90 K and afterwards reduced (field cooling)

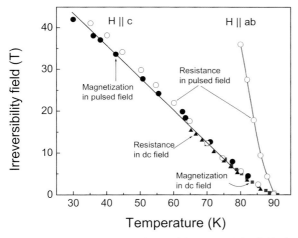

Figure 4.3 Temperature dependence of the irreversibility field of a bulk YBCO sample for $H\|c$ and $H\|ab$ obtained from different measurements in dc and pulsed magnetic fields as indicated in the figure

4.1.2
Upper Critical Fields

The upper critical field of bulk YBCO can be determined from the $M(T)$ dependence shown in Figure 4.2. In the temperature range in which the magnetization shows a reversible behavior, a linear $M(T)$ dependence is observed except at temperatures near T_c, where fluctuations dominate. This is in accordance with the *Ginzburg-Landau* theory, which predicts a linear decrease of the magnetization with increasing temperature into the normal state. Therefore, the temperature T_{c2} obtained by extrapolating the linear part of the reversible magnetization to $M = 0$ corresponds to the upper critical field $B_{c2} = 7$ T.

The procedure to determine H_{c2} is shown once more in Figure 4.4 for several applied magnetic fields. These static field measurements were compared with resistance measurements $R(T,H)$ performed both in static and pulsed field (see Figure 4.1). The B_{c2} data determined from $M(T)$ measurements were found to coincide with the field of the midpoint of $R(T)$ transition curves.

Upper critical field data of bulk YBCO are shown in Figure 4.5 in the field range up to 50 T for the two field orientations parallel and perpendicular to the c axis. From these data a temperature-independent B_{c2} anisotropy of $B_{c2}^{ab}/B_{c2}^{c} = (5.3 \pm 0.1)$ is obtained for the bulk YBCO in the investigated temperature range above 84 K. A similar anisotropy of $B_{c2}^{ab}/B_{c2}^{c} = 5.5$ was reported for a YBCO single crystal [4.4]. The slope $dB_{c2}^{ab}/dT = 8$ T/K found for the bulk YBCO sample for fields perpendicular to the c axis was observed to be smaller by about 20% than that of the single crystal for which $dB_{c2}^{ab}/dT = 10.5$ T/K [4.4].

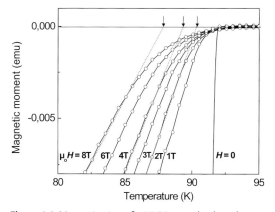

Figure 4.4 Magnetization of a YBCO sample plotted against the temperature at several dc fields $H\|c$. Shown is the reversible part of the magnetization data and the linear extrapolation of $M(T)$ to $M = 0$. The arrows mark the temperatures T_{c2} related to the B_{c2} values as indicated in the figure

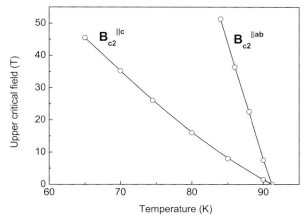

Figure 4.5 Upper critical field of a YBCO sample as a function of temperature for magnetic fields $H\|c$ and $H\perp c$.

4.1.3
Vortex Matter Phase Diagram

The vortex matter phase diagram of bulk YBCO obtained from the B_{irr} and B_{c2} data (see Figures 4.3 and 4.5, respectively) is shown in Figure 4.6 in the range of magnetic fields up to 50 T for the two field directions. For magnetic fields below B_{irr} the flux line lattice is plastically deformed by strong pinning and is called a disordered vortex glass. As mentioned above, the critical current density becomes zero in the field range between $B_{\mathrm{irr}}(T)$ and $B_{\mathrm{c2}}(T)$ of the vortex liquid which is dominated by thermal fluctuations.

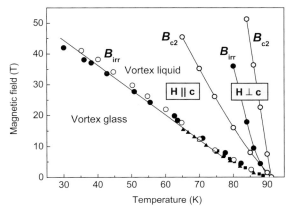

Figure 4.6 Vortex matter phase diagram of bulk YBCO for magnetic fields $H\|c$ and $H\perp c$ in the field range up to 50 T. $B_{\mathrm{irr}}(T)$ for $H\|c$ has been determined both from magnetization measurements in dc field (■) and pulsed field (●) and from resistance measurements in dc field (▲) and pulsed field (○). For details see text

With regard to applications, the most important feature of this phase diagram is the irreversibility line $B_{irr}(T)$ for fields parallel to the c axis. The irreversibility field parallel to the c axis increases from about 7 T at $T = 77$ K up to about 42 T at $T = 30$ K. Thus, there is a wide field range for applications, especially for temperatures below 77 K.

It is interesting to compare the obtained phase diagram for bulk YBCO with that of YBCO single crystals. The magnetic phase diagram reported by Nishizaki et al. [4.5] for clean, detwinned YBCO single crystals is shown in Figure 4.7. A typical feature, which can only be observed for very clean single crystals of YBCO, is a first-order transition line $B^*(T)$ dividing the vortex solid below the irreversibility line $B_{irr}(T)$ in a low-field *Bragg* glass and a high-field vortex glass.

The Bragg glass phase and the field-driven transition have been intensively studied in $Bi_2Sr_2CaCu_2O_y$ (Bi-2212) and $YBa_2Cu_3O_{6.6}$ single crystals with strongly reduced oxygen content using low-angle neutron scattering [4.6] and muon spin rotation [4.7–4.9]. These experiments revealed that the intensity of Bragg peaks observed at low magnetic fields suddenly decreases at a certain magnetic field B^*, indicating a field-driven transition from a nearly perfect into a strongly disturbed vortex lattice. At the transition field $B^*(T)$ an anomalous rise of j_c with increasing field has been observed.

The field-driven transition at B^* in the highly anisotropic $Bi_2Sr_2CaCu_2O_y$ has been explained by decoupling of flux lines into pancake vortices [4.8]. Decoupling of pancake vortices in different CuO_2 layers occurs at the decoupling or dimensional crossover field

$$B_{dec} \approx \frac{\Phi_o}{s^2 \gamma^2} \tag{4.2}$$

with Φ_o the flux quantum, s the interlayer distance between the CuO_2 layers, and γ the anisotropy parameter.

A similar vortex phase diagram as that for $Bi_2Sr_2CaCu_2O_y$ ($\gamma \approx 50 - 250$) has been found for $YBa_2Cu_3O_{6.6}$ [4.9]. Removal of oxygen from the CuO chains in YBCO is known to weaken the coupling between the CuO_2 planes and to increase the anisotropy parameter γ. This parameter is in the range of $\gamma \approx 5 - 7$ for optimally doped YBCO and increases up to $\gamma \approx 35.5$ for $YBa_2Cu_3O_{6.6}$. The order-disorder transition in $YBa_2Cu_3O_{6.6}$ observed at B^* has been interpreted in the same way as that for $Bi_2Sr_2CaCu_2O_y$ – by a dimensional crossover from a Bragg glass of flux lines to pancake vortices [4.9].

In optimally doped YBCO single crystals, the flux line lattice is much more stable with regard to plastic deformations than in the case of Bi-2212 or $YBa_2Cu_3O_{6.6}$. For optimally doped YBCO, one estimates from Eq. (4.2) a decoupling field of $B_{dec} \approx 77$ T using $\gamma \approx 6$ and the interlayer distance $s = 0.85$ nm. Thus, the dimensional crossover mechanism cannot be responsible for the order-disorder transition at the field B^* in Figure 4.7. For YBCO, the phase transition at B^* has been explained by entanglement of vortices, resulting in a strongly pinned state for fields above B^* [4.10, 4.11]. According to this model, the pinning and elastic energy of the vortex lattice are comparable at the field $B^*(T)$ of the order-disorder transition. Above this field, the pinning energy dominates and the stiffness of the flux lines becomes weaker because of the proliferation of dislocations giving rise to plastic deformations and a mechanically entangled configuration of flux lines with improved pinning properties. The entanglement field $B_E(0)$ of YBCO was estimated to be in the range 1.4–5.7 T, which is comparable with the experimental B^* data in Figure 4.7.

The vortex phase diagram of YBCO is determined by a competition between elastic, pinning, and thermal energies. Therefore, the vortex-matter phase of single crystals depends not only on magnetic field and temperature, but also on the level of disorder in the superconductor. Additional disorder can be introduced into YBCO single crystals by randomly distributed point defects produced by electron irradiation. It has been shown [4.5] that with increasing disorder due to point defects the characteristic field $B^*(T)$ shifts to a lower magnetic field together with the critical point at the melting line $H_m(T)$ at which $B^*(T)$ and $B_{irr}(T)$ of the Bragg glass meet each other (see Figure 4.7).

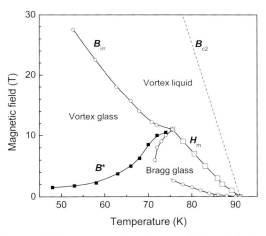

Figure 4.7 Vortex matter phase diagram of a YBCO single crystal for $H\|c$. Data from Ref. [4.5]. For details see text

The level of disorder can also be varied by changing the oxygen content and thus the density of clusters of oxygen vacancies. The evolution of the vortex phase diagram of $YBa_2Cu_3O_{7-x}$ single crystals in the range between x < 0.007 (overdoped YBCO, T_c = 89.8 K) and x = 0.1 (underdoped, T_c = 91.1 K) has been investigated by Küpfer et al. [4.12, 4.13]. Data for $B_{irr}(T)$ of the vortex glass phase and the melting line below the critical point are shown in Figure 4.8 for several selected oxygen concentrations. By reducing the oxygen content, the pinning energy rises in comparison to the elastic energy, with the result that the vortex lattice becomes more disordered at higher temperatures. Therefore, the critical point shifts to lower magnetic field along the melting line. At x = 0.1, the first-order melting transition and the order-disorder transition field $B^*(T)$ (not shown in Figure 4.8) completely disappear. Thus, the irreversibility line $B_{irr}(T)$ of the vortex glass phase starts to increase from zero-field as shown in Figure 4.8, similarly to $B_{irr}(T)$ in the case of bulk, melt-textured YBCO.

Thus, it can be concluded that the simple form of the vortex matter phase diagram in bulk YBCO is attributed to the strong disorder in this material.

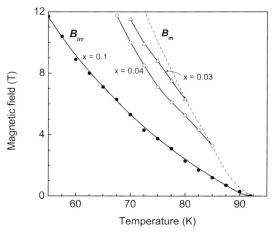

Figure 4.8 Evolution of the vortex matter phase diagram of a $YBa_2Cu_3O_{7-x}$ single crystal ($H||c$) with decreasing oxygen content between x <0.007 and x=0.1. Data from Ref. [4.13]

4.2
Critical Currents and Pinning Force

4.2.1
Transport Measurements

The critical current density j_c of type II superconductors can be determined from transport or from magnetic measurements. The most direct way for the determination of j_c is to apply a transport current and to measure the voltage U in the standard four-probe configuration. Then, j_c is defined at a certain value of the electric field $E_c = U/l$, where l is the distance between the voltage contacts. However, this method is restricted to superconductors in the form of wires or tapes. It is not difficult to prepare bar-like samples from bulk YBCO which could be used for transport measurements, but special techniques for preparing the current contacts are needed to ensure that high transport currents of more than 100 A, required already at 77 K, can flow without sample heating. Therefore, j_c data obtained from transport measurements are restricted to the temperature range around 77 K.

An example of the variation of j_c with magnetic field at $T = 77$ K using transport current is shown in Figure 4.9 [4.14]. The measurements were performed for two field directions ($H\|c$ and $H\perp c$) and for currents flowing within the ab planes and perpendicular to the applied field. The critical current density $j_c(H)$ is highly anisotropic, which is in accord with the strong anisotropy of YBCO found for $B_{c2}(T)$ and $B_{irr}(T)$ (see Section 4.1). For $H\|c$, j_c drops rapidly for applied fields $\mu_0 H > 5$ T, which is because of the irreversibility field of $B_{irr}^{\|}(77\ K) \approx 7$ T. For $H\perp c$, j_c is only weakly dependent on the applied field up to the highest measured field of 30 T, which is considerably smaller than the irreversibility field of $B_{irr}^{\perp}(77\ K) \approx 50$ T. In self-field, a critical current density of $j_c \approx 80$ kA cm^{-2} was found for this sample.

Transport critical current densities of $j_c = 70$ kA cm^{-2} in self-field and of $j_c = 11$ kA cm^{-2} at 20 T for $H\perp c$ were reported for a YBCO sample doped with BaSnO$_3$ [4.15].

4.2.2
Magnetization Measurements

Usually, the field and temperature dependence of the critical current density $j_c(H,T)$ of a melt-textured HTSC is determined from contact-free magnetization measurements. Such measurements are easy to perform, especially for applied magnetic fields directed along the c axis. In this case, which is most important for applications, supercurrents are induced in the ab planes, and the size of the current loops is given by the sample geometry. The critical current density $j_c(H,T)$ can be extracted from magnetization loops measured at several temperatures using one of the relations in Eqs. (1.14)–(1.16).

Typical $j_c(H)$ curves of a melt-textured YBCO sample obtained from magnetization measurements in the temperature range 77–59 K are shown in Figure 4.10 for magnetic fields applied along the c axis. At 77 K, the critical current density decreases monotonously from $j_c = 30$ kA cm^{-2} at $H = 0$ to 10 A cm^{-2} at $\mu_0 H = 7$ T (not shown in Figure 4.10), corresponding approximately to the irreversibility field at this temperature. For temperatures $T \leq 70$ K, a maximum of j_c is observed near zero magnetic field followed by a broad plateau where j_c remains almost constant. This field-independent j_c which is typical for bulk YBCO at these temperatures points to single vortex pinning where $R_c < a_o$, i.e. the transverse size R_c of the flux bundles becomes smaller than the flux line spacing a_o. This means that each flux line is pinned individually [4.16]. The j_c maximum near $H = 0$ is a geometrical effect due to substantial curvature of the flux lines. In this self-field dominated region, the critical state model cannot be applied, i.e. the field range of constant j_c is probably extended to even lower magnetic fields.

The critical current density in bulk YBCO can be substantially improved by the refinement of the Y211 precipitations. As an example, it was demonstrated that high j_c values up to 110 kA cm^{-2} at 77 K in self-field can be achieved by the use of very fine Y211 particles with diameters of 110 nm [4.17]. Other ways to improve the critical current density in bulk YBCO such as doping and irradiation techniques will be considered in Sections 6.1.1 and 6.1.3.

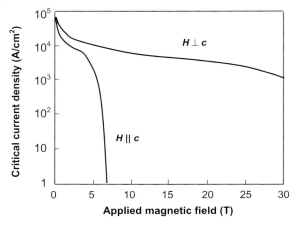

Figure 4.9 Field dependence of the critical current density of a bulk YBCO sample for $H\|c$ and $H\perp c$ obtained from transport measurements at 77 K. Data from Ref. [4.14]

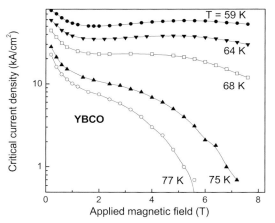

Figure 4.10 Field dependence of the critical current density of a bulk YBCO sample for $H\|c$ obtained from magnetization measurements at temperatures between 77 K and 59 K

The availability of j_c data over a wide field and temperature range allows us to test the scaling of the volume pinning force provided that the irreversibility field is known. The irreversibility field of YBCO at 77 K is typically $B_{irr} \approx 7$ T for $H\|c$. Magnetization measurements are restricted to applied fields below 15 T in most cases. Therefore, the scaling behavior of YBCO can usually be analyzed only in a limited temperature range above 70 K. In order to overcome this problem, which is connected with the lack of experimental data for B_{irr}, several authors have used the peak field at which the pinning force achieves its maximum value for the analysis of the scaling behavior of YBCO. However, the significance of this method is rather questionable, because even if the normalized pinning force does not scale with the reduced magnetic field, a scaling behavior will be enforced over the field range around the peak field.

Investigations of the irreversibility line in an extended field range up to 22 T have been reported [4.18]. A good scaling of the pinning force in these bulk YBCO samples has been found between 80 K and 46 K, as shown in Figure 4.11. It should also be noted that these YBCO samples had, with $B_{irr}(77\text{ K}) = 4$ T and $B_{irr}(46\text{ K}) = 20$ T, rather low irreversibility fields. This explains the wide temperature range over which the scaling behavior of this sample has been investigated.

The irreversibility line of the bulk YBCO sample whose $j_c(H)$ characteristics are shown in Figure 4.10 has been investigated in pulsed fields up to 50 T (see Figure 4.6). Using these data, the normalized volume pinning force was determined as a function of the reduced magnetic field H/H_{irr}, as shown in Figure 4.12. Deviations from the scaling which are visible in the low-field range for $T = 77$ K and 74.5 K may be attributed to contributions of different pinning sites to the pinning force. This explanation is supported by the rather good scaling behavior found for doped YBCO material (see, for instance Figure 6.2), in which one pinning mechanism becomes dominant.

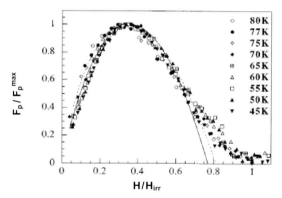

Figure 4.11 Normalized pinning force of a bulk YBCO sample (for $H||c$) plotted against the reduced magnetic field H/H_{irr}. Taken from Ref. [4.18]

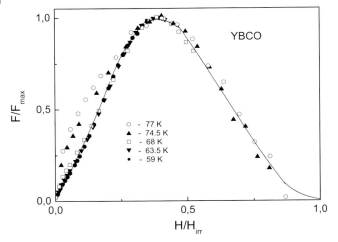

Figure 4.12 Normalized pinning force of a bulk YBCO sample (for $H \| c$) plotted against the reduced magnetic field H/H_{irr}. The $j_c(H,T)$ data for this sample are shown in Figure 4.10

4.3
Flux Creep

4.3.1
Flux Creep in Bulk YBCO

The magnetic relaxation $M(t)$ of HTSCs observed in a certain time window shows approximately a logarithmic time dependence. Usually, the temperature dependence of the relaxation rate $dM/d(\ln t)$ in HTSCs exhibits a peak at intermediate temperatures, which has been found to shift to lower temperature with increasing applied magnetic field. The unexpected decrease of the relaxation rate at high temperatures can be attributed to the fixed time window of the relaxation measurement [4.26].

It is convenient to consider the normalized flux creep rate $S \equiv 1/M \, dM/d \ln(t)$. Typical data for the temperature dependence of the normalized flux creep rate for YBCO are shown in Figure 4.13. Most surprising is the almost temperature-independent plateau of S at intermediate temperatures, which has been reported not only for bulk YBCO but also for single crystals and thin films [4.19, 4.20]. This plateau cannot be understood within the *Anderson-Kim* model, which predicts $S = -kT/U_o$, as can easily be derived from Eq. (1.21). According to this model, S is expected to curve upward with increasing temperature, because the barrier height U_o should decrease with increasing temperature. Within the collective creep model, one obtains from (1.23)

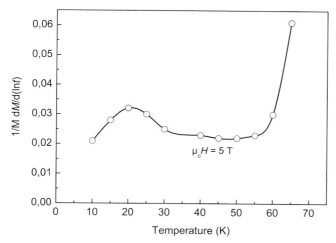

Figure 4.13 Temperature dependence of the normalized flux creep rate $S \equiv 1/M\, dM/d\ln(t)$ of a bulk YBCO sample measured at $\mu_o H = 5$ T for $H \| c$

$$S = \frac{kT}{U_o + \mu kT \ln t/t_o} \tag{4.3}$$

According to Eq. (4.3), with increasing temperature the relaxation rate approaches the limit

$$S = \frac{1}{\mu \ln t/t_o} \tag{4.4}$$

and thus becomes independent of temperature. Another prediction of this formula is that S decreases with time, which is in agreement with long-term relaxation data (see, for instance Figure 1.8). Experimental data for the height of the plateau of S in YBCO samples are in the range of several percent. From the expression (4.4), one obtains $S \approx 0.048$ and $S \approx 0.043$ for typical times of flux creep experiments of $t = 10^3$ s and $t = 10^4$ s, respectively, assuming $\mu = 1$ and $t_o = 10^{-6}$ s [4.21].

Surprisingly, a finite relaxation has been found at very low temperatures in several HTSCs including YBCO single crystals [4.22–4.24]. Thus, thermal activation does not freeze out at $T = 0$. This novel mechanism of flux creep at very low temperatures has been explained by quantum tunneling of vortices.

4.3.2
Reduction of Flux Creep

Thermally activated flux motion tends to reduce the field gradient and, thus, the maximum trapped field B_o in the superconductor or the levitation force of superconducting magnetic bearings. At 77 K, typical relaxation rates of $d(B/B_o)/d(\ln t/t_o)$

≈ — 0.05 were reported for melt-textured YBCO [4.25], i.e. the trapped field B_o reduces by about 5% per time decade. Hence, flux creep seems to be a serious problem for superconducting permanent magnets, and it is highly desirable to control and reduce the magnetic relaxation.

Flux creep depends on the pinning properties and is expected to become weaker by the introduction of strong pinning centers. This was confirmed by irradiation of YBCO single crystals with heavy ions [4.26]. The flux creep rate at 60 K has been found to decrease by a factor of about 2 after irradiation, which was caused both by the strong increase in j_c in the irradiated sample and by the higher activation energy for flux creep at the same level of j_c.

A very effective experimental procedure to reduce the magnetic relaxation was described by Beasley et al. [4.27]. Investigating the magnetic relaxation in the low-temperature superconductor PbTl at about 2 K, they found a significant reduction in the flux creep rate if the temperature was decreased after establishing the critical state in the superconductor. After reducing the initial temperature T_o by ΔT, the current density $j = j_c(T_o)$ of the supercurrents became less than the critical current density $j_c(T_o-\Delta T)$ at the lower temperature $T = T_o-\Delta T$. In this sub-critical state, flux creep is strongly reduced. The expression

$$\frac{\partial \Phi}{\partial t}\Big|_{T-\Delta T} \propto \exp(\Delta T \ \ln R) \ \frac{\partial \Phi}{\partial t}\Big|_{T} \tag{4.5}$$

was derived [4.27] for the time dependence of the total flux Φ in a cylindrical superconductor of radius R in the sub-critical state at the temperature $T = T_o-\Delta T$. According to Eq. (4.5), the flux creep rate at the lower temperature is expected to be attenuated exponentially in ΔT.

Using this procedure, a dramatic suppression of the magnetic relaxation was achieved in YBCO thin films [4.28], in aligned YBCO powders [4.29], and in YBCO single crystals [4.32]. It has also been found that the suppressed relaxation in the sub-critical state coincides with that observed after a long time in the critical state at the lower temperature when the critical current was decayed to the level of the lower j_c frozen in the superconductor at the higher temperature [4.32].

An example of the suppression of flux creep in bulk YBCO by this technique is shown in Figure 4.14 [4.33]. The sample, 5 mm in diameter, was magnetized in a superconducting magnet. In order to ensure that the sample was in the critical state, a complete hysteresis loop was passed through starting from $H = 0$ via $\mu_o H = 8$ T, $\mu_o H = -8$ T to $\mu_o H = 1$ T, before the time dependence of magnetization was measured in a vibration sample magnetometer. In order to suppress the flux creep, the sample was cooled from 77 K to 65 K shortly after the start of the relaxation measurement at 77 K. Afterwards, no flux creep effect was detectable with the magnetometer. For comparison, relaxation data obtained for $T = 77$ and 65 K are also shown in Figure 4.14.

The reduction of the relaxation rate has also been reported for a YBCO minimagnet consisting of two bulk YBCO disks [4.25]. A maximum trapped field of $B_{\sigma} = 4.3$ T has been obtained in this magnet at 65 K. After activating the YBCO magnet at this temperature, the magnetic relaxation was measured at 65 K and at several lower temperatures. The creep rate was found to be reduced by factors of 6, 197, and ≥ 1000 on lowering the temperature by $\Delta T = 2$ K, 4 K, and 6 K, respectively [4.25].

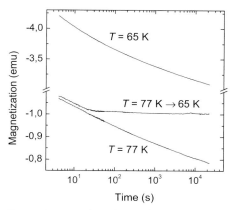

Figure 4.14 Time dependence of magnetization for a bulk YBCO sample measured at T = 77 K and T = 65 K. The relaxation has been found to be strongly suppressed on reducing the temperature from 77 K to 65 K (for details see text)

4.3.3
Pinning Properties from Relaxation Data

Relaxation data for bulk YBCO can be described by collective creep of flux bundles interacting with weak pinning sites (see Section 1.1.6). The non-logarithmic time dependence of the magnetization observed in bulk YBCO at given values of temperature and magnetic field can be described by Eq. (1.23). From these data, the parameter μ of the collective creep model can be extracted, as was demonstrated in Section 1.1.6.

The pinning potential U_o can be derived using relaxation data obtained at different temperatures following a procedure which was proposed by *Maley* et al. [4.29]. In order to fit the $U(j)$ relation (1.22) of the collective creep model to the experimental data, the activation energy for flux creep is written as

$$U = k_B T \left(C - \ln \left| \frac{dM}{dt} \right| \right) \qquad (4.6)$$

where the parameter C is a time-independent constant. Then, U is plotted against $M_{irr} \propto j$ using $k_B T \ln|dM/dt|$ from relaxation data obtained at different temperatures and a constant parameter C which has to be chosen so that all isotherms of the $U(j)$ plot fall on one smooth curve. It has been shown that the smoothness of this plot can be improved by using the scaling relation $U/g(T)$ (where $g(T) = [1-(T/T_c)^2]^{3/2}$) instead of U [4.30].

As an example, the current dependence of the activation energy $U/g(T)$ of a melt-textured YBCO sample is shown in Figure 4.15. It is clearly seen that $U/g(T)$ depends approximately logarithmically on j, which can be explained both by a logarithmic pinning barrier proposed by Zeldov et al. [4.31] and by the inverse power law barrier of the collective creep model [see Eq. (1.22)]. By using Eq. (1.22) to fit the experimental data, one obtains $U_0/k_B = 109$ K [4.34] for the height of the pinning barrier using standard relations [4.35] for the two quantities $j_{co}(T)$ and $U_0(T)$ which are not influenced by flux creep. Such low pinning energies, which are typically for HTSCs, justify the application of the flux creep model to these superconductors.

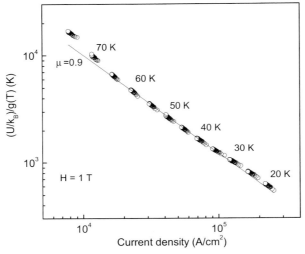

Figure 4.15 Current dependence of the activation energy $U/g(T)$ of a bulk YBCO sample obtained at $\mu_0 H = 1$ T in the temperature range between 20 K and 70 K

The collective flux creep model explains many relaxation data of HTSC samples. However, it should be noted that the collective creep can be destroyed if strong pinning centers are introduced into the superconductor. Then, two scenarios are possible: (a) the flux creep rate reduces to its low value typical for conventional superconductors if strong pins with a high density pin the vortex lattice individually, or (b) the flux creep rate increases because of plastic creep of weakly pinned flux lines along strongly pinned flux lines if the density of strong pins is low. Plastic vortex creep associated with the motion of dislocations in the vortex lattice has been observed in YBCO single crystals [4.36].

4.4 Mechanical Properties

High-T_c superconductors are brittle materials with poor mechanical properties. This significantly affects their application. Typically, at a critical strain the material fractures, which is caused by the propagation of microcracks, which are present in the superconductor. In bulk superconductors, large strains are generated by the application of magnetic fields, this being caused by the large field gradient arising in the critical state within the superconductor. Therefore, it is not surprising that the mechanical strength is extremely important with respect to the highest trapped fields achievable in these superconductors.

The mechanical properties may be of interest for another reason. If superconductivity in an HTSC is based on electron-phonon interaction, then lattice vibrations and the elastic properties of the lattice in the HTSC become relevant for the superconducting transition temperature.

In this chapter, several basic relations will be introduced illustrating the importance of the mechanical properties with regard to the understanding of the phenomenon of superconductivity and to its practical application. Then, data for the mechanical properties of bulk YBCO will be presented.

4.4.1 Basic Relations

Provided that the attractive interaction between the two electrons of a *Cooper* pair is mediated by phonons, then the superconducting transition temperature is closely related to the phonon spectrum and, in particular, to the *Debye* temperature Θ_D, which is connected to the highest frequency of the phonon spectrum considered in the *Debye* model. In this case, the superconducting transition temperature T_c can be expressed as

$$T_c = 1.13 \, \Theta_D \exp\left(-\frac{1+\lambda}{\lambda-\mu^*}\right) \tag{4.7}$$

where λ is the electron-phonon coupling constant and μ^* is the *Coulomb* pseudo potential.

For YBCO, high values of the *Debye* temperatures of $\Theta_D = 410$ K [4.37–4.39] and $\Theta_D = 460$ K [4.40] have been reported. The *Debye* temperature is proportional to the mean sound velocity, which can be replaced for HTSCs by the shear sound velocity v_s. Using the relation $v_s = (G/\rho)^{0.5}$ with G the shear modulus of the lattice and ρ the density, one obtains for T_c the simple relation

$$T_c \propto \Theta_D \propto G^{0.5} \tag{4.8}$$

i.e. in the case of a phonon mechanism of superconductivity, T_c should vary as the square root of the shear modulus.

The shear modulus G is defined by the relation

$$\tau = G\gamma \tag{4.9}$$

where τ is the shear stress and γ the shear strain. Other elastic moduli of the superconductor are the *Young's* (or elastic) modulus E which is defined by *Hooke's* law

$$\sigma = E\varepsilon \tag{4.10}$$

where σ and ε are tensile stress and strain, respectively, and the bulk (or compression) modulus

$$K = -V\frac{dp}{dV} \tag{4.11}$$

where p is the applied hydrostatic pressure and V is the volume of the superconductor.

The fracture strength of a brittle material is not an invariant quantity, but rather controlled by the microcracks within the material. The relation between fracture strength σ_{max}, fracture toughness K_{IC} and the critical length a of microcracks in a brittle material is given by

$$K_{IC} \sim \sigma_{max}(\pi a)^{0.5} \tag{4.12}$$

These microcracks set a limit for the fracture strength of the material because they start to propagate when the external tensile stress reaches the fracture toughness of the material. Taking into account that there is a distribution of microcracks, a corresponding distribution of fracture strengths should exist. According to *Weibull* [4.41], the statistics of fracture is described by

$$w = 1 - \exp[-V_s/V_o(\sigma_{max}/\sigma_o)^m] \tag{4.13}$$

where w is the probability of failure, σ_o is the *Weibull* characteristic strength, m is the *Weibull* modulus, V_o is a volume scale parameter, and V_s is the volume of the test specimen. An important consequence of this relation is that two samples with the same distribution of microcracks but of different size will have different fracture strengths. From Eq. (4.13), one obtains

$$\frac{\sigma_1}{\sigma_2} = \left(\frac{V_2}{V_1}\right)^{1/m} \tag{4.14}$$

i.e. the larger sample ($V_1 > V_2$) has a smaller mean fracture strength σ_1 than the smaller sample.

4.4.2
Mechanical Data for Bulk YBCO

The *elastic properties* of bulk YBCO have been investigated by mechanical and ultrasonic techniques. Data for ceramic, untextured YBCO obtained from ultrasonic measurements are collected in Table 4.1. The data for Young's modulus, shear modulus, and bulk modulus show a weak temperature dependence increasing slightly with decreasing temperature. The magnitude of the elastic moduli of these ceramic samples depends strongly on their porosity $p = 1-\rho/\rho_{max}$ or density ρ, which is less than the theoretical maximum density ρ_{max}. The ratio ρ/ρ_{max} is denoted the relative density in Table 4.1. The effect of porosity is to reduce the ultrasonic wave velocity and, hence, the elastic stiffness moduli. The elastic moduli in the last row of Table 4.1 were estimated [4.42] for a non-porous matrix from experimental raw data obtained on a ceramic sample at room temperature [4.42] by applying the wave-scattering theory in a porous medium [4.43].

Table 4.1 Elastic properties of ceramic YBCO

Young's modulus E (GPa)	T (K)	Shear modulus G (GPa)	T (K)	Bulk modulus K (GPa)	T (K)	Relative density $\rho/\rho_{max} \times 100$	Reference
92.4	288	38.5	291			89	4.44
92.4	252	38.5	252				"
93	200	38.9	207				"
94	150	39.5	139				"
95.1	87	39.9	79				"
95.8	11	40.2	15				"
				70.8	295	88	4.42
				72.5	200		"
				74.4	100		"
				75.3	20		"
95.2	295	37.3	295	70.8	295	88	4.42
125	295	48.5	295	98.7	295	100	"

Investigating the anisotropy of **Young's modulus** E at 300 K, Goyal et al. [4.45] found for a melt-textured YBCO sample a value of $E = 143$ MPa in the c direction and $E = 182$ MPa along the ab plane. Reported values of E for melt-textured YBCO measured at 300 K in the c direction vary between about 96 GPa [4.46] and 143 GPa [4.45]. The Young's modulus of a bulk YBCO sample (in the c direction) has been found to decrease from 128 to 100 GPa after a flux jump induced by applying a magnetic field of 6.5 T at 15 K [4.47]. This reduction in E was explained by additional microcracks introduced by the applied high magnetic field.

The **fracture toughness** (K_{IC}) of bulk YBCO was determined at 300 K by the indentation method and the single-edge notch beam technique. The values of the fracture toughness within the *ab* plane of melt-textured YBCO are in the range between 1.5 MPa√m and 2.1 MPa√m (see Table 4.1). A significantly lower fracture toughness (in the range 0.88–1.2 MPa√m) has been found in bulk YBCO without Y211 particles [4.48].

On studying the influence of thermal cycles (between 300 K and 77 K) on the mechanical properties of bulk YBCO, the fracture toughness was found to decrease with increasing number of thermal cycles. After 50 cycles, a reduction in K_{IC} by about 20% was observed. At the same time, the number of cracks per indentation increased [4.49]. Therefore, it was concluded that thermal cycling essentially contributes to the generation and evolution of macro-cracks. The temperature dependence of the fracture toughness of bulk YBCO has been studied in more detail by Yoshina et al. [4.50]. K_{IC} was found to decrease from 1.3 to 0.4 MPa√m on reducing the temperature from 300 K to 40 K. Cracks which were induced by the indentation process were observed to increase in length with decreasing temperature.

The **flexural** or **bending strength** σ_B of bulk YBCO was measured by a three-point bending test. Data for σ_B obtained for melt-textured YBCO at 300 K are collected in Table 4.1. The stress-strain relation of small YBCO samples cut from a large melt-textured YBCO disk has been investigated by the three-point bending test and the tensile test [4.51]. The **bending strength** (average value: 104 MPa) was found to be larger by almost a factor of two than the **tensile strength** (average value: 56 MPa), whereas coinciding average values of the **fracture toughness** (of 1.9 MPa√m) were determined by both methods. The main reason for the large difference in the values measured for fracture strengths is the different sample volume evaluated by these two methods. In particular, for the three-bending configuration, the region with the maximum strain is localized on that part of the sample on which the force is acting. It is well known (and has already been discussed in the previous section) that bigger samples are likely to fracture at lower external forces because the fracture always takes place at the weakest part in the sample. Therefore, tensile tests provide more reliable data for the evaluation of the fracture strength of the investigated small sample than bending tests. For the same reason, it is not surprising that the fracture strength of σ_{max} = 30 MPa estimated for big YBCO disks (25 mm in diameter) from cracking under the influence of electromagnetic forces (see Section 5.4.1) is significantly lower than the tensile strength of σ_{max} = 56 MPa obtained from tensile tests on small samples with dimensions of $3 \times 3 \times 4$ mm^3.

According to Eq. (4.12), microcracks inside the bulk YBCO start to propagate when the external tensile stress reaches the fracture toughness of the bulk YBCO. Using $K_{IC} \approx 1.5$ Mpa√m and $\sigma_{max} \approx 30$ MPa from cracking experiments, one estimates from Eq. (4.12) $a = 600$ μm for the size of the largest microcrack, which is confirmed by observations of the distribution of cracks in bulk YBCO.

The bending strength of samples cut from a bulk YBCO pellet has been measured at 300 K and 77 K. At 77 K, an increase of σ_B up to about 40% was observed (see Table 4.1) which was found to depend on the location of the sample in the YBCO pellet [4.52].

Improved values of the bending strength and fracture toughness at 300 K have been reported by the addition of silver [4.53–4.55]. In YBCO with 15 vol% Ag, the fracture toughness has been found to enhance from 1.6 up to 2.8 MPa√m or even 3.4 MPa√m in the case of small and uniformly distributed Ag particles [4.55]. This improved fracture strength is probably caused by the strongly reduced number of microcracks in the YBCO/Ag composite material [4.56] (see Section 6.2).

Table 4.2 Mechanical properties of bulk YBCO (at $T=300$ K unless otherwise noted; Ag addition in vol%)

Young's modulus (GPa)	Flexural strength (MPa)	Fracture toughness			Comment	Reference
		(001) (MPa√m)	(100/010) along ab (MPa√m)	(100/010) along c (MPa√m)		
128	77.3	1.48	1.48	2.05		4.47
95.9						4.46
			1.5 – 2.0	1.5 – 2.0		4.49
			0.88 – 1.2		without 211	4.48
			1.6 – 2.1		with 211	"
		1.9			without Ag	4.53
		2.4			+ 5.6% Ag	"
110	85	1.6			without Ag	4.55
103	103	2.1			+ 5% Ag[a]	"
97	120	2.8			+15% Ag[a]	"
103	110	2.6			+ 5% Ag[b]	"
101	135	3.4			+15% Ag[b]	"
	104	1.9			bending test[d]	4.51
	56[c]	1.9			tensile test[d]	"
		1.3			$T = 300$ K	4.50
		0.4			$T = 40$ K	"
	56.1				$T = 300$ K	4.52
	76.7				$T = 77$ K	"

a – average diameter of Ag particles 12 μm,
b – average diameter of Ag particles 2 μm,
c – tensile strength,
d – YBCO sample with 15 wt%Ag (or 8.4 vol.% Ag)

Enhanced values of the fracture strength of bulk YBCO have also been reported by resin impregnation [4.57]. This enhancement of σ_B was found to be stronger at 77 K than at 300 K, presumably because of the strong temperature dependence of the strength of the epoxy resin. On measuring the tensile strength of bulk YBCO at 300 K, an additional improvement was found on wrapping the bulk with carbon fiber fabric prior to resin impregnation [4.58]. The tensile strength of the bulk YBCO without resin impregnation was about 12 MPa, which was found to increase (a) to 18 MPa after resin impregnation and (b) to 29 MPa for the resin-impregnated YBCO wrapped with carbon fiber fabrics.

4.5
Selected Thermodynamic and Thermal Properties

In this section, selected thermodynamic and thermal properties such as the specific heat, the thermal conductivity, and the thermal expansion will be considered. Knowledge of specific heat and thermal conductivity data of HTSC is essential to understand the response of the superconductor to heat released due to variations of the applied magnetic field. Examples are the local heating of the superconductor if it is activated by pulsed fields and the phenomenon of flux jumps occurring at low temperatures. The consequence is in both cases a considerable limitation of the trapped field in the superconductor to a value below that expected from the critical current density at the given temperature.

In order to avoid cracking of YBCO permanent magnets, one can try to compensate the unavoidable magnetic tensile stress in the superconductor by coating the YBCO magnet with a metal casing, generating a compressive stress in the superconductor after cooling from 300 K to 77 K. In order to find a suitable material, one has to know the thermal expansion coefficients of YBCO.

Furthermore, thermodynamic and thermal properties of HTSC are of fundamental interest. In particular, specific heat and thermal conductivity measurements have been used to study the symmetry of the order parameter in HTSC. This aspect will be briefly considered in the next section. In the main part of this chapter, experimental data for YBCO will be presented.

4.5.1
Symmetry of the Order Parameter

Specific heat experiments on bulk superconductors are insensitive to the phase of the order parameter which is investigated in tunneling experiments. However, they can provide valuable information on the density of states $N(E)$ near the *Fermi* level because the electronic part of the specific heat c_{el} is proportional to $N(E)$.

The specific heat of a superconductor in the normal state in the temperature range immediately above T_c ($T < 0.1\Theta_D$ with Θ_D as the *Debye* temperature) is given by

$$c = c_{el} + c_{ph} = \gamma_N T + \beta T^3 \qquad (4.15)$$

where the first term is the electronic contribution to the specific heat, γ_N is the Sommerfeld parameter, and the second term is the phonon contribution. The phonon contribution resulting from the Debye model is cubic in temperature at temperatures above T_c and becomes constant at high temperatures ($T \gg \Theta_D$).

In the superconducting state, the electronic specific heat of a fully gapped s wave superconductor is exponentially activated and vanishes at low temperatures and in zero magnetic fields, i.e. $c_{el}/T \approx 0$. In contrast, in the case of d wave superconductivity, $c_{el}/T \propto T$ is expected at low temperatures and in zero magnetic fields. Also the field dependence of the electronic specific heat in the presence of flux lines in the mixed state is quite different for s and d wave superconductors. For isotropic s wave superconductors, $c_{el}(B)/T \propto B/B_{c2}$ increases proportionally to the area occupied by the vortex cores. In contrast, $c_{el}(B)/T$ scales as $(B/B_{c2})^{1/2}$ for d wave superconductors, i.e. $c_e(B)/T \propto (B/B_{c2})^{1/2}$.

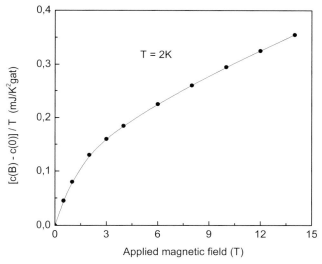

Figure 4.16 Specific heat difference $[c(H)-c(0)]/T$ of a YBCO single crystal versus magnetic field H applied parallel to the c axis at $T = 2$ K, showing a nearly square-root law. Data from Ref. [4.59]

Unfortunately, in the case of YBCO it is not easy to separate the electronic specific heat in the superconducting state from the total specific heat because of the presence of additional contributions. The specific heat of YBCO, which is dominated by the phonon term c_{ph}, contains several other ill-defined contributions arising from weakly interacting paramagnetic centers and a linear term $c_{lin} \propto T$ of uncertain origin. Nevertheless, the d wave character of YBCO has been verified by low-temperature specific heat measurements on high-quality YBCO single crystals containing a very low concentration of magnetic impurities [4.59]. The reported data show both the $c_e/T \propto T$ law at zero (and low) magnetic field and a $c_e(B)/T \propto B^{1/2}$ dependence at high magnetic fields (see Figure 4.16) expected in the case of $d_{x^2-y^2}$ pairing symmetry. Because a possible additive of a fully-gapped s wave component cannot be excluded from specific heat data for YBCO, they are compatible with both pure d wave and $(s + d)$ wave superconductivity.

Another experimental probe to explore the structure of the superconducting gap is to study the angular variation of the thermal conductivity for a magnetic field rotating in the CuO$_2$ plane. For an s wave superconductor in the high temperature regime below T_c, the thermal conductivity should be minimal when the magnetic field is perpendicular to the heat current and vice versa. Indeed, this twofold symmetry (for a variation of the angle between applied magnetic field and heat flow of 360°) has been observed in niobium [4.60]. However, a different behavior has been found for YBCO showing a four-fold symmetry of the thermal conductivity if the magnetic field rotates within the ab plane while the heat is injected along the b axis. This four-fold symmetry is consistent with a $d_{x^2-y^2}$ superconducting order parameter which has four nodes in the k space structure of the gap [4.60].

4.5.2
Specific Heat

At $T = T_c$, the electronic specific heat of a superconductor jumps abruptly from its normal state value $c_n = \gamma_N T_c$ (with γ_N as the *Sommerfeld* parameter) to the superconducting state value c_s. The BCS theory predicts (for s wave superconductivity, mean-field type transitions, and weak electron-phonon coupling) a ratio $(c_s - \gamma_N T_c) / (\gamma_N T_c) = \Delta c / \gamma_N T_c = 1.43$.

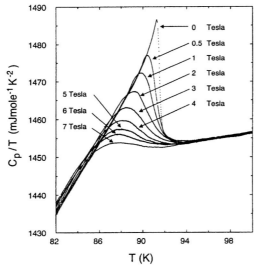

Figure 4.17 Specific heat anomaly of a Y-123 single crystal at the superconducting phase transition at various magnetic fields applied parallel to the c axis. Taken from Ref. [4.61]

In Figure 4.17, the specific heat for a Y-123 single crystal is shown [4.61] measured at temperatures around T_c for various magnetic fields oriented in the c direction. The jump of the specific heat at T_c cannot be described by the mean-field approximation, which is mainly because of strong fluctuation effects in Y-123. In addition, d wave superconductivity and strong electron-phonon coupling in Y-123 cause deviations from the BCS relation. It has been shown [4.61] that the specific heat data at zero field can be described on the basis of a 3D-XY fluctuation model for phase transitions [4.62].

The electronic contribution to the specific heat data in Figure 4.17 is, with $c_e/T_c = \gamma_N \sim 25$ mJ mol^{-1} K^{-2}, much smaller than the total specific heat measured at 94 K, which is completely dominated by the specific heat of the lattice.

4.5.3
Thermal Expansion

It is well known that the linear thermal expansion coefficient $a = 1/l\, \partial l/\partial T$ of a metal (or a superconductor in the normal state) can be described at low temperatures ($T < 0.05\Theta_D$) along the lines of the specific heat by

$$a = a_{el} + a_{latt} = a_1 T + a_2 T^3 \qquad (4.16)$$

where a_{el} and a_{latt} are the electron and lattice contributions to the thermal expansion, respectively. The ratio a/c_v from thermal expansion and specific heat at constant volume has been found to be temperature-independent over a wide temperature range and can be expressed by *Grüneisen's* equation [4.63]

$$\frac{a}{c_v} = \frac{\gamma_{Gr}}{K} \qquad (4.17)$$

where K is the compression modulus and γ_{Gr} is the *Grüneisen* parameter.

The linear temperature term in Eqs. (4.15) and (4.16) should be absent in the superconducting state. However, linear terms in specific heat and thermal expansion data have been observed for Y123 at low temperatures, and these have been attributed to non-superconducting phases. Reported values for the *Grüneisen* parameter of Y123 are $\gamma_{Gr} \approx 1.4$ [4.64–4.66] and $\gamma_{Gr} \approx 3$ [4.67, 4.68]. For the estimation of the lower value of γ_{Gr} from Eq. (4.17), a compression modulus of $K \approx 94$ GPa has been assumed, which is more realistic than the large value of $K \approx 180$ GPa used for the estimation of $\gamma_{Gr} \approx 3$.

Linear thermal expansion coefficients for Y-123 single crystals and melt-textured Y-123 have been measured using a high-resolution capacitance dilatometer [4.69]. At $T = 100$ K, the linear thermal expansion coefficient for both samples is in the range 10^{-5} to 4×10^{-6} K^{-1} along the c axis and in the ab plane, respectively, as shown in Figure 4.18. Thus, the thermal expansion coefficient is considerably larger along the c axis than along the a or b axis, which is in agreement with X-ray [4.70] and neutron diffraction measurements [4.71].

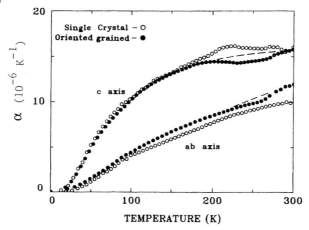

Figure 4.18 Temperature dependence of the linear thermal expansion coefficient for a Y-123 single crystal (open circles) and an oriented-grained YBCO sample (solid circles) along and perpendicular to the c axis. Taken from Ref. [4.69]

The second-order transition into the superconducting state at T_c is connected with a jump in the thermal expansion coefficient Δa which is closely related to the specific heat jump Δc according to the *Ehrenfest* relationship

$$\Delta a = \frac{\Delta c}{T_c} \frac{dT_c}{dp} \tag{4.18}$$

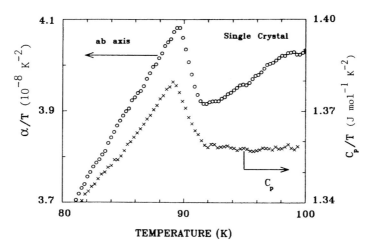

Figure 4.19 Discontinuities in the thermal expansion coefficient and specific heat curves at T_c for a Y-123 single crystal. Taken from Ref. [4.69]

where dT_c/dp is the pressure dependence of T_c. A jump Δa_{ab} in the thermal expansion coefficient a_{ab} (along the a or b axis) at T_c has been observed both for Y-123 single crystals (Figure 4.19) and melt-textured Y-123 samples [4.69]. Data for Δa_{ab} and Δc obtained for a YBCO single crystal [4.69] are shown in Figure 4.19. The relation between Δa_{ab} and Δc according to Eq. (4.18) has been used to estimate the uniaxial pressure dependence of T_c as $dT_c/dp_{ab} = 0.089$ K kbar^{-1} (for a Y-123 single crystal) and 0.036 K kbar^{-1} (for a melt-textured Y-123 sample). The larger value of dT_c/dp_{ab} for the single crystal has been attributed to a small amount of Al (about 2%) due to the method of processing it [4.69].

4.5.4
Thermal Conductivity

For several applications of bulk HTSC, its unusual thermal transport properties play a decisive role. For the application as bulk magnets at 77 K, the high heat conduction within the ab planes of bulk YBCO is essential to prevent thermomagnetic instabilities and flux jumps due to temperature rise. For the application as power current leads for superconducting coils working at 4.2 K, however, the low thermal conductivity of bulk YBCO at helium temperatures is indispensable.

The temperature dependence of the thermal conductivity κ (in polycrystalline YBCO or in the ab plane component of single crystalline and melt-textured YBCO) is characterized by a rapid increase in κ below T_c and a peak at approximately $T_c/2$ [4.72–4.77] as shown in Figure 4.20.

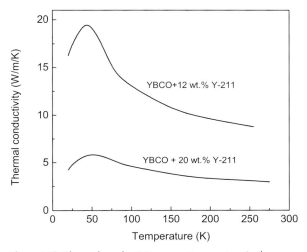

Figure 4.20 Thermal conductivity versus temperature in the ab plane for two melt-texured YBa$_2$Cu$_3$O$_{6.96}$ samples with 12 wt% Y-211 (upper curve) and 20 wt% Y-211 (lower curve). From Ref. [4.77]

To understand $\kappa(T)$, one has to take into account the phonon contribution κ_{ph} and the electron contribution κ_{el} to the total thermal conductivity κ. It is well known that at temperatures above T_c, $\kappa(T)$ in HTSC is dominated by κ_{ph}. Below T_c, electrons condensed into Cooper pairs cannot carry entropy and, thus, do not contribute to the thermal conductivity. However, one has to take into account the existence of normal electrons, so-called quasiparticles in the superconducting state. The scattering rate of these quasiparticles in YBCO has been observed to decrease rapidly with decreasing temperature. It is now widely accepted that the rapid increase of κ below T_c and the peak of κ are mostly due to the contribution of quasiparticles located in the CuO_2 planes [4.74, 4.76, 4.78]. However, a clear separation of the phonon and quasiparticle heat currents is difficult, and a number of authors have explained the peak in κ in YBCO as essentially due to a phonon contribution [4.79–4.81].

The absolute value of κ^* at the peak temperature strongly depends on the Y-211 content, as demonstrated in Figure 4.20. κ^* is significantly higher for the YBCO sample with the lower Y-211 content, and its value approaches that for good YBCO single crystals, which is in the range $\kappa^* \approx 20 - 28$ W m^{-1} K^{-1}. The reason for the lower thermal conductivity of the YBCO sample with the larger Y-211 content appears to be the scattering from the large number of Y-211 grain boundaries [4.77].

References

4.1 Y. Yeshurun, A. P. Malozemoff, Phys. Rev. Lett. **60**, 20202 (1988).

4.2 D. S. Fisher, M. P. A. Fisher, D. A. Huse, Phys. Rev. B **43**, 130 (1991).

4.3 J. L. O'Brien, H. Nakagawa, A. S. Dzurak, R. G. Clark, B. E. Kane, N. E. Lumpkin, N. Miura, E. E. Mitchel, J. D. Goettee, J. S. Brooks, D. G. Rickel, R. P. Starrett, Phys. Rev. B **61**, 1584 (2000).

4.4 U. Welp. W. K. Kwok, G. W. Crabtree, K. G. Vandervoort, J. Z. Liu, Phys. Rev. Lett. **62**, 1908 (1989).

4.5 T. Nishizaki, N. Kobayashi, Supercond. Sci. Technol. **13**, 1 (2000).

4.6 R. Cubitt et al., Nature **365**, 407 (1993).

4.7 S. L. Lee et al., Phys. Rev. Lett. **71**, 3862 (1993).

4.8 C. Bernhard, C. Wenger, Ch. Niedermayer, D. M. Pooke, J. L. Tallon, Y. Kotaka, J. Shimoyama, K. Kishio, D. R. Noakes, C. E. Stronach, T. Sembiring, E. J. Ansaldo, Phys. Rev. B **52**, R7050 (1995).

4.9 J. E. Sonier, J. H. Brewer, R. F. Kiefl, D. A. Bonn, J. Chakhalian, S. R. Dunsiger, W. N. Hardy, R. Liang, W. A. MacFarlane, R. I. Miller, D. R. Noakes, T. M. Riseman, C. E. Stronach, Phys. Rev. B **61**, R890 (2000).

4.10 B. Khaykovich, E. Zeldov, D. Majer, T. W. Li, P. H. Kes, M. Konczykowski, Phys. Rev. Lett. **76**, 2555 (1996).

4.11 B. Khaykovich, M. Konczykowski, E. Zeldov, R. A. Doyle, D. Majer, P. H. Kes, T. W. Li, Phys. Rev. B **56**, R517 (1997).

4.12 H. Küpfer, Th. Wolf, C. Lessing, A. A. Zhukov, X. Lancon, R. Meier-Hirmer, W. Schauer, H. Wühl, Phys. Rev. B **58**, 2886 (1998).

4.13 H. Küpfer, Th. Wolf, R. Meier-Hirmer, A. A. Zhukov, Physica C **332**, 80 (2000).

4.14 K. Salama and D. F. Lee, Supercond. Sci. Technol. **7**, 177 (1994).

4.15 M. Lepropre, I. Monot, M. P. Delamare, M. Hervieu, Ch. Simon, J. Prevost, G. Desgardin, B. Raveau, J. M. Barbut,

D. Bourgault, D. Braithwaite, Cryogenics **34**, 63 (1994).

4.16 A. I. Larkin, Yu. N. Ovchinnikov, J. Low Temp. Phys. **34**, 409 (1979).

4.17 S. Nariki, N. Sakai, M. Murakami, I. Hirabayashi, Supercond. Sci. Technol. **17**, S30 (2004).

4.18 B. Martinez, X. Obradors, A. Gou, V. Gomis, S. Pinol, J. Fontcuberta, H. Van Tol, Phys. Rev. B **53**, 2797 (1996).

4.19 A. P. Malozemoff, M. P. A. Fisher, Phys. Rev. B **42**, 6784 (1990).

4.20 L. Civale, A. D. Marwick, M. W. McElfresh, T. K. Worthington, A. P. Malozemoff, F. Holtzberg, J. R. Thompson, M. R. Kirk, Phys. Rev. Lett. **65**, 1164 (1990).

4.21 G. Blatter, M. V. Feigel'man, V. B. Geshkenbein, A. I. Larkin, V. M. Vinokur, Rev. Mod. Phys. **66**, 1125 (1994).

4.22 L. Fruchter, A. P. Malozemoff, I. A. Campbell, J. Sanchez, M. Konczykowski, R. Griessen, F. Holtzberg, Phys. Rev. B **43**, 8709 (1991).

4.23 G. T. Seidler, C. S. Carillo, T. F. Rosenbaum, U. Welp, G. W. Crabtree, V. M. Vinokur, Phys. Rev. Lett. **70**, 2814 (1993).

4.24 G. T. Seidler, T. F. Rosenbaum, K. M. Beauchamp, H. M. Jaeger, G. W. Crabtree, U. Welp, V. M. Vinokur, Phys. Rev. Lett. **74**, 1442 (1995).

4.25 R. Weinstein, J. Liu, Y. Ren, R. Sawh, D. Parks, C. Foster, V. Obot, in *Proceedings 10th Anniversary HTS Workshop on Physics, Materials and Applications*, edited by B. Batlogg, C. W. Chu, W. K. Chu, D. U. Gubser, and K. A. Müller (World Scientific, Singapore, 1996) p. 625.

4.26 Y. Yeshurun, A. P. Malozemoff, A. Shaulov, Rev. Mod. Phys. **68**, 911 (1996).

4.27 M. R. Beasley, R. Labusch, W. W. Webb, Phys. Rev. **181**, 682 (1969).

4.28 J. Z. Sun, B. Lairson, C. B. Eom, J. Bravman, T. H. Geballe, Science **247**, 307 (1990).

4.29 M. P. Maley, J. O. Willis, H. Lessure, M. E. McHenry, Phys. Rev. B **42**, 2639 (1990).

4.30 M. E. McHenry, S. Simizu, H. Lessure, M. P. Maley, J. Y. Coulter, I. Tanaka, H. Kojima, Phys. Rev. B **44**, 7614 (1991).

4.31 E. Zeldov, N. M. Amer, G. Koren, A. Gupta, M. W. McElfresh, R. J. Gambino, Appl. Phys. Lett. **56**, 680 (1990).

4.32 J. R. Thompson, Yang Ren Sun, A. P. Malozemoff, D. K. Christen, H. R. Kerchner, J. G. Ossandon, A. D. Marwick, F. Holtzberg, Appl. Phys. Lett. **59**, 2612 (1991).

4.33 G. Fuchs, S. Gruss, G. Krabbes, P. Schätzle, J. Fink, K.-H. Müller and L. Schultz, in *Advances in Superconductivity X*, ed. by K. Osamura and I. Hirabayashi, (Springer-Verlag, Tokyo, 1998), Vol. 2, p. 847.

4.34 S. Gruss, G. Fuchs, G. Krabbes, P. Schätzle, J. Fink, K.-H. Müller, L. Schultz, IEEE Trans. Appl. Superconductivity **9**, 2070 (1999).

4.35 J. R. Thompson, Yang Ren Sun, L. Civale, A. P. Malozemoff, M. W. McElfresh, A. D. Marwick, F. Holtzberg, Phys. Rev. B **47**, 14440 (1993).

4.36 Y. Abulafia, A. Shaulov, Y. Wolfus, R. Prozorov, I.. Burlachkov, Y. Yeshurun, Phys. Rev. Lett. **77**, 1596 (1996).

4.37 S. J. Collocott, R. Driver, E. R. Vance, Phys. Rev. B **41**, 6329 (1990).

4.38 A. Junod, in *Physical Properties of High Temperature Superconductors*, ed. by D. M. Ginsberg (World Scientific, Singapore, 1990), Vol. 2, Chap. 2.

4.39 S. E. Stupp. T. A. Friedmann, J. P. Rice, R. A. Schweinfurth, D. J. Van Harlingen, D. M. Ginsberg, Phys. Rev. B **43**, 13073 (1991).

4.40 H. J. Fink, Phys. Rev. B **58**, 9415 (1998).

4.41 W. Weibull, J. Appl. Mechan. **18**, 293 (1951).

4.42 M. Cankurtaran, G. A. Saunders, K. C. Goretta, Supercond. Sci. Technol. **7**, 4 (1994).

4.43 M. Cankurtaran, G. A. Saunders, J. R. Willis, A. Al-Kheffaji, D. P. Almond, Phys. Rev. B **39**, 2872 (1989).

4.44 H. Yusheng, X. Jiong, J. Sheng, H. Aisheng, Z. Jincang, Physica B **165**, 1283 (1990).

4.45 A. Goyal, P. D. Funkenbusch, D. M. Kroeger, S. J. Burns, J. Appl. Phys. **71**, 2362 (1992).

4.46 R. Ravinder Reddy, M. Murakami, S. Tanaka, P. Venugopal Reddy, Physica C **257**, 137 (1996).

4.47 Feng Yu, K. W. White, Ruling Meng, Physica C **276**, 295 (1997).

4.48 H. Fujimoto, M. Murakami, N. Koshizuka, Physica C **203**, 103 (1992).

4.49 A Leenders, M. Ullrich, H. C. Freyhardt, Physica C **279**, 173 (1997).

4.50 Y. X. Yoshino, A. Iwabuchi, K. Noto, N. Sakai, M. Murakami, Physica C **357-360**, 796 (2001).

4.51 T. Okudera, A. Murakami, K. Katagiri, K. Kasaba, Y. Shoji, K. Noto, N. Sakai, M. Muratami, Physica C **392-396**, 628 (2003).

4.52 M. Tomita, M. Murakami, K. Katagiri, Physica C **378-381**, 783 (2002).

4.53 P. Schätzle, G. Krabbes, S. Gruss, G. Fuchs, IEEE Trans. Appl. Supercond. **9**, 2022 (1999).

4.54 J. Joo, J.-G. Kim, W. Nah, Supercond. Sci. Technol. **11**, 645 (1998).

4.55 Jinho Joo, Seung-Boo Jung, Wansoo Nah, Jung-Yeul Kim, Tae Sung Kim, Cryogenics **39**, 107 (1999).

4.56 P. Diko, G. Fuchs, G. Krabbes, Physica C **363** 60 (2001).

4.57 M. Tomita, M. Murakami, Supercond. Sci. Technol. **15**, 808 (2002).

4.58 M. Tomita, M. Murakami, K. Yoneda, Supercond. Sci. Technol. **15**, 803 (2002).

4.59 Y. Wang, B. Revaz, A. Erb, A. Junod, Phys. Rev. B **63**, 094508 (2001).

4.60 H. Aubin, K. Behnia, M. Ribault, L. Taillefer, R. Gagnon, Z. Phys. B **104**, 175 (1997).

4.61 O. Jeandupeux, A. Schilling, H. R. Ott, A. van Otterlo, Phys. Rev. B **53**, 12475 (1996).

4.62 D. S. Fisher, M. P. A. Fisher, D. A. Huse, Phys. Rev. B **43**, 130 (1991).

4.63 E. Grüneisen, Ann. Phys. 2, 294 (1908).

4.64 S. J. Collocott, G. K. White, S. X. Dou, R. K. Williams, Phys. Rev. B **36**, 5684 (1987).

4.65 H. Ledbetter, M. Lei, A. Hermann, Z. Sheng, Physica C **225**, 397 (1994).

4.66 C. Haetinger, I. Abrega Castillo, J. V. Kunzler, L. Ghivelder, P. Pureur, S. Reich, Supercond. Sci. Technol. **9**, 639 (1969).

4.67 E. Salomons, H. Hemmes, J. J. Scholtz, N. Koeman, R. Brouwer, A. Driessen, D. G. de Groot, R. Griessen, Physica B **145**, 253 (1987).

4.68 C. A. Swenson, R. W. McCallum, K. No, Phys. Rev. B **40**, 8861 (1989).

4.69 C. Meingast, B. Blank, H. Bürkle, B. Obst, T. Wolf, H. Wühl, Phys. Rev. B **41**, 11299 (1990).

4.70 P. Horn, D. T. Keane, G. A. Held, J. L. Jordan-Sweet, D. L. Kaiser, F. Holtzberg, T. M. Rice, Phys. Rev. Lett. **59**, 2772 (1988).

4.71 J. J. Cappono, C. Chaillout, A. W. Hewat, P. Lejey, M. Marezio, N. Nguyen, B. Raveau, J. L. Soubeyroux, J. L. Tholence, R. Tournier, Europhy. Lett. **3**, 1301 (1987).

4.72 C. P. Popoviciu, J. L. Cohn, Phys. Rev. B **55**, 3155 (1997).

4.73 J. L. Cohn, E. F. Skelton, S. A. Wolf, J. Z. Liu, R. N. Shelton, Phys. Rev. B **45**, 13144 (1992).

4.74 R. V. Yu, M. B. Salomon, J. P. Lu, W. C. Lee, Phys. Rev. Lett. **69**, 1431 (1992).

4.75 G. A. Shams, J. W. Cochrane, G. J. Russel, Physica C **363**, 243 (2001).

4.76 B. Zeini, A. Freimuth, B. Büchner, R. Gross, A. P. Kampf, M. Kläser, G. Müller-Vogt, Phys. Rev. Lett. **82**, 2175 (1999).

4.77 G. A. Shams, J. W. Cochrane, G. J. Russel, Physica C **351**, 449 (2001).

4.78 K. Krishana, N. P. Ong, Y. Zhang, Z. A. Yu, R. Gagnon, L. Taileffer, Phys. Rev. Lett. **82**, 5108 (1999).

4.79 L. Tewordt, Th. Wölkhausen, Solid State Commun. **70**, 839 (1989).

4.80 J. L. Cohn, V. Z. Kresin, M. E. Reeves, S. A. Wolf, Phys. Rev. Lett. **71**, 1657 (1993).

4.81 A. V. Inyushkin, A. N. Taldenkov, Phys. Rev. B **54**, 13261 (1996).

5
Trapped Fields

5.1
Low-Temperature Superconductors

Trapped fields in low-temperature superconductors have been investigated at the temperature of liquid helium and below. As early as 1965 it was demonstrated that a magnetic field of 2.7 T can be trapped at 1.5 K in a bulk Nb_3Sn hollow cylinder [5.1]. Later, a hollow cylinder consisting of Cu-clad Nb_3Sn ribbons was used to trap a magnetic field of 2.2 T at 4.2 K [5.2]. However, the trapped field in bulk superconductors at low temperatures was found to be strongly limited by thermomagnetic instabilities, resulting in flux jumps as the external field exceeds a certain value of the applied magnetic field B_j. Under adiabatic conditions, the instability field B_j at which the first flux jump is observed decreases with the specific heat. Since the specific heat of superconductors is very low at helium temperatures, thermomagnetic instabilities are very pronounced at low temperatures.

The discovery of HTSC stimulated new activities in the field of superconducting permanent magnets in bulk material. Because of the large heat conduction of HTSC at higher temperatures, the released heat within the superconductor is removed in the surrounding cryogenic agent. Therefore, thermomagnetic instabilities have no influence on the trapped field in HTSC, at least at temperatures above 30 K. Very high trapped fields can be achieved in HTSC taking advantage of the increasing critical current density at temperatures below 77 K. However, the so enhanced trapped field is limited by the mechanical properties of the material. This is discussed in more detail in Section 5.4.

5.2
Bulk HTSC at 77 K

HTSC disks are usually characterized by mapping of the trapped field at 77 K using liquid nitrogen as coolant. It is convenient to use a superconducting coil with a room temperature bore in order to magnetize the HTSC disks. This can be done either by *field cooling* or by *zero-field cooling*. Both cases are illustrated in Figure 5.1. The applied magnetic field, which is directed perpendicular to the top sur-

High Temperature Superconductor Bulk Materials.
Gernot Krabbes, Günter Fuchs, Wolf-Rüdiger Canders, Hardo May, and Ryszard Palka
Copyright © 2006 WILEY-VCH Verlag GmbH & Co. KGaA, Weinheim
ISBN: 3-527-40383-3

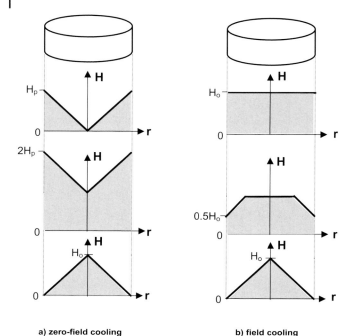

Figure 5.1 Field distribution within a bulk superconducting permanent magnet during and after the magnetizing procedure. **a** After zero-field cooling, magnetic fields H_p with H_p as the penetration field (upper panel) and $2H_p$ (middle panel) are applied. A trapped field $H_o = H_p$ remains in the superconductor after switching off the applied field (lower panel). **b** The superconductor is field-cooled applying a field H_o (upper panel). This field remains in the superconductor after switching off the applied field (lower panel)

face of the disks, should be at least twice as large as the maximum trapped field if the superconductor is cooled before applying the magnetic field (*zero-field cooling*). The advantage of the *field cooling* mode is the lower magnetizing field which should correspond approximately to the expected maximum trapped field in the superconductor.

In order to map the trapped field, a Hall sensor with a small active area is moved across the top face of the disk at a small, fixed distance above the surface of the sample. Typical equipment for field mapping is shown schematically in Figure 5.2.

The field profile of a YBCO disk 25 mm in diameter is shown in Figure 5.3. The curved shape of the distribution of the trapped field in bulk YBCO superconductors reflects the field dependence of the critical current density. A maximum trapped field of $B_o = 1.2$ T was found for this YBCO disk, this being a typical value for unirradiated YBCO of this size. In most cases, the reported values of B_o are not corrected with respect to (a) the distance between sample and active area of the Hall probe and (b) the time between magnetization procedure and field mapping. The trapped field distribution shown in Figure 5.3 was measured about 5 min after the magnetization procedure. The correction for a shorter time of 10 s and zero distance from the active area of the Hall probe would enhance the maximum trapped field by about 15%, i.e. $B_0 = 1.4$ T.

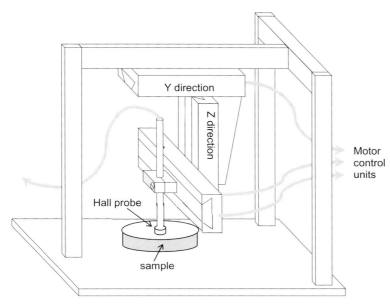

Figure 5.2 Equipment for field mapping. Not shown is the container with liquid nitrogen for cooling the sample to 77 K

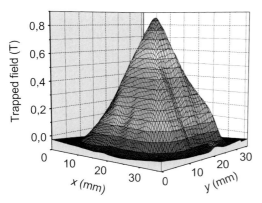

Figure 5.3 Typical distribution of the trapped field above a YBCO sample 25 mm in diameter measured at 77 K

Recently, a series of round robin tests of the flux-trapping ability of two high-quality, epoxy-reinforced melt-processed Sm123 samples was performed [5.3] over a 20 month period in four European laboratories within the EFFORT consortium (European Forum for Processors of Large Grain (*RE*)BCO Materials – see www.rebco-effort.net) in collaboration with SRL-ISTEC. These tests were initiated by the international scientific community under the auspices of the Versailles Project on Advanced Materials and Standards (VAMAS) and the Technical Committee 90 (TC90) of the International Electrotechnical Commission (IEC), and was organized by the IRC in Superconductivity (Cambridge, UK). The aim of these tests was to check the reliability and reproducibility of field profile measurements as a standard method to characterize bulk superconductors.

The Sm123 samples used for the round robin tests were fabricated in the Superconductivity Research Laboratory (ISTEC, Japan) by a top-seeded melt growth technique. After polishing the top and bottom surfaces of the melt-grown pellets, the mechanical properties of the samples were improved by a resin impregnation technique. Finally, the samples were coated with a 0.5 mm thick CFRF layer in order to protect the surface of the superconductor. The two pellets, 50 mm in diameter and 19 mm in thickness, were investigated by the following laboratories: IPHT Jena (initial and final measurement), IFW Dresden, ZFW Göttingen, and ATI Vienna.

The trapped-field distribution of the upper and lower surfaces of both samples have been measured in liquid nitrogen using the locally standard measurement conditions in each laboratory. Each measurement involved cooling the sample to 77 K in an applied magnetic field, removing the magnetizing field to generate a trapped field in the sample, and scanning with a Hall probe across the sample surface in order to measure the distribution of the trapped field. The obtained field profiles were found to be particularly sensitive to the distance between sample and *Hall* probe and to flux creep effects within the time window between magnetizing the samples and scanning the trapped field. Both effects have to be taken into account in order to get comparable results. The data obtained for the maximum trapped field were found to agree within 10%, if the same conditions were realized in the participating laboratories, i.e. (a) a 15 min relaxation time prior to scan and (b) a distance of 0.8 mm between Hall probe and sample.

It was observed that one sample retained its field trapping ability throughout the tests, whereas a slight deterioration in the properties was found for the other sample. It was also found that the flux-trapping ability of the top surface of each sample was around twice as good as that of the bottom surface. This is not surprising, and is consistent with data from ATI Vienna. On studying the flux-trapping properties in various layers of YBCO pellets, it was found that the top 4 mm of a superconductor (diameter $\approx$ 26 mm, thickness $\approx$ 10 mm) contributes up to 70% of the trapped field at the surface of the magnetized pellet [5.4]. The corresponding numbers for the samples investigated in the round robin tests indicate that the top 6 mm contribute ca. 50% to the maximum trapped field.

5.3
Trapped Field Data at 77 K

Trapped field data of HTSC disks reported by several groups at 77 K are collected in Table 5.1. This table also contains results for YBCO materials improved by doping and irradiation, which will be considered in Chapter 6.

In Table 5.1, the ratio $\mu_o^{-1} B_o/d$ is shown in order to take the different sample sizes into account. This ratio can be considered as an "average" critical current density of the bulk HTSC sample, averaged not only over the sample volume but also over the field-dependent critical current densities within the field range up to B_o. Therefore, this average $j_c^{av} = \mu_o^{-1} B_o/d$ is much smaller than the critical current density measured on a small part of the bulk HTSC sample in self-field.

Table 5.1 Maximum trapped field B_o of bulk HTSC at 77 K

Bulk HTS	B_o (T)	Size d (diameter or square edges) (mm)	$\frac{B_o}{\mu_o}/d$ (kA cm^{-2})	Remarks on the preparation	References
YBCO	1.2	25	3.8	Zn doped	5.5
	1.2	26	3.7	Li doped	5.6
	1.15	38 × 38	2.4	CeO$_2$ addition, batch [a]	5.7
	1.4	38 × 38	2.9	CeO$_2$ addition	5.7
	1.3	38 × 38	2.7	CeO$_2$ addition	5.8
	1.3	60	1.7	Batch [a]	5.9
	1.6	33	3.9	fine Y211, Ag add.	5.10
	0.5	9 × 9	4.4	YBaCuO precipitates	5.11
	2.1	20	8.4	U/n [b]	5.12
	2.1	26	6.4	Neutron irradiation	5.5
Sm-123/Ag	2.1	36	4.6	OCMG [a]	5.13
	1.6	60	2.1		5.14
Gd-123/Ag	1.4	24	4.6	fine Gd211, OCMG [c]	5.15
	2.7	50	4.3	fine Gd211, OCMG [c]	5.15
	3.0	65	3.7		5.16
	2.1	65	2.6		5.9
Dy-123	1.4	32	3.5	Dy211	5.17
	1.7	48	2.8	in air	5.17
Nd-123	1.4	30	3.7	pre-sintering in oxygen	5.18

a representative values for batches of up to 16 pieces
b uranium-doped and neutron-irradiated
c OCMG: processed under low oxygen partial pressure

For bulk YBCO, several groups have achieved trapped fields B_o (77 K) in the range between 1.2 and 1.3 T. The highest trapped fields have been reported by neutron irradiation. In this context, it should be noted that even higher trapped fields, up to 3.2 T, have been obtained in a mini-magnet consisting of a stack of irradiated YBCO tiles [5.12]. Furthermore, the improvement of B_o achieved by the use of very fine Y211 particles resulting in high critical current densities up to 1.1×10^5 A cm^{-2} [5.10] should be emphasized (see Chapter 3).

In bulk R-123 with R = Sm and Gd, trapped fields B_o (77 K) above 2 T have been achieved even without irradiation [5.13, 5.15, 5.16]. In these HTSCs, not only has j_c been improved, but also the size of the single-domain material has been successfully enlarged.

5.4
Limitation of Trapped Fields in Bulk YBCO at Lower Temperatures

Trapped fields in YBCO can be drastically increased by taking advantage of the higher critical current density and pinning force at temperatures below 77 K. The so enhanced trapped field is limited by the mechanical properties of the material. Usually, cracking is observed as a result of the large tensile stresses acting on the superconductor during the magnetizing procedure. If cracking can be avoided, then thermomagnetic instabilities are observed in the temperature range below about 20 K.

The different limitations of the trapped field of a melt-textured YBCO sample are illustrated in Figure 5.4. In the case of weak pinning (see Figure 5.4a, the trapped field $B_o(T)$, which is determined by critical current density (and sample size), increases with decreasing temperature. Thermomagnetic instabilities

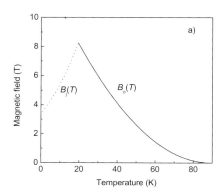
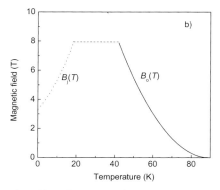

Figure 5.4 Illustration of the different limitations of the maximum trapped field $B_o(T)$ in different temperature ranges. **a** Weak pinning: limitation due to j_c and sample size (solid line) for $T > 20$ K as well as due to flux jumps (dotted line) for $T < 20$ K. **b** Strong pinning: additional limitation of $B_o(T)$ due to magnetic tensile stress (dashed line) in the temperature range between 20 and 40 K

become predominant at temperatures below 20 K, and in this region the trapped field is strongly reduced by flux jumps [5.19]. For stronger pinning (see Figure 5.4b), the trapped field between 20 K and 40 K is limited by the mechanical properties of the superconductor. Below 20 K, again thermomagnetic instabilities become predominant.

5.4.1
Magnetic Tensile Stress and Cracking

The shape of the field profile of superconducting permanent magnets (shown, for instance, in Figure 5.3) is responsible for the large forces acting on the superconductor during and after the magnetizing procedure. The strong flux gradient of the trapped field generates a tensile stress which is transferred to the crystal lattice by the pinned flux lines via the pinning centers which are embedded in the crystal lattice of the superconductor. In general, field gradients within the superconductors result in body forces and macroscopic mechanical deformations.

Therefore, if a zero-field cooled superconductor undergoes a magnetization loop, a compressive force acts in the case of increasing applied magnetic fields and an expansive one with decreasing fields. Magneto-elastic effects in HTSC have been reviewed by Johansen [5.20]. The distribution of the stresses in the ab plane of a bulk superconductor during the magnetizing procedure was calculated by Ren et al. [5.21] within a 2D approach, neglecting the stress along the c axis. Later, this model was extended to the 3D case taking into account also the stress along the c axis [5.22]. According to these models, trapped fields in superconducting permanent magnets are exposed to large tensile stresses. The highest tensile stress occurring at the center of the sample depends on the applied magnetic field. In the field-cooling mode, the maximum tensile stress is given by

$$\sigma = \frac{B_o^2}{2\mu_o}\left[b_m^2 - b^2 - \frac{1-2v}{12(1-v)}(b_m - b)^2(b^2 + 4b + 2b^2 b_m + 8b_m - 3b_m^2)\right] \quad (5.1)$$

where $b = B/B_o$ and $b_m = B_m/B_o$. Here, B_m is the maximum applied field, B_o is the maximum trapped field, and v is the Poisson ratio. The field dependence of σ is shown in Figure 5.5 for $b_m = 1$ and $b_m = 1.5$ using $v = 0.3$ for the Poisson ratio of YBCO. For $b_m = 1$, one obtains

$$\sigma(0) = 0.76 \frac{B_o^2}{2\mu_o} \quad (5.2)$$

in the remnant state at $b = 0$ and a slightly higher tensile stress of $\sigma = 0.78 \, B_o^2 / (2\mu_o)$ at $b = 0.127$. If B_o is quoted in T and σ in MPa, then $\sigma(0)$ reads $\sigma(0) = 0.30 \times B_o^2$. Thus, at $B_o = 9$ T the tensile stress becomes 24.3 MPa, which is in the range of the tensile strength $\sigma_{max} = 20 - 30$ MPa of bulk, melt-textured YBCO.

In practice, it is difficult to realize the case $b_m = 1$, i.e. to choose the activating field B_m exactly equal to the maximum trapped field B_o of the superconductor, which is *a priori* not known. If the activating field B_m exceeds B_o, then the maximum tensile stress strongly increases during the magnetizing procedure. This is shown in Figure 5.5 for $b_m = 1.5$. The tensile stress increases in this case up to about $\sigma = 1.6\ B_o^2\ /\ (2\mu_o)$ at $b \approx 0.52$. For fields $0.52 > b > 0$, the tensile stress decreases linearly with the applied magnetic field and arrives at the same value of $\sigma(0)$ in the remnant state for $b = 0$ as obtained for $b_m = 1$. The linear part of the $\sigma(b)$ relation in Figure 5.5 is given by

$$\sigma = \frac{B_o^2}{2\mu_o}\left[\frac{2}{3}\frac{2-v}{1-v}b + \frac{7-2v}{12(1-v)}\right] \tag{5.3}$$

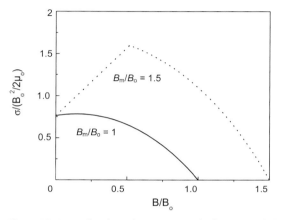

Figure 5.5 Normalized tensile stress vs applied magnetic field normalized by the maximum trapped B_o for two values of the ratio B_m/B_o, where B_m is the magnetizing field applied to activate the superconducting permanent magnet

Hence, the maximum tensile stress acting on the superconductor for $b_m = 1.5$ is about twice as large as that for $b_m = 1$. Therefore, in order to minimize the maximum tensile, it is important to choose a magnetizing field B_m that is not too high.

Tensile stresses in the superconductor cause cracking if the tensile strength σ_{max} of the material is reached. Cracking of the material is indicated by a sudden decrease in the signal of the Hall probe used for measuring the trapped field during the magnetizing procedure. A visual inspection of damaged samples shows a crack on the surface of the sample in some cases, but sometimes no crack is visible. The crack within the superconductor can easily be detected by a field profile measurement at 77 K. A typical example is shown in Figure 5.6.

The crack divides the disk into two superconducting domains. Currents can flow only within the superconducting domains and not across the boundary between the two domains.

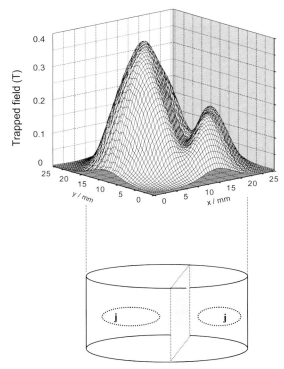

Figure 5.6 Field profile of a damaged YBCO sample divided by a crack. The sketch in the bottom panel shows the two superconducting domains of the YBCO disk

The relatively low tensile strength of $\sigma_{max} = 20 - 30$ MPa found for bulk YBCO originates from small cracks in melt-textured YBCO arising from the sample preparation [5.23, 5.24]. It is evident that cracks preferably parallel to the c axis ("c cracks") start to propagate when the tensile stress developing under the influence of electromagnetic forces reaches the fracture toughness K_{IC} of the material. The length of the largest c cracks represents the flaw size a in the relation $\sigma_{max} \approx K_{IC} (\pi a)^{-0.5}$ [see Eq. (4.12)], which determines – together with the fracture toughness K_{IC} – the fracture strength σ_{max} of brittle ceramics. Using $K_{IC} \approx 1.5$ MPa/$\sqrt{m}$ and $\sigma_{max} \approx 30$ MPa, one estimates from Eq. (4.12) a flaw size $a \approx 600$ μm, which is in agreement with observations of the distribution of cracks in bulk YBCO.

5.4.2
Thermomagnetic Instabilities

The nonuniform field distribution of the trapped field in bulk superconductors does not correspond to an equilibrium state. Therefore, it is not surprising that flux jumps arise in the critical state under certain conditions, resulting in a flux redistribution toward the equilibrium state, which is accompanied by heating of the superconductor. Whereas a local flux jump is restricted to a small fraction of the sample, a global flux jump is characterized by the motion of vortices in the entire volume of the sample. The character of the flux jump depends on the dynamics of the flux and temperature redistribution. The ratio

$$\tau = \frac{t_m}{t_t} = \mu_o \frac{\kappa \sigma}{c} \tag{5.4}$$

of the magnetic (t_m) and thermal (t_t) diffusion time constants has been used [5.25] to describe these dynamics. The thermal diffusion time constant is defined as $t_t = c/\kappa$, where κ is the heat conductivity and c the heat capacity per volume, whereas the magnetic diffusion time constant $t_m = \mu_o \sigma = \mu_o \, \partial j/\partial E$ (σ – electric conductivity) is defined by the slope of the E-j curve of the superconductor.

For $\tau \ll 1$ (adiabatic approach), there is not enough time to remove the heat released due to flux motion, and therefore the flux motion is accompanied by an adiabatic heating within a small fraction of the superconductor. For $\tau \gg 1$ (dynamic approach), the redistributed field profile remains fixed during the rapid heating of the entire volume of the superconductor. Monofilamentary low-temperature superconductors are in the adiabatic limit. The dynamic limit is realized in low-temperature superconducting composites, as, for instance, in multifilamentary wires or tapes consisting of a matrix of normal metal with superconducting filaments embedded in it [5.25].

In the adiabatic case, the released magnetic energy changes instantly into thermal energy as the external field exceeds a certain value of the applied magnetic field B_j which depends on the specific heat and on dj_c/dT according to

$$B_j = \left(\varepsilon \frac{\mu_o c j_c}{|dj_c/dT|} \right)^{1/2} \tag{5.5}$$

This adiabatic stability criterion was derived by Swartz and Bean [5.26], who obtained for the parameter ε the value $\varepsilon = \pi^2/4$. Later, slightly modified values, such as $\varepsilon = 3/2$ [5.27], $\varepsilon = 2$ [5.28] and $\varepsilon = 3$ [5.29] have been obtained for this parameter.

In the dynamic case, the relation

$$B_j = \left(\frac{2\mu_o^2 j_c h}{n(dH/dt)} \right)^{1/2} \tag{5.6}$$

has been derived for the first flux-jump field in a superconducting slab of thickness $2d$ subjected to a magnetic field H parallel to the sample surface [5.30]. Here, dH/dt is the field ramp rate at which the external magnetic field is increased, h is the heat transfer coefficient to the coolant, and n is given by the exponent of the power law $j \propto E^{1/n}$ describing the E-j curve of the superconductor.

Experimental data for the temperature dependence of the first flux-jump field of a bulk YBCO sample 4 mm in diameter are shown in Figure 5.7 [5.19]. The disk-shaped sample was cooled in flowing helium gas. B_j was found to increase with temperature up to 8.5 T at 20 K, which is the point of intersection with $B_o(T)$.

Because of the small sample diameter, the level of B_o was low enough to avoid cracking of the sample, similarly to the case illustrated in Figure 5.4a. Comparable time constants have been found in these experiments for the magnetic flux and the thermal diffusion. Therefore, the data could not be analyzed within either the adiabatic or the dynamic approach.

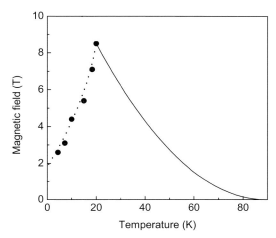

Figure 5.7 Temperature dependence of the maximum trapped field B_o in a small YBCO sample 4 mm in diameter. Below about 20 K, B_o is limited by flux jumps (after [5.19]).

This case, i.e. $\tau \approx 1$, has been considered for a superconducting slab [5.31] and for a superconducting cylinder of radius R [5.32]. Using the model proposed by Takeo et al. [5.32], a satisfactory description of the experimental data shown in Figure 5.7 has been achieved [5.19].

5.5
Magnetizing Superconducting Permanent Magnets by Pulsed Fields

A crucial problem which has to be solved in order to utilize the high trapped fields of bulk YBCO in superconducting magnets, superconducting magnetic bearings, or rotating superconducting machines is to develop an efficient method to magnetize them. Commonly, superconducting magnets are used for the magnetization of bulk YBCO. However, such big magnets are not suitable to magnetize bulk YBCO magnets installed in one of the applications mentioned above. High magnetic fields can also be generated in pulsed-field magnets, which are compact and relatively simple devices. However, a serious problem with this promising technique is the heating of the bulk superconductor under the influence of the pulsed field. Electric fields which are induced in a surface layer of the bulk superconductor by the rapid motion of magnetic flux penetrating the superconductor are directed parallel to the supercurrents. The resulting losses locally heat the superconductor. Therefore, the critical current density is suppressed in the warmer outer region of the YBCO disk and the field gradient in this region becomes smaller, resulting in a lower maximum trapped field than that obtained by magnetizing the superconductor in a static field. The temperature and field distribution during pulsed-field activation of superconducting permanent magnets has been calculated within a critical state model [5.33]. The results, shown in Figure 5.8, are obtained for a superconducting slab of width $2w$ = 1 cm, applying a triangular pulse of magnitude $B_M = 2B_p$ = 5.6 T (where B_p is the penetration field) and using typical values for the specific heat, the heat conductivity, and the critical current density of bulk YBCO at 77 K. The model predicts that in the case of a short rise time of the pulsed field, the outer warmer region of the YBCO disk hardly contributes to the trapped field in the YBCO disk. Therefore, the maximum trapped field is strongly reduced.

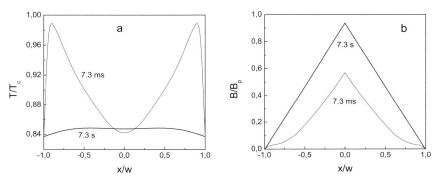

Figure 5.8 Calculated temperature (**a**) and flux density profiles (**b**) for a slab of width $2w$ in the remanent state after applying a pulsed magnetic field $2B_p$ for two different rise times of the pulsed field (after [5.33])

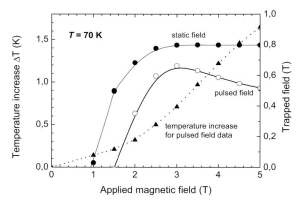

Figure 5.9 Trapped field (and temperature increase for pulsed field data) vs applied static and pulsed field in a bulk YBCO sample (diameter 25 mm) at T = 70 K

In Figure 5.9, trapped field data obtained in a YBCO sample (diameter 25 mm) at 70 K in static and pulsed fields are compared. The trapped field was measured with a Hall probe mounted at the center of the sample. Shown is the dependence of the trapped field on the applied magnetic field after cooling the superconductor at zero magnetic field. If pulsed fields are used, only this magnetization procedure is possible. A special feature of the zero-field cooling mode is that the trapped field in the sample center remains zero for applied fields $H < H_p$, where H_p is the penetration field. This is illustrated in Figure 5.1. From the static field data in Figure 5.9, a penetration field B_p of about 1 T is derived. The trapped field increases for fields $\mu_o H > B_p$ and saturates at about 0.8 T for applied dc fields $\mu_o H > 2.5$ T. If the sample is magnetized by pulsed fields, then higher fields are needed to get the same trapped fields as those obtained with static fields because of heating effects in the bulk YBCO. The temperature increase during the pulsed field experiments has been measured (see Figure 5.9). Because of heating, the highest trapped field reduces from $B_o = 0.8$ T (in static field) to $B_o^{puls} = 0.65$ T (in pulsed field). Heating is also responsible for the maximum in the $B_o^{puls}(H)$ dependence observed at $\mu_o H = 3$ T as well as for the reduction of B_o^{puls} for higher applied fields. The average temperature of the YBCO disk was found to increase with the magnitude of the pulsed field. A temperature rise of about 1.5 K was observed after applying a pulsed field of 5 T.

Data for the temperature dependence of the trapped field in a YBCO sample which was magnetized in static and pulsed field (rise time 5 ms) are compared in Figure 5.10 [5.34]. The trapped fields B_o^{puls} and B_o obtained in pulsed and static field, respectively, are comparable only at high temperatures $T > 75$ K. With decreasing temperature, the ratio B_o^{puls}/ B_o strongly reduces. The pulsed field data have been obtained by applying a single field pulse. Slightly higher trapped fields can be achieved by applying several field pulses (IMRA method). In this case, the magnitude of the pulsed field was reduced stepwise in order to avoid too strong heating of the sample. Nevertheless, for the applied pulse fields with a rise time of 5 ms the heating effect at 30 K is so strong that B_o^{puls} is only about half as large as the trapped field B_o obtained in static field.

A multi-pulse technique with successive cooling steps (MPSC) has been proposed by Sander et al. [5.35, 5.36]. The magnetizing procedure was started, for example, at $T = 75$ K by applying three pulses. Then, the magnetized sample was cooled to 70 K and three additional pulses were applied. The same procedure can be repeated to obtain lower temperatures. The advantage of this method is that only relatively small portions of magnetic flux have to be introduced into the superconductor in order to achieve the critical state at the lower temperature. However, experimental data for the MPSC procedure also confirmed that the rise time of the pulsed field is extremely important. In particular, higher trapped fields have been obtained in a YBCO sample with 30 ms pulses (rise time 15 ms) than with 3 ms pulses (rise time 1.5 ms) after the MPSC procedure [5.36].

It has been demonstrated [5.37] that the heating of the bulk superconductor can be reduced and even avoided by applying pulsed fields with a long rise time. This is shown in Figure 5.11, where trapped field data for a YBCO disk are compared.

This sample was magnetized in static field and in pulsed fields with two different rise times of 10 ms and 300 ms. For the short pulse, a similar suppression of the trapped field B_0^{puls} has been found, as reported by Yanaga et al. [5.34] (see Figure 5.10). The rise time of 300 ms turned out to be long enough to avoid heating of the YBCO disk. No difference between pulsed field and static field data has been found. The trapped field of 5T obtained in this way at 30 K is remarkably high, taking into account the relatively small diameter of 10 mm of the YBCO disk.

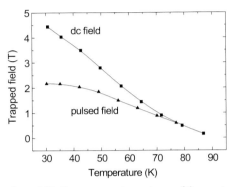

Figure 5.10 Temperature dependence of the maximum trapped field of a YBCO sample magnetized in static and in pulsed field (after [5.34]).

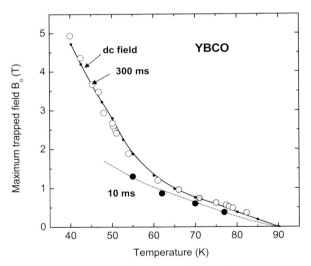

Figure 5.11 Temperature dependence of the maximum trapped field B_0 of a YBCO sample (diameter 10 mm) magnetized in static field (small filled symbols, solid line) and pulsed field of different rise time: 10 ms (large filled symbols, dotted line) and 300 ms (open symbols)

Furthermore, novel ideas for the use of pulsed fields for magnetizing bulk HTSC have been proposed including a temporary opening of the superconducting loop in order to allow flux penetration [5.38]

5.6
Numerical Calculations of the Local Critical Current Density from Field Profiles

The profiles of the magnetic field measured 2 mm above the surface of YBCO samples have extremely different shapes depending on the real structure in the samples.

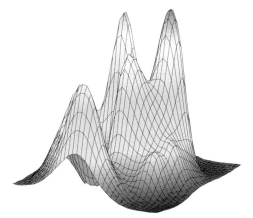

Figure 5.12 The magnetic field distribution for an HTSC bulk composed of several superconducting sub-domains. (Same plane and component as shown in Figure 5.3.)

The profile shown in Figure 5.12 indicates a sample consisting of several small sub-domains in contrast to the perfect single domain and high quality superconductor which was characterized in Figure 5.3. Only the single domain sample can react properly to external magnetic fields in any practical application. The identification of the positions, dimensions, and orientations of the superconducting sub-domains together with the determination of the corresponding j_c values is of the greatest importance. In the case of single-domain material, this seems to be, at first glance, no longer relevant. However, it should be pointed out that also single-domain HTSC is, on a microscopic scale, strongly inhomogeneous, not only because of the formation of defects, but also because of the growth conditions. Weak links in the bulk superconductor can disturb the symmetry of the patterns of supercurrents circulating in the superconductor and of the corresponding field profiles. This is illustrated in Figure 5.13.

In order to improve the sensitivity and significance of field profiles, several modifications of the magnetizing procedure have been proposed, which are discussed in Section 5.7 below. Using these techniques, inhomogeneities in single-domain HTSC can be much better visualized than by the usual method.

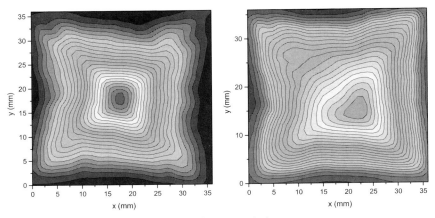

Figure 5.13 Contour maps of trapped fields for two single-domain YBCO samples ($35 \times 35 \times 15$ mm^3). **a** Undisturbed field distribution with 4 edges corresponding to the 4 grain boundaries between the growth sectors. Maximum trapped field $B_0 = 1.2$ T. **b** Disturbed field distribution with 3 edges. Maximum trapped field $B_0 = 0.8$ T

At this stage, it is highly desirable to calculate the currents from the magnetic image, which is essentially an inversion of *Biot-Savart's* law. Several attempts to solve this problem for cylindrical bulk samples were reported [5.39] shortly after the discovery of high-T_c superconductivity. Later, models for the calculation of current maps in thin films were proposed [5.45–5.47]. These models have been extended to 3D samples [5.48–5.53] assuming that the field distribution $\mathbf{B}_z(x,y,z)$ is generated by circulating currents $\mathbf{J}(x,y)$ (Figure 5.15), which are confined to the basal *ab*-planes of the HTSC and uniform in the z direction. Furthermore, these models have been applied successfully to calculate the current distribution from magneto-optical flux imaging experiments [5.49–5.51] and from field profiles of bulk HTSC samples obtained by a Hall probe scanning technique [5.52, 5.54]. Whereas the algorithms used in the above-mentioned models are based on the Fourier transformation, the inversion problem has also been solved by a finite-element procedure [5.42, 5.43] or using *Biot-Savart's* law directly [5.44]. These methods and their applications to bulk HTSC are described in the following sections.

5.6.1
Inverse Field Problem: Two-Dimensional Estimation

The determination of the current distribution within a magnetized HTSC from field profiles measured above a superconductor is a typical inverse problem and belongs to the class of the improperly posed tasks. That is, its solution may neither be unique nor continuous to the input data. Thus, the standard numerical

5.6 Numerical Calculations of the Local Critical Current Density from Field Profiles

field calculations cannot be applied directly to the identification of the current density distribution within the HTSC.

To solve this inverse problem, the finite element method has been applied. The fundamental finite element equation set defining the magnetic field distribution in the whole region has been extended by equations describing the magnetic field in the regions of measurements. This leads finally to an over-determined, linear, and ill-conditioned set of equations with two unknowns: the vector potential values in the whole region and the current density values within the superconductor:

$$\mathbf{MZ} = \mathbf{U} \tag{5.7}$$

Since the coefficient matrix $\mathbf{M}$ is of the size m × n (with m > n), the solution of the equation set exists in the sense of the least squares:

$$\mathbf{Z_0} = \min\|\mathbf{U} - \mathbf{MZ}\|^2 \tag{5.8}$$

This solution can be written as:

$$\mathbf{Z_0} = \mathbf{RU} \tag{5.9}$$

where $\mathbf{R}$ denotes the Moore-Penrose pseudo-inverse [5.40]. The minimal least squares solution of this problem, i.e. the vector of minimum Euclidean length which minimizes Eq. (5.8) can be computed by the singular values decomposition of $\mathbf{M}$. The details of this method have been described in [5.42]. In this way the current density distribution within the HTSC can be determined.

It should be noted that knowledge of the exact distribution of the current density influences only the force density distribution and not the total levitation force. Thus, from the practical point of view it is necessary to define the average current densities of some subregions of the HTSC only. This simplifies the calculation process. According to the required accuracy, the bulk superconductor has been artificially divided into rings of identical cross-section, and the mean j_c for each of these rings has been determined. By this approach, the number of unknown j_c values can be much reduced (e.g., to 16, 25, or 36). This cuts down at the same time the extension of the measurement area, finally leading to much smaller equation sets, which stabilizes the calculation process.

As an example the j_c distribution within a magnetized HTSC (radius 30 mm, height 15 mm) is presented in Figure 5.14. The bulk superconductor has been divided into 36 superconducting rings.

The proposed algorithm enables a stable and physically correct solution of the analyzed inverse problem (determination of the current density distribution within the HTSC), and thus it can be applied for the detection of flaws that could exist within the HTSC (regions with small j_c values). The solution of the problem differs greatly according to the number of the assumed subdivisions of the HTSC region. For practical applications, it is sufficient to examine the mean j_c distribution in a maximum of 36 HTSC rings. The mean value of the determined j_c in each sub-region can be called the "engineering current density", as only this value is responsible for the force interacting between the HTSC and any external magnetic field with local gradients.

For several applications of HTSC, as for field excitation units within electrical machines, knowledge of the mean value of j_c in a reduced number of sub-domains is fully adequate. However, for other applications, in which the shielding properties of the superconductor become important (as in reluctance machines or magnetic bearings), the distribution of the current density has to be determined more precisely.

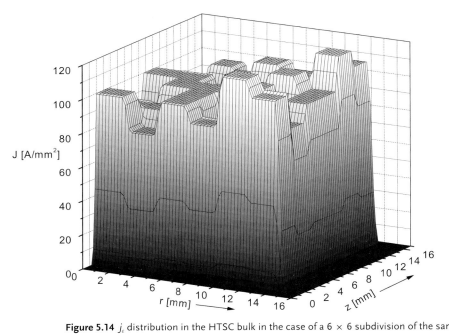

Figure 5.14 j_c distribution in the HTSC bulk in the case of a 6 × 6 subdivision of the sample

5.6.2
Three-Dimensional Estimation

To define the mathematical 3D-model for the evaluation of the current distribution, the superconducting bulk has been artificially divided into N identically shaped sub-cubes with three individual current sheet values and directions s_c surrounding each cube. This is based on the critical-state model of an HTSC, which is applied to these small sub-elements (Figure 5.15).

The magnetic field distribution of such cubes can be calculated by the Biot-Savart law (the magnetic permeability of all areas is equal to μ_0). The two components of the magnetic field generated by the critical current sheet s_{yc} at $z = e$ shown in Figure 5.15 are given by:

$$H_z = |s_{yc}| \ln \frac{y - d + \sqrt{(x - x_0)^2 + (y - d)^2 + (z - e)^2}}{y - c + \sqrt{(x - x_0)^2 + (y - c)^2 + (z - e)^2}} \Big|_{x_0 = a}^{x_0 = b} \quad (5.10)$$

and

$$H_x = |s_{yc}| \left[\arctan \frac{(x - x_0)(y - d)}{(z - e)\sqrt{(x - x_0)^2 + (y - d)^2 + (z - e)^2}} + \\ - \arctan \frac{(x - x_0)(y - c)}{(z - e)\sqrt{(x - x_0)^2 + (y - c)^2 + (z - e)^2}} \right]_{x_0 = a}^{x_0 = b} \quad (5.11)$$

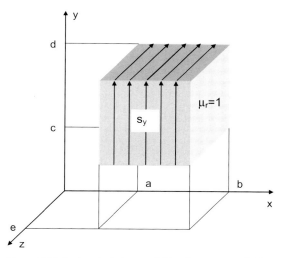

Figure 5.15 Basic calculation model for the magnetic field caused by the currents of a small HTSC subdomain. As a simplification, only the current sheets s_y in the x-z and x-y planes are shown.

Using Eqs. (5.10) and (5.11), the components of the magnetic field at each point of the outer area of the HTSC can be determined as the sum of the magnetic fields of each sub-domain (Figure 5.15). The calculation of the current density distribution within the HTSC from the magnetic field distribution in any region leads to an over-determined, linear, and ill-conditioned set of $3P$ equations (P: number of evaluation points) with $3N$ unknown surface current densities s_c ($P \gg N$). The minimal least squares solution of this problem can be again obtained by singular values decomposition of the main matrix, as described in Ref. [5.44].

The above method has been applied for the calculation of the j_c distribution within a multi-domain YBCO sample using the magnetic field distribution in a plane 3 mm above the surface of the superconductor (see Figure 5.16). The current densities within the superconductor – determined by means of the proposed method – generate a magnetic field distribution which matches the measured one of Figure 5.16 precisely. This proves that regions with small magnetization (of poorer quality) have been found accurately.

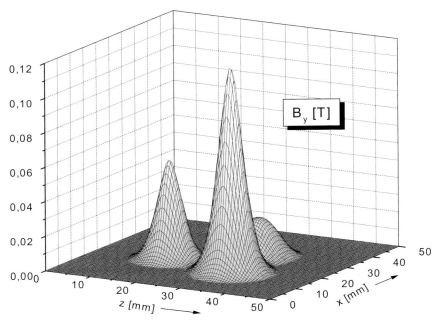

Figure 5.16 Flux density distribution at a distance of 3 mm above the HTSC surface (normal component of a multi-domain probe)

The above algorithm has been used for quality checking of YBCO bulks used for superconducting bearings or motors [5.7]. The YBCO bulks, 10 mm in height with a cross-section of 36 × 36 mm were magnetized in an external magnetic field of >2 T at 77 K. For the quality survey, the distribution of the normal component of the magnetic flux density was measured in five planes lying between 0.6 mm and 8.6 mm above the HTSC sample and parallel to it. Each plane contained 29 field points in both directions. The HTSC sample itself was divided into 10 × 5 × 10 identical parts representing the assumed subdomains. This leads to an over-determined system of 4205 equations with 1500 unknown current densities. Numerical calculations indicated that the current densities within the individual sub-parts are almost identical, which means that their distribution within the entire HTSC bulk is advantageously homogeneous: the equivalent current density for each sub-domain exhibits values in the range $j_c = 10$ kA cm^{-2}.

From the practical point of view, the measurements of the flux density can be restricted to one plane near the HTSC surface only and to not so many points. It is also possible to scan the field distribution for a cluster of several magnetized HTSC bulks simultaneously. In both cases, the proposed method enables the quality of the individual HTSC probes to be evaluated. This algorithm may serve – even in the assembled state – as a non-destructive quality control measurement for HTSC bulks used for magnetic bearings or the excitation of electrical machines.

5.7
Visualization of Inhomogeneities in Bulk Superconductors

Melt-processed HTSC are, on a microscopic scale, strongly inhomogeneous, which is not only because of the formation of defects, but also because of the growth conditions resulting in different growth sectors which are separated by grain boundaries. Two different methods, based on the Hall probe scanning techniques, have been proposed [5.56, 5.57] to visualize local variations of the critical current density in bulk YBCO. In both methods, local supercurrents are induced by modified magnetizing procedures.

Using pulsed fields with a short rise time of about 1 ms, regions with weak links have been shown to be heated more than those without weak links [5.56]. By applying pulsed fields exceeding the maximum trapped field of the YBCO superconductor, trapped fields have been found to remain preferably in the grain boundaries between the a growth sectors within the ab plane, forming a characteristic "cross" on the top of the bulk superconductor (see Figure 2.22). Therefore, it has been concluded that weak links are distributed mainly inside the a growth sectors, whereas the grain boundaries separating these growth sectors are almost free of weak links.

The other technique [5.57] uses a small permanent magnet to magnetize small regions of the bulk superconductor instead of magnetizing the whole sample. Local shielding currents are induced by the permanent magnet, which has been fixed near the Hall probe during the scanning process. The penetration depth of the magnetic field of this magnet is in the region of 1 mm. Two features have been observed in the ab plane using this so-called "magnetoscan" technique:

1. The best shielding is found near the grain boundaries between the a growth sectors. That is to say, in these regions the critical current density is particularly high, which is in agreement with the results mentioned above.
2. Only weak shielding is detected near the seed crystal.

References

5.1 F. Lange, Cryogenics **5**, 143 (1965).
5.2 M. Rabinowitz, H. W. Arrowsmith, S. D. Dahlgren, Appl. Phys. Lett. **30**, 607 (1977).
5.3 D. A. Cardwell, M. Murakami, M. Zeisberger, W. Gawalek, R. Gonzales-Arrabal, M Eisterer, H. W. Weber, G. Fuchs, G. Krabbes, A. Leenders, H. C. Freyhardt, N. Hari Babu, Physica C **412-414**, 623 (2004).
5.4 R. Gonzales-Arrabal, M. Eisterer, H. W. Weber, J. Appl. Phys. **93**, 4734 (2003).
5.5 G. Fuchs, G. Krabbes, K.-H. Mueller, P. Verges, L. Schultz, R. Gonzalez-Arrabal, M. Eisterer, H. W. Weber, J. Low Temp. Phys **133**, 159 (2003).
5.6 L. Shlyk, G. Krabbes, G. Fuchs, K. Nenkov, P. Verges, Appl. Phys. Lett. **81**, 5000 (2002).
5.7 W. Gawalek, T. Habisreuther, M. Zeissberger, D. Litzkendorf, O. Surzhenko, S. Kracunovska, T. A. Prikhna, B. Oswald, L. K. Kovalev, W. Canders, Supercond. Sci. Technol. **17**, 1185 (2004).
5.8 H. Walter, M. P. Delamare, B. Bringmann, A. Leenders, H. C. Freyhardt, J. Mat. Research **15**, 1231 (2000).
5.9 Nippon Steels, Company Prospectus.
5.10 S. Nariki, N. Sakai, M. Murakami, I. Hirabayashi, Supercond. Sci. Technol. **17**, S30 (2004).
5.11 A. Cardwell, private communication, see also: N. Hari Babu, E. S. Reddy, D. A. Cardwell, A. M. Campbell, C. D. Tarrant, K. R. Schneider, Appl. Phys. Lett. 83, 4806 (2003).
5.12 R. Weinstein, R. Sawh, Y. Ren, D. Parks, Mater. Sci. Eng. B **53**, 38 (1998).
5.13 H. Ikuta, A. Mase, U. Mizutani, Y. Yanagi, M. Yoshikawa, Y. Itoh, T. Oka, IEEE Trans. Appl. Supercond. **9**, 2219 (1999).
5.14 Dowa Chemicals, Company Prospectus.
5.15 S. Nariki, N. Sakai, M. Murakami, Physica C **378-381**, 631 (2002).
5.16 S. Nariki, N. Sakai, M. Murakami, Supercond. Sci. Technol. **18**, S126 (2005).
5.17 S. Nariki, N. Sakai, M. Murakami, Physica C **357-360**, 814 (2001).
5.18 M. Matsui, S. Nariki, N. Sakai, M. Murakami, Supercond. Sci. Technol. **15**, 781 (2002).
5.19 C. Wenger, Thesis, TU Dresden (1999).
5.20 T. H. Johansen, Supercond. Sci. Technol. **13**, 830 (2000).
5.21 Y. Ren, R. Weinstein, J. Liu, R. P. Sawh, C. Foster, Physica C **251**, 15 (1995).
5.22 T. H. Johansen, Phys. Rev. B **60**, 9690 (1999).
5.23 P. Diko, Supercond. Sci. Technol. **13**, 1202 (2000).
5.24 P. Diko, G. Fuchs, G. Krabbes, Physica C **363**, 60 (2001).
5.25 R. G. Mints, A. L. Rakhmanov, Rev. Mod. Phys. **53**, 551 (1981).
5.26 P. S. Swartz, C. P. Bean, J. Appl. Phys. **39**, 4991 (1968).
5.27 T. Akachi, T. Ogasawara, K. Yasukochi, Jpn. J. Appl. Phys. **20**, 1559 (1981).
5.28 K.-H. Müller, C. Andrikidis, Phys. Rev. B **49**, 1294 (1994).
5.29 S. L. Wipf, Cryogenics **31**, 936 (1991).
5.30 R. G. Mints, Phys. Rev. B **53**, 12311 (1996).
5.31 K. Yamafuji, M. Takeo, J. Chikaba, N. Yano, F. Irie, J. Phys. Soc. Jpn. **26**, 315 (1969).
5.32 M. Takeo, J. Phys. Soc. Jpn. **30**, 697 (1971).
5.33 S. Bræck, D. V. Shantsev, T. H. Johansen, Y. M. Galperin, J. Appl. Phys. **92**, 6235 (2002).
5.34 Y. Yanagi, Y. Itoh, M. Yoshikawa, T. Oka, A. Terasaki, H. Ikuta, U. Mizutani, in *Advances in Superconductivity*, edited by K. Osanura and I. Hirabayashi (Springer-Verlag, Tokyo, 1998), Vol. 2, p. 941.
5.35 M. Sander, U. Sutter, R. Koch, M. Kläser, Supercond. Sci. Technol. **13**, 841 (2000).
5.36 M. Sander, U. Sutter, M. Adam, M. Kläser, Supercond. Sci. Technol. **15**, 748 (2002).
5.37 Unpublished results of the authors.

5.38 M. Sander, Physica C **392-396**, 704 (2003).
5.39 L. W. Conner, A. P. Malozemoff, Phys. Rev. B **43**, 402 (1991).
5.40 M. Däumling, D. C. Larbalestier, Phys. Rev. B **40**, 9350 (1989).
5.41 C. L. Lawson, R. J. Hanson, Solving least squares problems, Prentice-Hall, 1974.
5.42 R. Palka, *Synthesis of magnetic fields by optimization of the shape of areas and source distributions*, Archiv für Elektrotechnik **75**, 1-7 (1991).
5.43 R. Palka, H. May, W.-R. Canders, *Nondestructive quality testing of high temperature superconducting bulk material used in electrical machines and magnetic bearings, Optimization and Inverse Problems in Electromagnetism*, Kluwer Academic Publishers, 2003.
5.44 R. Palka, H. May, W.-R. Canders, Przeglad Elektrotechniczny **2003** (No10: Special issue XII International Symposium on Theoretical Electrical Engineering ISTET'03), p. 329.
5.45 P. D. Grant, M. W. Denhoff, W. Xing, P. Brown, S. Govorkov, J. C. Irwin, B. Heinrich, H. Zhou, A. A. Fife, A. R. Cragg, Physica C **229**, 289 (1994).
5.46 W. Xing, B. Heinrich, Hu Zhou, A. A. Fife, A. R. Cragg, J. Appl. Phys. **76**, 4244 (1994).
5.47 H. Kamijo, K. Kawano, IEEE Trans. Appl. Supercond. **7**, 1228 (1997).
5.48 E. H. Brandt, Phys. Rev. B **46**, 8628 (1992).
5.49 R. J. Wijngarden, H. J. W. Spoelder, R. Surdeanu, R. Griessen, Phys. Rev. B **54**, 6742 (1996).
5.50 R. J. Wijngarden, K. Heeck, H. J. W. Spoelder, R. Surdeanu, R. Griessen, Physica C **295**, 177 (1998).
5.51 Ch. Joos, A. Forkl, R. Warthmann, H. Kronmuller, Physica C **299**, 215 (1998).
5.52 J. Amoros, M. Carrera, X. Granados, J. Fontcuberta, X. Obradors, *Applied Superconductivity 1997*, Inst. Phys. Conf. Ser. No. 158, ed. by H. Rogalla et al. (IOP Publishing, Bristol, 1997) p. 1639.
5.53 G. K. Perkins, Yu. V. Boguslavsky, A. D. Caplin, Supercond. Sci. Technol. **15**, 1140 (2002).
5.54 E. Mendoza, M. Carrera, E. Varesi, A. E. Carillo, T. Puig, J. Amoros, X. Granados, X. Obradors, *Applied Superconductivity 1999*, Inst. Phys. Conf. Ser. No 167, ed. by X. Obradors et al. (IOP Publishing, Bristol, 2000) p. 127.
5.55 H. Ikuta, A Mase, U. Mizutani, Y. Yanagi, M. Yoshikawa, Y. Itoh, T. Oka, IEEE Trans. Appl. Supercond. **9**, 2219 (1999).
5.56 A. B. Surzhenko, S. Schauroth, D. Litzkendorf, M. Zeisberger, T. Habisreuther, W. Gawalek, Supercond. Sci. Technol. **14**, 770 (2001).
5.57 M. Eisterer, S. Haindl, T. Wojcik, H. W. Weber, Supercond. Sci. Technol. **16**, 1282 (2003).

6
Improved YBa$_2$Cu$_3$O$_{7-\delta}$-Based Bulk Superconductors and Functional Elements

When attempting the improvement of superconductors for various applications, the parameters taken into consideration must be relevant to the desired properties for a specific engineering concept. In this context, we shall distinguish between superconducting properties, such as flux pinning, affecting the "intragrain" critical current, and the "effective" critical current, which, together with the single domain size, determines trapped fields. Improvement in mechanical properties is indispensable for applications in high fields. Furthermore, technological aspects – the economic production of large and complicated shapes – are also important. As superconductors are on the threshold of commercial application, a sizable and economic technology must be developed which is closely linked to advanced ceramics technology and characterized by batch processing rather than continuous line processing. Functional elements consist of the superconductor in the appropriate shape and specification together with such non-superconducting items as are necessary for its function as a part of a facility or machine. In this chapter, functional elements based on *bulk BSCCO* material will also be taken into account.

6.1
Improved Pinning Properties

6.1.1
Chemical Modifications in YBa$_2$Cu$_3$O$_7$

The *influence of lattice defects and site substitutions in YBa$_2$Cu$_3$O$_{7-\delta}$* has been discussed with respect to T_c in Chapter 1.3. Depending on the affected site, the change of charge by heterovalent substitution results either in carrier doping directly or by charge transfer if carriers are itinerant or localized in holes (polarization near the charged defect) or the excess charge is compensated by another type of defect ("defect chemistry", non-stoichiometric oxygen). Here we shall consider in detail the influence of chemical alterations on critical currents, and the pinning behavior will be highlighted.

High Temperature Superconductor Bulk Materials.
Gernot Krabbes, Günter Fuchs, Wolf-Rüdiger Canders, Hardo May, and Ryszard Palka
Copyright © 2006 WILEY-VCH Verlag GmbH & Co. KGaA, Weinheim
ISBN: 3-527-40383-3

An approximate idea of the *compatibility with lattice sites* for potential substituting ions can be obtained from the ionic radii. Whereas a "tolerance factor" can be clearly defined for the perovskite lattice, the criterion is weaker in the 123 structure; nevertheless in most cases the ratio $r_{substituent}/r_{host}$ is in the range between 0.85 and 1.15.

Table 6.1 compiles ionic radii of important substituents and hosts. Processing is similar to that for Y123 unless the degree of substitution is high. Then, heating regime and oxygenation have to be adjusted.

Table 6.1 Ionic radii r^{n+} (nm) of potential substituents and Cu^{n+} ions

Shannon's parametric system (see also Table 2.1). $T(m_1)$ is the invariant peritectic temperature at $p(O_2) = 0.21$ bar.
CN is the coordination number, * for CN = 9 (if data for CN10 not available)

Ionic radii for alkaline earth elements and certain rare earth ions Shannon's parameters

	Ca^{2+}	Sr^{2+}	Ba^{2+}	Yb^{3+}	Y^{3+}	Nd^{3+}	Ce^{3+}	Ce^{4+}
CN 8	0.112	0.126	0.142	0.099	0.102	0.111	0.114	0.097
CN 10	0.123	0.136	0.152	0.105*		0.116*	0.125	0.107

Ionic radii of potential substituents on Cu sites; *t* tetrahedral, *s* square

	Li^+	Mg^{2+}	Al^{3+}	Co^{2+}	Ni^{2+}	Pd^{2+}	Zn^{2+}	Cd^{2+}	Ag^+	Ag^{2+}
CN 4	0.059	0.057 t	0.039 t	0.057	0.049 s	0.064 s	0.060	0.078	0.102	
CN 6	0.076	0.072	0.054	0.065	0.055 t					0.079

	Cu^{3+}	Cu^{2+}	Cu^+ CN2: 0.056 "dumbbell"
CN 4		0.057 s	0.060
CN 6	0.054	0.073	0.077

Substitution on Y sites

The most relevant substituents on Y sites so far are Ca^{2+} ions, which increase the density of holes, thus driving the material from an underdoped to an overdoped state [6.1]. The shift of T_c passing a maximum is important, affecting j_c remarkably. Surprisingly, investigations on Ca-doped thin films demonstrated an unexpectedly strong increase in the transport currents across a [001]-tilt grain boundary as large as 24°. This result was explained by band bending near the grain boundary resulting in a local depletion of hole density, which is compensated by holes from Ca doping [6.2]. The behavior of melt-textured bulk materials is characterized by the influence of both the dopant and the oxygen nonstoichiometry, since – under identical oxygenation conditions – δ is different for Ca-doped and undoped $YBa_2Cu_3O_{7-\delta}$. In this case, the major contribution to a peak appearing in the magnetization curve is not generated by the substituent but by oxygen disordering, which disappears under appropriate annealing [6.3]. On the other hand, j_c in a sample containing 0.1wt% Ca and doped nearly to the optimal carrier concentration exceeds the undoped reference material by a factor of 1.6 at 2 T. This is significantly less than in thin films and has to be attributed to the typical mosaic structure in bulk material, which is characterized by a high density of low-angle grain boundaries and their scattering to wider angles. For comparison, j_c increased after Ca doping by a factor of 2 in the artificial 8° [100] twist grain boundary of a bilayered thin film. The factor decreases asymptotically and no significant improvement was recognized for a 45° [001] tilt boundary [6.4].

Substitution on Cu sites
Although Cu(1) and Cu(2) sites apply to an octahedral coordination sphere (orthogonal bonding axes), a preferable occupancy either in chains, Cu(1), or in planes, Cu(2), is observed for a number of elements. Fe, Co, V, Al, Ga [6.5] and also Pd and Cd [6.6] were found to have a preference for chain sites, whereas Zn ions prefer the Cu(2) plane sites. Substitution by Ni ions is an example of occupying both Cu sites in varying partition ratios. The ratio of occupancy was found to be controlled by the preparation conditions [6.19]. In-plane substitution was observed to increase in less-oxidizing atmospheres. As an additional complication of Ni substitution, the precipitation of an NiO-rich phase was reported [6.20].

Substitution on Cu(1) site in Y123
The incorporation of Al^{3+} on the Cu(1) position has an important influence on properties of bulk materials since alumina is widely used as a reasonably priced material for crucibles and supports for Y and other RE-based superconductors. Thereby, Al was found to break the chains by branching, but retaining the 4-fold coordination to oxygen in a strongly distorted tetrahedron [6.5, 6.7]. The consequences are multiple:

1. Al^{3+} traps a hole via apical oxygen.
2. It blocks the path for oxygen diffusion.
3. Although one might expect a defect structure promoting flux pinning, experience shows a remarkable decrease in T_c.

Therefore, avoiding Al pollution is a rule for improving the properties in bulk materials.

In-plane substitution on Cu(2) site in Y123
Here, the influence of Zn and Li doping on the superconducting properties of bulk YBCO and, in particular, on the flux pinning has to be considered. The optimum concentration of the active dopant with regard to optimal pinning is always less than 1 wt%; therefore, processing remains nearly unchanged and the modi-

fied melt crystallization growth MMCG or MMCP with a precursor of doped Y123, yttria, and Pt has also been applied [6.8–6.11]. Intensive investigation of Zn doping was initialized by the surprisingly strong suppression of T_c of YBCO by Zn doping, with a suppression rate of about –10 K/at% [6.12]. Thus, the T_c of $YBa_2Cu_3O_7$ is much more strongly suppressed by non-magnetic Zn than, for instance, by magnetic Ni, for which a suppression rate of only – 4 K/at% was found. This unexpected behavior can be explained at least qualitatively taking into account the completely filled d-shell corresponding to S = 0 for Zn^{2+} in contrast to S = $1/2$ for $Cu^{2+}(d^9)$. The doped holes in the standard CuO_2 plane compensate the copper spin by forming the Zhang-Rice singlet quasi-particle [6.13], which is accepted to contribute essentially to the pairing process. The Zhang-Rice singlet cannot develop with a doped hole in the neighborhood of the zinc impurity, thus leaving an induced magnetic moment, or in other words: zinc impurities in a spin singlet wave function induce local moments [6.14]. The resulting disturbance by a magnetic moment within the CuO_2 planes has been pictured in $Bi_2Sr_2CaCu_2O_8$ superconductors by scanning tunneling microscopy and its size was estimated to amount 1.5 nm [6.15]. Induced local moments have been revealed by NMR in underdoped [6.14] and slightly overdoped [6.5] $YBa_2(Cu_{1-x}Zn_x)O_{7-\delta}$, and it was found that they reside on the nearest neighboring copper orbitals around the zinc site.

In Figure 6.1 the disturbed region in the neighborhood of a zinc site is shown schematically. The coupling of the local moments with the conduction band is sufficient to induce pair breaking of the superconducting carriers, which explains the strong suppression of T_c by Zn doping.

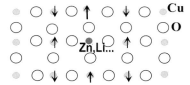

Figure 6.1 Schematic representation of a local magnetic moment resulting from uncompensated spins due to a nonmagnetic impurity in the superexchange network of a CuO_2 plane.

Surprisingly, the pinning properties of YBCO were found to improve by Zn doping for low Zn concentrations. This has been reported for a $YBa_2Cu_3O_7$ single crystal [6.16] and later also for bulk YBCO materials [6.8]. Improved pinning in Zn-doped YBCO can be related to the disturbed regions around the Zn impurities. Their in-plane size is comparable with the coherence length of YBCO. Therefore, these normal conducting regions around the Zn impurities can be considered as effective pinning defects for low Zn concentrations.

The same mechanism is assumed to explain the strongly enhanced pinning properties found recently for Li-doped YBCO. Lithium occupying Cu sites in the CuO_2 planes such as Zn suppresses T_c of YBCO at a rate of 6 K/at% Li. Li^+ is nonmagnetic, like Zn^{2+}, but has a different valence. Therefore, not only a local magnetic moment, but also an additional hole is induced when a Cu atom in a CuO_2 plane is substituted by a Li atom. The volume pinning force was found to increase with the Zn or Li concentration at low impurity content. For higher impurity concentrations, the pinning centers induced by Zn or Li doping start to overlap, and T_c and the pinning force are strongly suppressed. Impurity concentrations of 0.36 at% Zn and 0.6 at% Li were found to be optimal for the flux pinning in bulk YBCO, which corresponds to the distance of mean spacing between dopants of roughly 2 nm [6.17], which is comparable with the coherence length in YBCO [6.18].

Li doping has even been favored to generate more effective pinning properties; however, very recent comparative studies indicate quite similar results for Zn- and Li-doped YBCO. The volume pinning force $F_p = j_c \cdot B$ of Li-doped YBCO is compared in Figure 6.2 with that of undoped YBCO at $T = 75$ K. The samples were prepared with their optimum doping concentrations under equivalent thermal treatment and at the same final oxidizing temperature. The pinning force which is plotted in Figure 6.2 versus the reduced magnetic field $b = B/B_{irr}$ is strongly enhanced for Li-doped YBCO [6.11].

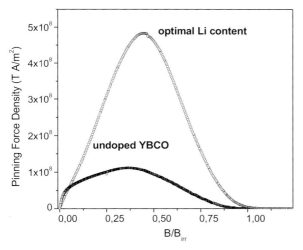

Figure 6.2 Development of volume pinning force $F_p = j_c B$ for undoped and doped samples of bulk YBCO prepared under optimized conditions: 1: undoped, 2: Li-doped. $T = 75$ K after [6.11]

The $j_c(H)$ dependencies are represented in Figure 6.3 at 77 K for two bulk YBCO samples doped with optimum Li concentration and undoped, respectively. Zn and Li doping affects the pinning properties in a similar way. The magnitudes of j_c of the two materials, but also the field dependence of j_c are very similar. A distinct peak effect develops in both cases at temperatures below 77 K, which will be discussed in more detail in Chapter 8.

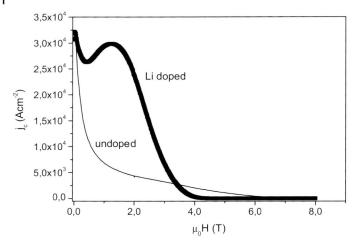

Figure 6.3 Critical current density vs field, $j_c(H)$, at 77 K for undoped and Li-doped YBCO [6.67]

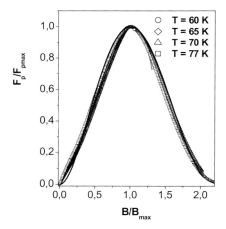

Figure 6.4 Scaling of F_p vs normalized field $b=B/B_{irr}$ [6.11]

The experimental $j_c(H,T)$ data of doped YBCO can be collected in the form of a scaling law for the volume pinning force $F = j_c B$ as a function of the reduced field $b = B/B_{irr}$. This is shown in Figure 6.4 using the data of Li-doped YBCO presented in Figure 6.3. There is a common $F(b)/F_{max}$ dependence in the investigated range between 55 and 85 K. The maximum volume pinning force F_{max} appears at reduced fields $b_m \approx 0.33 - 0.4$.

The scaling behavior of the pinning force in this Li-doped YBCO sample is much better pronounced than that of undoped YBCO, where deviations from the scaling behavior were found especially at low magnetic fields. The improved scaling in Zn- or Li-doped YBCO is attributed to the predominance of the pinning centers induced by Zn or Li doping, whereas in undoped YBCO various types of pinning sites contribute to the pinning force.

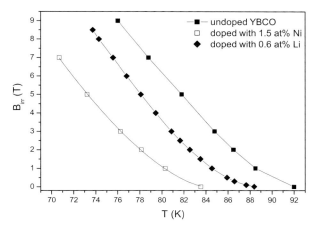

Figure 6.5 Irreversibility lines for bulk Li, Ni, and undoped YBCO, after [6.68]

The irreversibility lines for different dopants are shown in Figure 6.5. The shift of the irreversibility line to lower values observed for the Ni- and Li-doped YBCO in comparison to the undoped material can be explained by the lower T_c in doped materials.

As a consequence of the improved pinning, high critical currents j_c and trapped fields up to 1.2 T on top of cylinders of 25 mm diameter have been obtained at 77 K for Zn- and Li-doped bulk YBCO samples.

6.1.2
Sub-Micro Particles Included in Bulk YBCO

Defects generated by second-phase inclusions in melt-processed bulk materials are an important medium for flux pinning. It is now a standard procedure to use an excess of the peritectic 211 phase and Pt additives dissolving in the melt which support the particle refinement. Like Pt, additives of CeO_2 have been proven to reduce the critical radius of 211 particles which can be trapped (see Section 2.4.3). This is the reason for improved pinning observed with CeO_2 additives [6.71, 6.72, 6.73]. The influence of 211 grain size on j_c has been widely reported, although a quantitative relationship cannot be stated because of a number of influencing parameters. Fine particles have a preference to be pushed. Furthermore, very fine particles – if added *ex situ* – cannot easily be distributed homogeneously. Promising results were recently obtained for particles which are precipitated *in situ* during melt processing. Samples with inclusions of small particles of the uranium-containing phase $U_{1.2}Pt_{0.8}Y_2Ba_4O_{12}$, originally prepared as a target for the U-n fission products method (see Section 6.1.3) revealed a significantly increased critical current already in the as-prepared state [6.21].

Later, the "double perovskite structure" of this phase was identified and a series of "nanoscaled" precipitations in melt-grown YBCO have been developed with the composition $Y_2Ba_4CuMO_y$ (M = U, W, Nb, Ta, ...) related to cubic "double perovskite" structures [6.22]. The very fine particles (10 nm) can be assumed to precipitate from an oversaturated liquid or in the solid phase. Since a significant change of T_c has not been reported, the degree of substitution in Y123 remains small.

Very recently, a chemical route to extended defects with nanosized cross section was reported resulting in enhanced pinning properties [6.75]. This may give a chance to develop a less expensive alternative to irradiation techniques which so far have been used to generate linearly elongated pinning active defects.

6.1.3
Irradiation Techniques

Flux pinning can be improved by radiation-induced columnar defects using heavy ions of high energy. These defects are limited on a relatively thin layer of tens of microns near to the surface of the sample (e.g., 52 μm for 2 GeV ^{197}Au ions), which was found to affect flux pinning in slabs of 1 mm thickness [6.23]. Increasing beam energy will not only increase the penetration depth but also the degree of irradiation damage in the sample (causing j_c to decrease), and it will change the mechanisms of damaging. A continuous linear damage appears if the absorbed energy per track length unit S_e exceeds a threshold value of 35 GeVmm^{-1}, otherwise Multiple-In-Line-Damage (MILD) is formed as a linear arrangement of numerous potential pinning centers [6.24]. Current experimental investigations promise an increase in j_c even at magnetic fields of 1 T by factors of more than 10 in slabs of thickness up to 1 mm.

In contrast, neutrons have significant larger penetration depth. Irradiation of bulk YBCO with fast neutrons results in a large variety of defects, which are generated by nuclear interactions between the incident neutrons and the lattice atoms of the superconductor. The size of the defects depends on the energy of the incident neutrons. Fast neutrons with energies $E > 0.1$ MeV create so-called collision cascades. A cascade is a spherical normal conducting region with a diameter of about 6 nm including the strain field around the defect. The cascade density is proportional to the fast neutron fluence. A neutron fluence of 2×10^{21} m^{-2} corresponds, for instance, to a cascade density of 1×10^{22} m^{-3} [6.25]. Neutrons with smaller energies ($E < 0.1$ MeV) are able to produce point defects, primarily in the oxygen sublattice. Isolated point defects do not play an important role for pinning at high temperatures. However, they may cluster during irradiation and create larger defects with an effective size appropriate for pinning, or may be able to increase the effective size of the collision cascades or the as-grown defects and, therefore, contribute effectively to pinning [6.26].

In Figure 6.6, $j_c(H)$ data for a Zn-doped YBCO sample before and after neutron irradiation are shown for three temperatures. A strong increase in the critical currents is observed after neutron irradiation. At 77.5 K, the critical current density increases up to $j_c \approx 100$ kA/cm^2 at zero magnetic field and becomes 2.5 times higher than before irradiation. An enhancement by a factor of about 8 was observed at 60 K at an external field of 0.8 T. The enhancement factors become lower with increasing external field, but a factor of ≈ 5 still remains at 2 T.

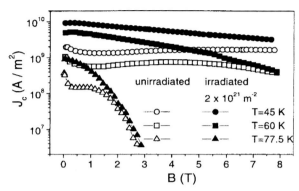

Figure 6.6 $j_c(H)$ of a Zn-doped YBCO sample prior to and after (full symbols) neutron irradiation (semilogarithmic scale, Zn concentration 0.1 wt%, [6.27])

The collision cascades produced by neutron irradiation are strong pinning centers, which are most effective at low magnetic fields, leaving the high-field region of the $j_c(H)$ dependence and the irreversibility field almost unchanged. The low-field plateau of j_c observed before irradiation at 77.5 K completely disappears because of the huge enhancement of j_c at low applied fields after irradiation. The corresponding critical current density monotonously decreases with increasing applied field [6.27].

An alternative and effective approach to introducing strong pinning defects into bulk superconductors is the so-called U/n method, which is based on a nuclear reaction inside the bulk material. In a first step, ^{235}U-doped YBCO material is melt processed. Then, the bulk material is irradiated with thermal neutrons, initializing nuclear fission events in the sample [6.28]. Fission products leave columnar defects on their tracks through the material of almost optimum size with a diameter of about 5 nm. High critical current densities up to 8.5×10^4 A/cm^2 have been reported at 77 K and trapped fields up to 2.1 T have been obtained on the surface of irradiated YBCO samples 20 mm in diameter at 77 K [6.29]. For comparison, the best trapped field results at 77 K for non-irradiated bulk YBCO of comparable size are about 1.3 T (Table 5.1).

6.2
Improved Mechanical Properties in $YBa_2Cu_3O_{7-\delta}$/Ag Composite Materials

Although grown by primary crystallization from the melt and behaving in some details like single crystals, the superconducting bulk materials are composite materials consisting of the superconducting matrix with Y211 particles included. Nevertheless, the materials remain brittle because of the restricted fracture toughness, as is typical of ceramic materials, and this became a limiting factor for applications under the influence of high forces. Tensile stresses larger than 25 MPa appear while magnetizing the bulk materials in high fields, which causes conventional YBaCuO bulk materials to crack spontaneously (Chapter 8). The demand for mechanical strength was the motivation to develop composite materials combining the brittle ceramics with ductile silver metal.

6.2.1
Fundamentals of the Processing and Growth of YBCO/Ag Composite Materials

The solidus temperature is decreased remarkably by the addition of silver. The lowest (eutectic) invariant melting temperature at 0.21 bar O_2 is reduced to 870 °C (from 899 °C for pure YBaCuO) [6.30].

Despite the sketchy knowledge about the four-component phase diagram $YO_{1.5}$-BaO-CuO_n-Ag, the sections in Figures 6.7 and 6.8 can explain the relevant features of the process. The situation along the section in Figure 6.7 is representative of a conventional precursor (Y123 or 123 + 211)-Ag, whereas Figure 6.8 represents the modified precursor 1/6 ($YBa_2Cu_3O_7$ + 0.24 Y_2O_3)-Ag. The monotectic behavior indicates the appearance of a silver-rich ("metallic") melt and the cuprate melt, with Ag dissolved in both the "oxide" melt and the superconducting phase.

The Y123 phase forms a solid solution $YBa_2(Cu_{1-y}Ag_y)_3O_{7-\delta}$ (Y123ss) by the solid state reaction (ssr)

$$ssr: \quad YBa_2Cu_3O_7 + 3y\,Ag + 1.5y\,O_2 \rightarrow YBa_2(Cu_{1-y}Ag_y)_3O_7 + 3y\,CuO \tag{6.1}$$

and it will also appear during heating of the green body, accompanied by small amounts of CuO. Note that the composition of this ss does not belong to the sections considered in Figure 6.7. The maximum Ag concentration in the subsolidus Y123ss corresponds to $y = 0.02$, whereas the Y123ss phase grown from the $YO_{1.5}$-BaO-CuO_n-Ag system in the melt-texturing process contains less Ag ($y' = 0.016$).

The lowest temperature at which a melt in a $YBa_2Cu_3O_7$-Ag mixture was observed is $T = 944$ °C, i.e. 10 K below the melting point of Ag in air. The appearing melt L″ consists mainly of Ag and should be attributed to a pseudo-eutectic reaction (e′)

$$e': \quad Y123ss + CuO + Ag = L'' + O_2 \tag{6.2}$$

6.2 Improved Mechanical Properties in YBa$_2$Cu$_3$O$_{7-\delta}$/Ag Composite Materials

This reaction appears invariant with respect to temperature at the considered section since the amount of CuO resulting from *ssr* [Eq. (6.1)] does not exceed the solubility limit of CuO in the Ag-rich melt – L″ in Figure 6.7. Between 944 and 980 °C, the remaining Y123ss coexists with this "metallic" melt L″. At concentrations higher than x_m (2.5 mol% Ag at 0.21 bar O$_2$), the complete decomposition of Y123ss takes place at 980 °C in a reaction of the monotectic type (*m*)

$$m: \quad Y123 + L'' = 211 + L' + O_2 \quad (6.3)$$

L′ and L″ designate the YBa$_2$Cu$_3$O$_7$-rich and the Ag-rich composition of the melt, respectively, appearing along the immiscibility line according to Figure 6.7, where x_m stands for the monotectic composition, which is about 2.5 mol% in 0.21 bar O$_2$ (corresponding to 5 mol% Ag in the L′ liquid or 2.6 wt% Ag$_2$O addition to YBa$_2$Cu$_3$O$_7$).

A peritectic decomposition reaction m_1, as known from the ternary YO$_{1.5}$-BaO-CuO system, proceeds at concentrations of Ag less than 2.5 mol%, though Y123ss coexists here with L′ in the range between 980 °C and 1020 °C since m_1 is not invariant with respect to *T* in the quaternary system.

$$m_1: \quad Y123 = L' + 211 + O_2 \quad (6.4)$$

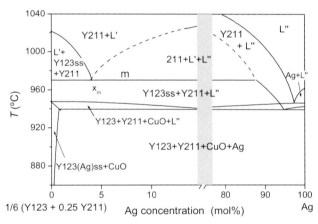

Figure 6.7 The vertical section 1/6YBa$_2$Cu$_3$O$_7$-Ag at 0.21 bar oxygen. Note that the composition of the solid solution generated according to Eq. (6.1) does not belong to this section.

The phase relationships become more complicated for the Y123 + *n* Y$_2$O$_3$ + *x* Ag composition: Y$_2$O$_3$ does not appear in equilibrium and it is not placed at the considered section. During slow heating in the subsolidus region, Y$_2$O$_3$ reacts with a part of Y123 to form Y211 + CuO, which appear together with Y123ss in the appropriate section, Figure 6.8. The main alterations are
   1. The monotectic temperature *T(m)* decreases (for *n* = 0.24 to 970 °C).

2. The monotectic point moves, for $n = 0.24$ to 5 mol% Ag (which corresponds to 10.7 mol% Ag in L' or 5.2 wt% Ag_2O addition to $YBa_2Cu_3O_7$), and the preferred precursors have a higher than monotectic Ag content, i.e. they have hypermonotectic composition.
3. In the Ag-rich part, L" appears in equilibrium with Y211 + Y123 + Ag or Y211 + Ag + CuO, respectively, depending on composition.

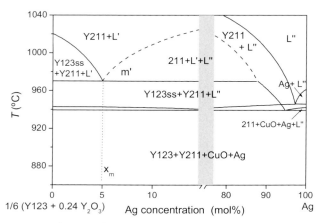

Figure 6.8 The vertical section $1/6(YBa_2Cu_3O_7 + n\,Y_2O_3)$-Ag at 0.21 bar oxygen and $n = 0.24$

6.2.2
Processing and Results

Systematic experimental investigations have been performed with the modified precursor compositions $YBa_2Cu_3O_7 + n\,Y_2O_3$ with n preferably between 0.2 and 0.4 and varying additions of Ag powder or Ag_2O (between 8 and 20 wt% Ag).

Mechanism of solidification and the generated microstructure depend on concentration of Y_2O_3 (typical: $0.2 < n < 0.4$) and especially Ag. Solidification proceeds according to the monotectic reaction m, [Eq. (6.2)] between 8 and 20 wt% Ag. The mean size of the Y211 particles was found to be about 1.7 µm and independent of the Ag concentration. Typical Ag inclusions of about 10 µm are formed, sometimes filling pores. Apart from the outer sphere, the Ag particles are homogeneously distributed (Figure 6.9). At lower Ag concentration, the melt remains in the supersaturated state (rather than separating into L' + L"), from which the solid solution $YBa_2Cu_{3-y}Ag_yO_{7-x}$ will crystallize in a monotectic reaction m, leaving the residual melt L' [6.30–6.32].

The composition of the Ag particles was determined by EDAX to be $Ag_{0.88 \pm 0.02}$ $Cu_{0.02 \pm 0.005}\,O_{0.10 \pm 0.03}$, roughly in accordance with the high-temperature phase diagram of the ternary system Ag-Cu-O, whereas the superconducting phase $YBa_2M_3O_7$ is represented by a solid solution containing silver with M = ($Ag_{0.016\pm0.004}Cu_{0.984\pm0.004}$).

6.2.3
Properties of Bulk YBaCuO/Ag Composite Materials

Microcracks appear in melt-grown YBCO both parallel to the crystallographic c-axis and within the a-b plane. The more relevant cracks in the [001] direction (c-cracks) are generated by the strong contraction of the c-axis during cooling and oxygenation after the tetragonal-to-orthorhombic transformation in the as-grown samples, when remarkable stresses appear around the 211 inclusions in the 123 matrix.

Furthermore, c-cracks reduce the cross section of the current path parallel to the a-b plane, and consequently the superconducting current of the current loop perpendicular to c as well as the trapped field parallel to c will decrease.

In a first approach, a significant increase of fracture toughness was expected by ductile Ag particles acting as "stoppers" of propagating cracks under the influence of stress in strong magnetic fields. Indeed, this seemingly has been verified in samples with very fine Ag particles distributed in a polycrystalline sintered sample [6.33]. K_{IC} however was found to increase only from 1.9 to 2.4 MPa m$^{-0.5}$ in melt-textured single-grain material [6.32]. More careful analysis of microstructure indicated a significantly decreased density of c microcracks in the as-grown and oxidized bulk Ag-YBCO composite material compared with standard YBCO (Table 6.2). This fact supports the concept assuming the reduced flaw density will cause improved strength in high magnetic fields. Thus, Ag particles play an active role in the temperature range between solidification and room temperature.

Because of their high plasticity, silver particles should be, in the first approximation, considered to behave mechanically like a liquid rather than a rigid solid. Then, the total volume change due to temperature change is responsible for the appearing microstresses induced by Ag particles. The relative volume change during cooling down from processing temperature to room temperature is $\Delta V_{Ag}/V_{Ag} = 8.87 \times 10^{-6}$ for Ag and $\Delta V_{comp}/V_{comp} = 2.46 \times 10^{-6}$ for the composite of 123 + 21 vol% 211. The larger volume change of Ag compared with the composite causes compressive tangential stresses σ_θ of $\sigma_\theta \approx 265$ MPa around Ag particles [6.34]. These compressive stresses around a Ag particle can compensate tangential tensile stresses around 211 particles as it is illustrated in Figure 6.10, thus suppressing a-b and c-microcracking. This interpretation is in accord with a considerably increased spacing between the microcracks in the presence of Ag particles [6.34] and explains the improved resistance of bulk samples against damage during the magnetizing process.

On the other hand, microcracks promote oxygen diffusion. Therefore, oxygenation of YBaCuO/Ag composites proceeds even more sluggishly and the appearance of tetragonal islands in some cases is an unfavorable aspect of the addition of silver, which has to be compensated by a controlled oxygenation regime. Another source of deteriorated superconducting properties in YBaCuO/Ag composites is the partial substitution of Ag for Cu [6.35]. However, the silver concentration in $RE\text{Ba}_2(\text{Cu}_{1-y}\text{Ag}_y)_3\text{O}_{7\pm\delta}$ grown in equilibrium with the peritectic melt is only $y \cong 0.016$. Therefore, its deteriorating influence turned out to be small and can be compensated by partial doping with Zn^{2+} ions.

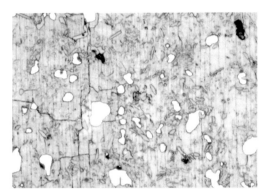

Figure 6.9 Ag particles distributed in the 123 matrix. Note the suppression of c microcracks in the Ag-rich region. Reprinted from [6.34] with permission from Elsevier

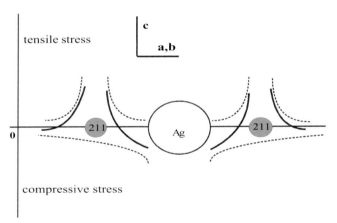

Figure 6.10 Tensile stresses around 211 particles (dashed) are suppressed in the compressive stress field around Ag particles [6.70]

6.3
Near Net Shape Processing: Large Sized Bulk Superconductors and Functional Elements

In order to increase the performance of devices such as superconducting motors, bearings, current leads and superconducting permanent magnets, the size and shape of the single domain YBCO superconductors have to be adapted to the engineering concept (as will be introduced in later Chapters 11–13). Furthermore, aspects of large-scale production have to be taken into account. Near net shaping is an alternative to conventional finishing by a cut and grind approach.

6.3.1
Finishing and Shaping

The simplest approach to an appropriate *size of superconducting elements* is the lateral stacking of single domain blocks. A stable connection between rectangular blocks can be obtained using appropriate epoxy resins. The resulting elements can be cooled down to cryogenic temperature without serious problems. The joints, however, are insulating, not permitting any electrical contact between adjacent grains.

On the other hand, cylindrically shaped green bodies are most appropriate to melt processing using the top seeding method. Also, hexagonal bodies have been successfully manufactured to increase the packaging of arrays. Even "hot seeding" has been accomplished, i.e. placing the seed in a central position with the *c*-axis perpendicular to the top face of the partially molten sample outside or inside the hot furnace. This becomes more complicated for rectangular or multiseeded samples for which additionally a (110) orientation must be achieved and consequently the lateral seed position becomes important. According to our present knowledge, hot seeding is indispensable for certain $RE_{1+y}Ba_{2-y}Cu_3O_7$ materials.

In summary, these facts indicate that the appropriate technology relates rather to single-crystal manufacturing than to a ceramic technology, but in any case nothing like machining metallic alloys. Pre-shaping, however is limited to simple forms. Whereas for simple levitator applications precise finishing is not necessary, the application in electrical machines, high-speed high-load bearings etc. needs precise mechanical finishing. The dense non-porous material is brittle, and therefore cutting and polishing are the preferred techniques. For high precision and complicated shapes, numerically controlled machines have been successfully tested, also including ultrasonic tools.

Further, one should bear in mind that the materials are thermodynamically metastable compounds at room temperature. In the presence of H_2O and CO_2 (both already present in the environmental air atmosphere), RE_2O_3 or $RE_2O_3 \cdot nH_2O$, Ba_2CO_3 or $Ba(OH)_2$, and CuO or basic copper carbonates are the decomposition products. Experience shows that high-quality and dense samples can be handled at room temperature in dry air without restrictions because the diffusion into a dense bulk material with a low ratio of surface to volume is sluggish. Even a short contact with wetting media during machining has impact only on the surface from which products of deterioration (e.g. the green Y211 phase) can easily be removed by polishing. Samples of lower quality, however, can be seriously affected by treatment under moist conditions. Porosity, grain boundaries, grain size, and purity are important factors.

The same may happen for permanent use of the bulk superconductors in a wet environment. Therefore, special care has to be taken for warming up samples from cryogenic temperatures.

Finishing to the final shape is a time- and cost-consuming procedure, especially if curved surface segments are necessary in functional elements of machines. Optimal design considering both functionality and machining, near net shaped processing and optimized finishing have to merge into a proper technology of bulk functional elements for power technology.

6.3.2
The Multi-Seed Technique

Several strategies have been introduced to increase the critical current density in an applied magnetic field. However, the size of the samples which can be achieved by the modified melt crystallization process seems to have been brought near to the limits with respect to economic processing. Indeed, a diameter of 10 cm has been obtained [6.36], but the mean effective critical current density of the current loop is significantly decreased in comparison to that of smaller samples (Table 5.1). This behavior is related to the mosaic structure resulting from the appearing low-angle grain boundaries. Moreover, the preparation of such large pellets is time consuming (about one week for samples of diameter 70 mm). Therefore, the multi-seeded melt growth process (MSMG) and welding techniques are promising alternatives to the conventional top-seeded melt growth process (TSMG) to prepare larger samples with good superconducting properties.

Indeed, tiles composed of several well-oriented single domains were developed when an appropriate number of SmBCO or NdBCO seeds were placed with parallel orientation on the top of the sample [6.37]. Thus, the machining time to grow plates of 35×110 mm² is reduced by 75% if four seeds are set at equal distances on top of the green body.

6.3.3
Rings of 123 Bulk Materials

The multi-seed method was tested for differently shaped materials including rings with various orientations of the *c*-axis of the seeds with respect to the ring axis [6.62, 6.63]. If the *c*-axis of the seeds is parallel to the ring axis, then the rings look quite regular. However, strong tangential stresses appear in rings seeded with their *c*-axes in a radial direction, and this initializes tangential cracks. These strong stresses result from the extension of the crystal lattice during oxygenation.

The authors are well aware of both the expectations of engineers and the spread of misunderstanding in parts of the community, and would like to emphasize here that it is not possible to prepare a ring-shaped crystal with *continuously radial* orientation of a particular axis (e.g. a *c*-axis perpendicular to the inner circumference of the ring)! This becomes obvious if we bear in mind the fact that a hypothetical ring which merges the low-angle grain boundary limit (i.e. < 4°) must consist of at least 360/4 = 90 single grains grown from 90 seeds. The ring shown in Figure 6.12 consists of 16 segments and the corresponding angle is 22.5°. Therefore, there is no advantage, for physical reasons, in a multi-seeded monolithic block with the *c*-axis of Y123 in the radial direction. Nevertheless, for any practical case, alternatives have to be taken into account: multi-seeded, multi-segmental monolithic rings or rings constructed of an appropriate number of shaped (monodomain or multi-seeded) segments. Functional aspects as well as engineering and manufacturing costs have to be taken into account. Cutting or milling, e.g., with a supersonic tool, remains a proper option for rings with an outer diameter less than about 80 mm.

Figure 6.11 presents a selection of functional elements from single-domain and multiseeded YBCO material. Cylinder segments in the figure consist of multi-seeded tiles, which – if necessary – were assembled using an appropriate resin. The contour of the trapped field of the multi-seeded ring (with the *c*-axis of the seeds parallel to the ring axis) is shown in Figure 6.12.

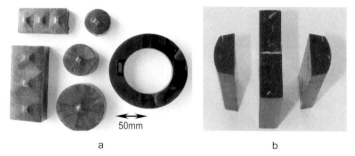

Figure 6.11 a Collection of single-domain and multiseeded materials a, and b finished elements for use in the rotor of a reluctance motor (Chapter 12) [6.63]

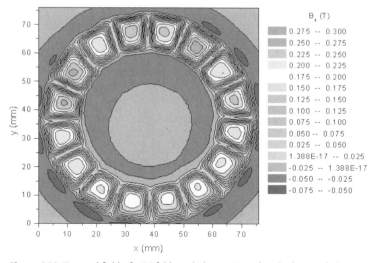

Figure 6.12 Trapped field of a 16-fold seeded ring. Note that the low angle limit is not matched, after [6.63]

The intergrain critical current density across the boundary between adjacent grains prepared by the multiseeding technique was found to be about 10^3 A cm^{-2} at a field of 20 mT as determined by local magnetization measurement [6.47]. This is about 10% of the average intragrain j_c. The difference between intra- and intergrain values results from the angle misfit at the grain boundary and the *segregation* of non-superconducting phases at the interface, which is accompanied by *voids* and the formation of *cracks*. Segregation may be caused by pushing 211 and residual (incongruent) melt solidifying in that region. Furthermore, one should consider that the *mosaic structure* appearing inside each grain itself causes a significant decrease in the intragrain j_c near the edge of the grains compared with the average j_c used as the reference.

6.3.4
Joining of Separate Single Grains

Joining of separately grown single-grain YBCO blocks is an alternative approach to obtain large bulk superconductors. At least four methods are currently being investigated.

Direct contact welding has been realized for small contact areas to investigate the influence of misorientation of "artificial" grain boundaries on j_c which have been formed by rejoining two parts obtained after cutting a single grain [6.38]. The trapped field was completely "reconstructed" after immediately rejoining the two sections at their interface by annealing in a pure O$_2$ atmosphere under the directional pressure of their own load (~10^{-3} bar). The corresponding current density across the welded grain boundary was $j_c \approx 10$ kA cm^{-2}. Applying the same procedure to weld two independently grown single grain bulks, the current density j_c in the joint strongly decreased to about 0.4 kA cm^{-2}, i.e. to about 40% of the bulk value, indicating the strong influence of the crystallographic orientation in the adjacent sections [6.39].

Welding by use of intrinsic filler materials (i.e. using only components which also form single grains) is related to the liquidus phase relationships [6.40–6.42]. An example is the rejoining of YBCO bulk tiles by means of a composition "Ba$_3$Cu$_5$O$_x$" which corresponds (nearly) to the composition of the peritectic melt. This enables joining below the peritectic melt temperature; however, it may leave nonsuperconducting components in the interface.

Joining by use of a soldering material relies on the lower peritectic temperatures of certain LnBa$_2$Cu$_3$O$_7$ compounds of heavy lanthanides such as Ln = Er, Tm, Yb. The soldering material forms a superconducting link between the properly oriented superconducting blocks. Potential solders should fit the epitaxy limitation for the bulk host, and their solidus temperature should be less than that of the superconductor composition. Local critical current densities up to 10 kA cm^{-2} have been obtained so far in such joints [6.43–6.45].

Welding by use of an YBCO/Ag composite is a promising alternative, since the matrix material also appears in the joint [6.46]. The cross section of the joint, presented in Figure 6.13, indicates the dense structure of the joint, which appears mostly free of Ag particles. However, Ag particles are enriched in the adjacent part of the matrix if a hypermonotectic spacer composition with Ag concentration higher than the monotectic composition is used. During heating, first the Ag-rich liquid L″ will appear at 940 °C according to the phase diagram (Figure 6.7), resulting from a pseudo-eutectic reaction (6.2) and distributing also over voids in the adjacent matrix. The Ba- and Cu-rich liquid L′ appears above the pseudo-monotectic temperature 970 °C because of the monotectic reaction m [Eq. (6.3)]. The crystallization proceeds from the depleted melt with low Ag contents during cooling, thus leaving Ag-rich melt in the pores of the adjacent matrix. Measurements of the critical current density in small specimens taken from the matrix and the joint have proven that j_c in the joint can excel critical current densities in parts of the matrix. This is because of the mosaic structure in certain parts of the matrix and is true for dense joints without voids. On the other hand, any defects such as cracks, voids, pores, and large-sized inclusions in the joint can be sensitively detected by irregularities in trapped-field mapping on top of the bulk [6.47].

Joining is a promising method which now is to be brought to a practicable technology. Problems to be solved concern appropriate temperature fields with respect to the applied solder and fixing the single tiles tightly during the process.

Recently, a *"repair and weld" method* was proposed starting from multi-seeded YBCO blocks. In a first step, material about 0.6 mm around the grain boundary (which contains the injurious defects) was removed by cutting, thus leaving a small stage for mechanical fixing. Then, a dense slice of the solder (YBCO/15 % Ag) was introduced before the thermal treatment was applied. The trapped field after "repairing" retains in the region of the joint up to 80% of the value of the bulk [6.47].

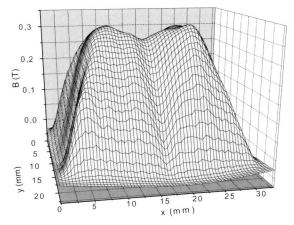

Figure 6.13 Trapped field of 2-fold seeded YBCO tile prepared by the "repair and weld" method [6.47]. Note that a "roof" (instead of a "cone") for the rectangular sample shape matches Bean's model.

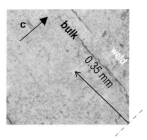

Figure 6.14 The micrograph proves the dense structure in the joint, after [6.47]

Another interesting alternative is to prepare large single-domain bulk HTSC by the so-called "MUltiseeded Seam-Less" (MUSLE) technique [6.48]. In the example given, a composed cylindrical green sample was prepared by stacking two layers of different REBaCuO materials. The lower part had a lower peritectic temperature than the upper part. Then, four seed crystals were placed on top of the stack, which thereafter was melt processed. Surprisingly, the four-seed structure in the upper part coincided, thus transforming the information of orientation to the lower part, which was found to consist of a single domain after this procedure.

6.4
Bulk Materials and Processing Designed for Special Applications

Here, materials and technology development will be considered as they are required for current leads and fault current limiters. In both cases, the total transport current and the quench behavior become significant properties, whereas j_c and the trapped fields become less important (see Chapter 13).

6.4.1
Infiltration Technique and Foams

The idea of an infiltration of the Ba-Cu oxide-rich melt into *fibers or fabrics* of yttria was initially proposed to produce melt-textured long-length conductors and large-area sheets [6.49]. It was found that commercial ceramic fibers and fabrics are easily penetrated by the melt, thereby saturating them with Y_2O_3. During subsequent slow cooling, the Y123 phase crystallizes according to the same chemistry as in MMCP. Surprisingly, highly textured materials have been obtained after seeding as usual for bulk YBCO bodies. Advantagously pre-shaped elements even of complex forms can easily be prepared by the method. Again, it should be kept in mind that the crystal orientation cannot follow any curved surface!

The infiltration of *bulk bodies* was initially applied to pre-shaped pressed or pre-sintered bodies made from the peritectic a phase, i.e. Y211 in the case of YBCO. It can also be applied to a porous 3-dimensional framework ("*foam*") of sintered 211 ceramics with pores of diameter 10–100 μm. Under appropriate process conditions, the near-peritectic $Ba_{0.45}Cu_{0.65}O_{0.97}$ melt wets and penetrates the frame without filling the pores completely. After seeding and cooling, a textured porous body is formed. The porosity enables oxygenation to proceed easily. From magnetization loops $M(H)$ at 77 K, an intergrain current density close to 40 kA cm^{-2} has been obtained at zero magnetic field [6.50]. The real current path and intra-grain current density in a superconducting framework is currently the object of theoretical analysis.

The preparation of the porous 211 foam was achieved with the aid of an advanced ceramic technology: a slurry with fine 211 grains suspended in water and polyvinyl alcohol was infiltrated into a body of polyurethane foam with a definite pore size. Then, the organic matrix was burned out in a thermal pre-reaction, and finally a seeded melt-texturing process was applied [6.51].

6.4.2
Long-Length Conductors and Controlled-Resistance Materials

Processing

Rods or tapes of length more than 10 cm are necessary for conducting elements in power applications to provide high-current-consuming facilities (e.g., superconducting magnets) or as a switching element in fault current limiters (see Chapter 13). Directional solidification in high-temperature gradients has successfully been applied to grow highly oriented Y123/211 composites of different 211 contents, proceeding according to the same principles as the melt texturing process. Preferably, a Bridgman technique or zone melting are applied. For *current leads* [6.52], the high j_c in the CuO_2 planes is used, and the bulk HTSC is therefore grown parallel to the *a*-axis. For resistive *fault current limiters* (see Chapter 13) [6.53], the low j_c of *c*-axis-grown material and its high resistance in the normal conducting state are used. The high resistance of *c*-axis-grown material prevents overheating and crash of the bulk HTSC during the transition from the superconducting to the normal state. However, the low mechanical stability due to cracking parallel to the planes is a serious disadvantage.

A promising alternative (in the case of resistive *fault current limiters*) is to prepare meander-shaped elements by cutting large disks from a single-grain YBCO material (Figure 6.15). In this way, a long current path and a restricted cross section can be realized. Such elements are advantageously directly contacted with a metal alloy shunt parallel to the superconductor [6.54] (Chapter 13). A melt-textured YBCO body with artificially patterned holes was presented as a possible 3-dimensional equivalent to meander-shaped fault current limiting elements with a more complicated current path containing branches also in the *c*-direction [6.55].

Fault current limiters and current leads are the most significant examples of the use of Bi2212-type superconductors in the form of bulk BSCCO tubes.

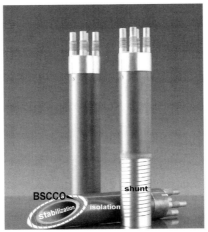

Figure 6.15 **a** Meander-shaped YBCO element with contacts (meander length 1 m [6.57]) and **b** a BSCCO tube [6.69] for use in resistive fault current limiters.

Contacting

Applications of superconductors in the high current range need advanced contacts between superconducting ceramics and metallic conductors. Soldering is the usual solution to the problem. Special care has to be taken in the wetting of the solder on the ceramics, and consequently the superconductor is covered with a metallic (Cu or Ag) layer which then is the base for the soldered contact. Electrolytic deposition is an appropriate method (see, e.g., Ref. [6.56]).

6.4.3
Bi2212 Bulk Materials and Rings

The application of BSCCO materials at 77 K in magnetic fields is limited by strong flux creep effects due to the decoupling of flux lines into pancake vortices in this nearly 2-dimensional material. Nevertheless, for applications at low magnetic fields as current leads, fault current limiters, and magnetic shielding, bulk BSCCO is often the better choice. For such applications, less strictly oriented material can be used. The demands on the texture depend on the ratio between weak link and flux creep behavior limiting the critical current density in BSCCO.

A very effective "Melt Cast Process" has been developed for bulk Bi2212, enabling processing of even large tubes of diameter 25–200 mm and length up to 1000 mm [6.58]. This process consists of successive melting, centrifugal casting, heat treatment, and oxygenation. For centrifugal casting, the melt of a composition appropriate to the subsequent solidification of Bi2212 is heated in a rotating metal cylinder (~400 rpm), thereby becoming distributed over the inner wall. Solidification in air proceeds from a temperature near 900 °C under controlled cooling. Finally, the precise dimension is obtained by ceramics machining [6.59, 6.60].

Bifilar functional elements for fault current limiters (FCL) are obtained from such tubes by cutting a spiral nearly to the end of the tube. A sophisticated technology is used to produce a 3-layer compound element containing also the metallic shunt (by soldering a thin metal tube onto the Bi2212 tube) and additionally a thin tube of fiber-reinforced plastics. A 1 mm slot is cut precisely, with computer control, using either a disc saw or high-speed milling. The following data give an idea of the size of a Bi2212 FCL element: total spiral length ~5400 mm, tube length 300 mm, wall thickness 4 mm, internal diameter 42 mm, height of winding 6 mm [6.58]. It should be noted that only Bi2212 material – the solidification of which is much easier to control than that of Bi2223 – satisfies the parameters for certain applications of the bulk materials [6.61].

6.4.4
Batch Processing of 123 Bulk Materials

Processing of 123 materials has been introduced, up until now, in scalable technology, which is characterized by a batch regime with the crystallization process as the central point.

Typical batch sizes – limited by the layout of the furnace – have been reported in the order of 3–20 kg. A typical technological scheme is represented in Figure 6.16. Ingot materials are available in appropriate quality from a number of providers. Here one should again recognize that the central technological step is not a ceramic process, hence "ceramic" powder criteria are at least not sufficient [6.62–6.65].

Some authors propose a final resin impregnation of the "as-grown" bulk materials to improve mechanical properties (Chapter 9), which also is claimed to prevent corrosion under the influence of moisture [6.66]. Whether or not such treatment is necessary depends on properties such as compactness (porosity), flaw density etc. of the "as-grown" bulk material as well as on the conditions during its use (frequent changes of temperature and contact with moisture etc.). For YBCO tiles grown with high mechanical quality, embedding in paraffin both gives sufficient protection and fixes the element in the machine for use.

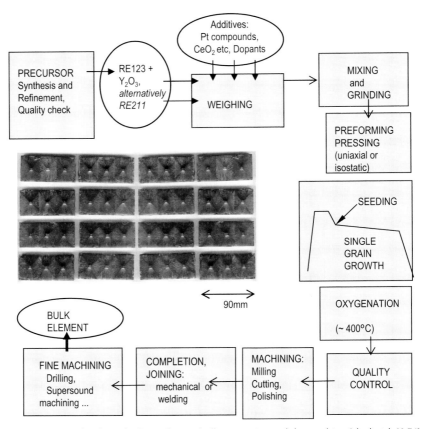

Figure 6.16 Technological scheme for 123 bulk processing and the resulting 8 kg batch [6.74]

References

6.1 J. L. Tallon, J. W. Loram, G. V. M. Williams, J. R. Cooper, I. R. Fisher, J. D. Johnson, M. P. Staines, C. Bernhardt, Physica Status Solidi B **215**, 531 (1999).

6.2 C. W. Schneider, R. R. Schulz, B. Goetz, A. Schmehl, H. Bielefeldt, H. Hilgenkamp, J. Mannhart, Appl. Phys. Lett. **75**, 850 (1999).

6.3 L. Shlyk, G. Krabbes, G. Fuchs, K. Nenkov, Physica C **383**, 175 (2002).

6.4 A. Schmehl, B. Goetz, R. R. Schulz, C. W. Schneider, H. Bielefeldt, H. Hilgenkamp, J. Mannhart, Europhys. Lett. **47**, 110 (1999).

6.5 R. Dupree, A. Gencten, D. M. Paul, Physica C **193**, 81 (1992).

6.6 L. Shlyk, G. Krabbes, G. Fuchs, G. Stöver, S. Gruss, K. Nenkov, Physica C **377**, 437 (2002).

6.7 M. Scavini, L. Mollica, R. Bianchi, G. A. Costa, M. Ferretti. P. Mele, A. Ubaldini, A. Ghigna, L. Malavasi, P. Mustarelli, Int. J. Mod. Phys. B **17**, 936 (2003).

6.8 G. Krabbes, G. Fuchs, P. Schätzle, S. Gruß, J. W. Park, F. Hardinghaus, G. Stöver, R. Hayn, S.-L. Drechsler, T. Fahr, Physica C **330**, 181-190 (2000).

6.9 Y. X. Zhou, W. Lo, T. B. Tang, K. Salama, Supercond. Sci. Technol. **15**, 723 (2002).

6.10 M. T. Gonzalez, N. H. Babu, D. A. Cardwell, Supercond. Sci. Technol. **15**, 1372 (2002).

6.11 L. Shlyk, G. Krabbes, G. Fuchs, K. Nenkov, P. Verges, Appl. Phys. Lett. **81**, 5000 (2002).

6.12 S. Zagulaev, P. Monod, J. Jegoudez, Phys. Rev. B **52**, 1074 (1990).

6.13 F. C. Zhang, T. M. Rice, Phys. Rev. B **37**, 3759 (1988).

6.14 A. V. Mahajan, H. Alloul, G. Collin, J. F. Marucco, Phys. Rev. Lett. **72**, 3100 (1994).

6.15 S. H. Pan, E. W. Hudson, K. M. Lang, H. Eisaki, S. Uchida, J. C. Davis, Nature **403**, 746 (2000).

6.16 M. Hussain, S. Kuroda, K. Takita, Physica C **297**, 176 (1998).

6.17 G. Krabbes, G. Fuchs, P. Verges, P. Diko, G. Stöver, S. Gruss Physica C **378-381**, 636 (2002).

6.18 H. A. Mook, P. Dai, S. M. Hayden, G. Aeppli, T. G. Perring, F. Dogan, Nature **395**, 580 (1998).

6.19 S. Adachi, C. Kasai, S. Tajima, K. Tanabe, D. S. Fujihara, T. Kimura, Physica C **351**, 323 (2001).

6.20 Y. Zhao, C. H. Cheng, J. S. Wang, Supercond. Sci. Technol. **17**, S34 (2004).

6.21 R. P. Sawh, R. Weinstein, D. Parks, A. Gandini, Y. Ren, I. Rusakova, Physica C **383**, 411 (2003).

6.22 N. H. Babu, E. S. Reddy, D. A. Cardwell, A. M. Campbell, Supercond. Sci. Technol. **16**, L44 (2003).

6.23 E. Mezzetti, R. Gerbaldo, G. Ghigo, L. Gozzelino, B. Minetti, G. Krabbes, A. Rovelli, Physica C **354**, 289 (2001).

6.24 R. Weinstein, A. Gandini, R-P. Sawh, D. Parks, B. Mayes, Physica C **387**, 391 (2003).

6.25 H. W. Weber, H. Böck, E. Unfried, L. R. Greenwood, J. Nucl. Mater. **137**, 236 (1986).

6.26 F. M. Sauerzopf, Phys. Rev. B **57**, 10 959 (1997).

6.27 R. Gonzalez Arrabal, M. Eisterer, H. W. Weber, G. Fuchs, P. Verges, G. Krabbes, Appl. Phys. Lett. **81**, 868 (2002).

6.28 Y. Ren, R. Weinstein, R. P. Sawh, J. Liu, Physica C **282-287**, 2301 (1997).

6.29 R. Weinstein, R. Sawh, Y. Ren, D. Parks, Mater. Sci. Eng. B **53**, 38 (1998).

6.30 P. Diko, G. Krabbes, C. Wende, Supercond. Sci. Technol. **14**, 486 (2001).

6.31 Y. Nakamura, K. Tachibana, H. Fujimoto, Physica C **306**, 259 (1998).

6.32 P. Schätzle, G. Krabbes, S. Gruß, G. Fuchs, IEEE Transact. Appl. Supercond. **9**, 2022 (1999).

6.33 J. Joo, S-B. Jung, W. Nah, J-Y. Kim, T. S. Kim, Cryogenics **39**, 107 (1999).

6.34 P. Diko, G. Fuchs, G. Krabbes Physica C **363**, 60 (2001).

6.35 F. Y. Chuang, D. J. Sue, C. Y. Sun, Mater. Res. Bull. **30**, 1309 (1995).

6.36 R. Tournier, E. Beaugnon, O. Belmont, X. Chaud, D. Bourgault, D. Isfort, L. Porcar, P. Tixador, Supercond. Sci. Technol. **13**, 886-895 (2000).

6.37 P. Schätzle , G. Krabbes, G. Stöver, G. Fuchs and D. Schläfer, Supercond. Sci. Technol. **12**, 69 (1999).

6.38 K. Salama, V. Selvamanickan, Appl. Phys. Lett. **60**, 898 (1992).

6.39 L. Chen, H. Claus, A. P. Paulikas, H. Zheng, B. W. Veal, Supercond. Sci. Tech. **15**, 672 (2002).

6.40 D. Shi, Appl. Phys. Lett. **66**, 2573 (1995).

6.41 D. A. Cardwell, A. D. Bradley, N. H. Babu, M. Kambara, W. Lo, Supercond. Sci. Tech. **15**, 639 (2002).

6.42 H. Zheng, M. Jiang, R. Nikolova, L. Chen, U.Welp , B. W. Veal, Physica C **322**, 1 (1999).

6.43 T. Prikhna, W. Gawalek, V. Moshchil, A. Surzhenko, A. Kordyuk, D. Litzkendorf, S. Dub, V. Melnikov, A. Plyushchay, N. Sergienko, A. Koval, S. Bokoch, T. Habisreuther, Physica C **354**, 333 (2001).

6.44 Ch. Jooss, B. Bringmann, H. Walter, A. Leenders, H. C. Freyhardt, Physica C **341–348**, 1423 (2000).

6.45 S. J. Manton, C. Beduz, Y. Yang, IEEE Trans. Appl. Supercond. **9**, 2089 (1999).

6.46 C. Harnois, G. Desgardin, I. Laffez, X. Chaud, D. Bourgault, Physica C **383**, 269 (2002).

6.47 Th. Hopfinger, R. Viznichenko, G. Krabbes, G. Fuchs, K. Nenkov, Physica C **398**, 95 (2003).

6.48 M. Sawamura, M. Morita, H. Hirano, Supercond. Sci. Tech. **17**, S418 (2004).

6.49 E. S. Reddy, J. G. Noudem, M. Tarka, G. J. Schmitz, J. Mater. Res. **164**, 955 (2001).

6.50 E. S. Reddy, G. J. Schmitz, Supercond. Sci. Tech. **15**, L21 (2002).

6.51 E. S. Reddy, M. Herweg, G. J. Schmitz, Supercond. Sci. Tech. **16**, 608 (2003).

6.52 Y. Imagawa, K. Kakimoto, Y. Shiohara Physica C **280**, 245 (1997).

6.53 R. F. Tornier, D. Isfort, D. Bourgault, X. Chaud, D. Buon, E. Floch, L. Porcar, P. Tixador, Physica C **386**, 467 (2003).

6.54 M. Morita, O. Miura, D. Ito, Physica C **357**, 870 (2001).

6.55 F. G. Noudem, S. Meslin, C. Harnoir, Supercond. Sci. Tech. **17**, 931 (2004).

6.56 M. Ueltzen, I. Martinek, F. Syrowatka, U. Floegel-Delor, T. Riedel, Physica C **372**, 1653 (2002).

6.57 F. N. Werfel, U. Floegel-Delor, R. Rothfeld, D. Wippich, T. Riedel, B. Göbel, *Schmelztexturierte YBCO Komponenten in Supraleitung und Tieftemperaturtechnik*, ed. VDI Technologiezentrum Düsseldorf, 2003 (CD attachment), ISBN 3-93-138444-6M

6.58 S. Elschner, F. Breuer, M. Noe, T. Rettelbach, H. Walter, J. Bock, IEEE Trans. Appl. Supercond. **13**, 1980 (2003).

6.59 Y. M. Park, G. E. Jang, Cryogenics **41**, 169 (2001).

6.60 J. Bock, S. Elschner, P. F. Herrmann, IEEE Trans. Appl. Supercond. **5**, 1409 (1995).

6.61 T. Lang, D. Buhl, D. Schneider, S. Al-Wakeel, L. J. Gauckler J. Electroceram. **1**, 133 (1997).

6.62 H. Walter, S. Arsac, J. Bock, S. O. Siems, W. R. Canders, A. Leenders, H. C. Freyhardt, H. Fieseler, M. Kesten, IEEE Trans Appl. Supercond. **13**, 2150 (2004).

6.63 G. Krabbes, Processing of High Performance (RE)BaCuO Superconductor Bulk Materials: A Thermodynamic Approach in Advances in Condensed Matter and Materials Research Vol. 4, edited by F. Gerard (Nova Science Publishers Hauppauge N.Y. 2003) pp. 179-219.

6.64 D. Litzkendorf, T. Habisreuther, J. Bierlich, O. Surzhenko, M. Zeisberger, S. Kracunovska, W. Gawalwek, Supercond. Sci. Tech. **18**, S206 (2005).

6.65 S. Sengupta, J. Corpus, J. R. Gaines, V. R. Todt, X. F. Zhang, D. J. Miller, C. Varanasi, P. J. Mc Ginn, Appl. Supercond. **7**, 1723 (1997).

6.66 M. Tomita, M. Murakami, K. Itoh, H. Wada, Supercond. Sci. Tech. **17**, 78 (2004).

6.67 L. Shlyk, G. Krabbes, G. Fuchs, K. Nenkov, B. Schüpp, J. Appl. Phys. **96**, 3371 (2004).

6.68 L. Shlyk, G. Krabbes, K. Nenkov, G. Fuchs, *Chemical doping in YBCO bulk material and its influence on pinning properties* in Applied Superconductivity, edited by A. Andreone, G. P. Pepe, R. Cristiano, G. Masullo, IOP Conf. Ser. 181, IOP Publ. Bristol and Philadelphia 2003, p. 1401.

6.69 J. Bock, F. Breuer, H. Walter, M. Noe, R. Kreutz, M. Kleinmaier, K. H. Weck, S. Elscher, Supercond. Sci. Tech. **17**, S122 (2004).

6.70 P. Diko, Supercond. Sci. Tech. **17**, R45 (2004).

6.71 C. J. Kim, H. W. Park, K. B. Kim, G. W. Hong, Supercond. Sci. Tech. **8**, 652 (1995).

6.72 S. Pinol, F. Sandiumenge, B. Martinez, N. Vilalta, X. Granados, V. Gomis, F. Galante, J. Fintcuberta, X. Obradors, IEEE Trans. Appl. Supercond. **5**, 1549 (1995).

6.73 P. Diko, C. Wnde, D. Litzkendorf, T. Klupsch, W. Gawalek, Supercond. Sci. Tech. **11**, 49 (1998).

6.74 G. Stöver, P. Schätzle, G. Krabbes, Research Report IFW Dresden (not published).

6.75 L. Shlyk, G. Krabbes, G. Fuchs, K. Nenkov, Appl. Phys. Lett. **86**, 092503 (2005).

7
Alternative Systems

Remarkably improved critical current density has been reported for bulk $Nd_{1+y}Ba_{2-y}Cu_3O_{7\pm\delta}$-based material in pioneering work by Yoo et al. [7.1]. The irreversibility line is shifted and the irreversibility field at 77 K can much exceed 10 T [7.1], [7.2]. Exceptionally high j_c values, even in applied fields of 2 T or more, are characteristic for $(RE)Ba_2Cu_3O_7$-based bulk materials with the large Ln^{3+} ions of the light lanthanide series with a certain homogeneity range $Ln_{1+y}Ba_{2-y}Cu_3O_{7\pm\delta}$ ("solid solution-like behavior"). Not nearly all questions have been answered concerning the mechanisms of processing and control of the properties of these materials. Nevertheless, the available material now allows us to extract the fundamental orientations, and this chapter will throw light on the cornerstones.

Increasing interest in processing $LnBaCuO$ materials in combination with Ag is based on the influence of silver on the phase equilibria, resulting in a remarkable decrease in the peritectic temperatures, thus compensating for the increase in the melting (decomposition) temperatures of $LnBa_2Cu_3O_7$ compared with $YBa_2Cu_3O_7$.

7.1
Impact of Solid Solutions $Ln_{1+y}Ba_{2-y}Cu_3O_{7\pm\delta}$ on Phase Stability and Developing Microstructure

T_c decreases with increasing substitution by Ln^{3+} for Ba^{2+}. This is expected from Section 1.2, taking into account that an excess of Ln gives rise to underdoping, which can be compensated by oxidizing only within a certain limit far below the optimum (Figure 1.14). T_c of fully oxygenated $Nd_{1+y}Ba_{2-y}Cu_3O_{7\pm\delta}$ as a function of y achieves 96 K ($y \cong 0$), and superconductivity disappears for $y > 0.35 (T_c = 0)$ [7.3]. A wide range of T_c and j_c values and wide variation in the dependency of j_c vs H was reported for different melt-grown samples from different groups and preparation routes. High-quality samples are characterized by $j_c > 20$ kA cm^{-2} at applied fields $\mu_0 H \cong 2 - 4$ T or a peak in this field (Chapter 8).

On the other hand, the broad homogeneity range of $Ln123$ phases is a reason for severe problems in processing of high quality materials in a reproducible way.

High Temperature Superconductor Bulk Materials.
Gernot Krabbes, Günter Fuchs, Wolf-Rüdiger Canders, Hardo May, and Ryszard Palka
Copyright © 2006 WILEY-VCH Verlag GmbH & Co. KGaA, Weinheim
ISBN: 3-527-40383-3

Processing in 0.01 bar oxygen partial pressure turned out to be a preferable condition to prepare bulk Nd123 materials, and this has now also been applied successfully to other materials of the light lanthanide series [7.1], [7.3].

The exceptionally high critical current density in superconducting compositions ($y \ll 0.3$), even in an applied field of several Tesla, had been attributed to the solid-solution behavior permitting fluctuating T_c in appropriate samples. On the other hand, decomposition of the Nd123 solid-solution phase into alternating $Nd_{1+y_1}Ba_{2-y_1}Cu_3O_{7\pm\delta}$ ($y_1 \approx 0.8$) and $Nd_1Ba_2Cu_3O_{7-\delta}$ compositions was concluded from energy-dispersive X-ray spectral analysis of traveling solvent-grown single crystals performed in TEM [7.4].

X-ray diffraction and X-ray absorption spectra near the Cu edge as well as Raman spectroscopy indicate ordered orthorhombic superstructures of "Nd213" at the Nd-rich edge of $Nd_{1+y}Ba_{2-y}Cu_3O_{7\pm\delta}$ single crystals ($y > 0.7$), whereas a continuous field of a "cation-disordered phase" between 0.3 and 0.7 might be explained by stages of a transformation (2nd-order type) between 123 and "213" compositions (not superconducting and not explicitly shown in Figure 7.1) [7.5]. The hypothesis of fluctuating T_c by alternating y seems to conflict with the experience that high j_c in applied fields or a peak in j_c vs H is exclusively obtained in samples the composition of which is close to the nominal 1:2:3 composition!

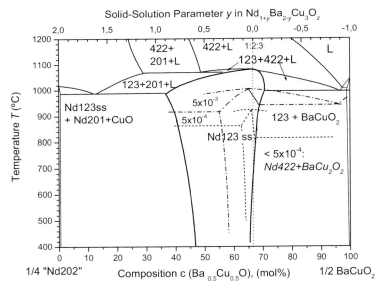

Figure 7.1 The section "NdCuO$_2$" – BaCuO$_2$ of the phase diagram Nd-Ba-Cu-O for p(O$_2$) = 0.21 bar with superposition of the $Nd_{1+y}Ba_{2-y}Cu_3O_{7\pm\delta}$ phase fields at 5×10^{-3} and 5×10^{-4} bar, respectively

A further piece in the puzzle was contributed by tuning the anomalous peak effect (Chapter 8) in nearly stoichiometric Nd123 single crystals in a series of annealing experiments under controlled $p(O_2)$/temperature conditions [7.6]. A

7.1 Impact of Solid Solutions $Ln_{1+y}Ba_{2-y}Cu_3O_{7\pm\delta}$ on Phase Stability and Developing Microstructure

typical peak near 2 T appears in the j_c vs. H curve of Nd123 crystals after a 2-step annealing in oxygen at 500 °C and 340 °C, or by slow continuous cooling from 600 °C followed by oxygenation at 340 °C. In contrast, low j_c results in applied fields if the same material is quenched to the oxygenation temperature 340 °C. $Ln_{1+y}Ba_{2-y}Cu_3O_{7\pm\delta}$/Ag composite materials with Ln = Nd, Sm behave similarly, the relevant annealing temperature being between 300 and 400 °C [7.7].

At this point a review of more recent investigations of phase relationships at the temperatures of sample preparation is helpful. Sections Ln101 – Ba011 for Ln = Sm and Nd in Figures 2.10 and 7.1 indicate the broad existence region for 123 solid solutions at $p(O_2)$ = 0.21 bar, whereas the extension of the 123 phase fields for 0.05 and 0.005 bar O_2 partial pressure is much smaller (indicated by the superimposed phase fields for Nd123 in Figure 7.1) [7.5], [7.8]. The "cation-disordered" region is not explicitly referred to in both figures.

Three features distinguish Ln123 bulk superconductors from Y123:
1. For both Nd123 and Sm123, regions of retrograde solubility, i.e. change of sign for dy'/dT at a certain temperature are observed at the Ba-rich and Nd-rich edges of the 123 phase field, respectively (y' represents the composition at the phase field boundary).
2. For both Nd123 and Sm123, the width of the solidus homogeneity ranges strongly expands toward higher values of y with increasing $p(O_2)$.
3. Crystals of the solid solution are usually expected to be inhomogeneous because of thermodynamic and kinetic distribution coefficients if special means are not applied.

The retrograde solubility causes phase separation due to decomposition of the solid solution near its Ba-rich composition edge at reduced T'. A spinodal decomposition mechanism is favored when the as-grown composition (idealized, the as-grown composition y_s at the solidus temperature T_s) becomes unstable with respect to the thermodynamic stability criterion ($\partial G/\partial y$), whereas the limit for irreversible diffusion (criterion $\partial^2 G/\partial y^2$) has not been exceeded. Spinodal decomposition is controlled by the degree of undercooling and by the activation energies for nucleation and diffusion [7.9]. The dotted lines in Figure 7.2 indicate schematically the extension of the metastable region between the (real or hypothetical) thermodynamic stability field of 123 and the corresponding spinodal line (representing the stability limit for diffusion). The decomposition according to Eq. (7.1) into $BaCuO_2$ + 123 solid solution (with the Ba-depleted equilibrium composition at $T' < T_s$) represents the thermodynamic boundary at the Ba-rich edge:

$$a\ Ln_{1\pm y_s}Ba_{2\pm y_s}Cu_3O_{7-\delta} \rightarrow b\ Ln_{1+y'}Ba_{2-y'}Cu_3O_{7-\delta'} + c\ BaCuO_2 + d\ O_2 \qquad (7.1)$$

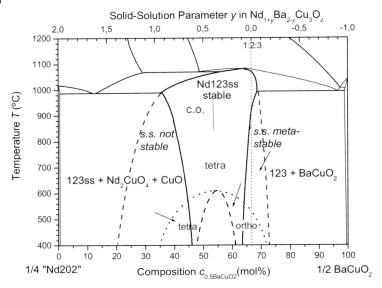

Figure 7.2 Sketch of the $Nd_{1+y}Ba_{2-y}Cu_3O_{7\pm\delta}$ phase field on the section "$NdCuO_2$" – $BaCuO_2$ indicating the metastable regions between equilibrium phase boundaries and spinodals; ortho, tetra, and a. o. indicate the stability regions of the orthorhombic, tetragonal and concentration controlled orthorhombic II phases [7.5], [7.12]. An arrow indicates the stability limit with respect to spinodal decomposition [7.6], [7.35]

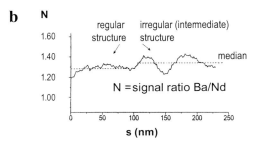

Figure 7.3 TEM image of irregular structure in the $Nd_{1+y}Ba_{2-y}Cu_3O_{7\pm\delta}$ phase and fluctuating Nd/Ba ratio in melt-textured and annealed material [7.10]

Indeed, TEM investigations indicate a number of defects in samples of appropriate composition containing typical regions of nonregular structure, in which alternating values of the Ba/Nd ratio have been determined by high-resolution microprobe analysis [7.10]. The periodic length of 50 nm shown in Figure 7.3 is too large to be explained by defects on the unit cell length scale; however, it agrees with an intermediate or metastable state of spinodal decomposition. In additional to these regions, nano-sized second-type grains have also been observed. The physical behavior of the melt-textured material was found to be in fair agreement with annealed single crystals in Ref. [7.6]. It became evident that the relevant decomposition proceeded at the Ba-rich edge of the homogeneity range.

A remarkable shift was also found at the Ln-rich limit of the solid solution field from $y \cong 0.7$ (950 °C) to $y \leq 0.2$ for T < 700 °C for $Sm_{1+y}Ba_{2-y}Cu_3O_{7\pm\delta}$ (Ln=Sm) in air, and this tendency is also indicated for Nd [7.11]. Here the stability limit in the subsolidus region is determined by reaction (7.2)

$$Ln_{1+y_s}Ba_{2-y_s}Cu_3O_{7-\delta} \to b\ Ln_{1+y''}Ba_{2-y''}Cu_3O_{7-\delta'} + c\ Ln_2CuO_4 + d\ O_2 \qquad (7.2)$$

enabling a similar spinodal mechanism to proceed in the Ln-rich part also, thus leaving a superconducting phase with y no larger than 0.2 (Figure 7.1).

On the other hand, several solid-state reaction routes from different precursors lead to Ln-rich solid solutions in which either cation ordering or oxygen ordering or both have been identified (maybe in a metastable state) [7.12], [7.13]. Thus, time-temperature transition (TTT) diagrams may be a more appropriate approach to controlling solid-state synthesis of these solid solutions, which become thermodynamically unstable in the subsolidus temperature field [7.11].

Although both ordered and clustered defects as well as structures resulting from spinodal decomposition are potential pinning centers, the defect structure of practical relevance should not cause a significant decrease of T_c, as would be the case for Ln-rich compositions. This conclusion is in accordance with the appearance of non-superconducting phase(s) already for $y > 0.3$. No indication of spinodal decomposition of the Nd123 solid solution into Nd-rich and Ba-rich 123 has been found in Ref. [7.13]. Oxygen ordering has also been verified in melt-textured Nd123 near the stoichiometric 123 composition, which can be correlated to the beginning of cationic disorder [7.14]. Really, it might already appear in the as-grown state as a consequence of antisite defects and Ba excess, although one should bear in mind that the Ba-rich composition becomes unstable during cooling or annealing below 900 °C.

Oxygen reordering however has also to appear in the course of spinodal decomposition and cationic disordering. Therefore, both types of defects should be expected to be present *simultaneously* in melt-grown materials. Additionally, deviations in cation stoichiometry and ordering have a controlling influence on the low-temperature oxygenation, which becomes apparent in pinning properties and the j_c vs H relationship [7.15].

Here it should be emphasized that a number of processes may proceed simultaneously because of a lack of any deep minimum at the free enthalpy hyperface, especially in the subsolidus region. Consequently, different types of defects appear in the same sample, and its microstructure is sensitively influenced by kinetic and thermodynamic parameters. This situation may explain the wide scattering of properties and features in the magnetization curve. Despite a number of open questions, the present situation already allows us to give a prognosis for the generation of a certain type of defects. Their relevance to pinning capability will be discussed in Chapter 8.

Growth-induced inhomogeneity results from the wide homogeneity ranges of Ln123 phases in contrast to Y123. Thus, for large-size Ln^{3+} ions, the peritectic reaction m_1' [Eq. (2.15)] proceeds *invariantly* only with one fixed solidus composition in air, near to (but not necessarily equal to) the stoichiometric composition 1:2:3 for Nd123 or Sm123, respectively. Note that this composition is not the optimum with respect to high critical currents.

Neglecting the excess of 211 (or 422) particles in the "semisolid liquid", the phase diagram Figure 7.1 implies, for an Nd-rich ingot [left hand from the composition at $T(m_1)$], that the *solidus* composition of the Ln123 phase is enriched with Ba compared with the liquid. In the course of further processing, the solidus composition moves continuously, resulting in Nd-richer crystals, which therefore appear inhomogeneous. In contrast, the shift to Ba-rich compositions is expected for Ba-rich ingots ("thermodynamic distribution coefficient"). An additional concentration shift has to be expected due to growth kinetics.

Stability range dependent on $p(O_2)$. The sketch of Figure 7.1 indicates that the stability range of Nd123 and Sm123 is drastically reduced at the Nd-rich and Sm-rich sides, respectively, if chemical potentials of oxygen [i.e. $p(O_2)$] are reduced. This is the physico-chemical background for cation stoichiometry control in Ln123 by the so-called "oxygen-controlled melt growth" OCMG process. The *reduced* oxygen partial pressure is the most important feature of this method [7.3].

7.2
Advanced Processing of Ln123

Growth-induced inhomogeneity and the strong influence of the actual composition on superconducting properties are the main problems for processing 123 solid solutions. The range of optimal compositions, however, is narrow and localized near the Ba-rich limit.

Oxygen potential-controlled processing and concentration-controlled processing are the two principal approaches to be considered.

7.2.1
Oxygen Potential Control

The extension of the solid solution to high Nd (Sm) concentrations is restricted in reduced oxygen pressure, whereas the temperature window remains almost unaffected. Therefore, stoichiometric materials are stabilized under these conditions, which is underlined when we compare the areas of stability of Nd123ss from 0.21 bar to 0.0005 bar oxygen in Figure 7.1.

7.2.2
Oxygen-Controlled Melt Growth Process (OCMG)

Low oxygen pressure processing of NdBaCuO bulk materials, which was proposed by Murakami et al. [7.3], can meanwhile be considered to be the "classical" tool to improve the critical current and irreversibility field in NdBaCuO superconductors. In the OCMG process, an appropriate heat–dwell–cool down cycle proceeds at a reduced oxygen pressure, which is kept constant in the course of solidification.

7.2.3
Isothermal Growth Process at Variable Oxygen Partial Pressure (OCIG)

An alternative approach to controlling growth and cation stoichiometry of the resulting $Ln123$ matrix is an isothermal low-pressure process in which $p(O_2)$ instead of T is altered dependent on time [7.16]. Three steps of this "Oxygen-Controlled Isothermal Growth" Process (OCIG) are visualized in the log $p(O_2)$ vs T^{-1} phase diagram for the NBCO system in Figure 7.4. First, the sample is heated under reduced but constant $p(O_2)$ until a melt is formed in a peritectic-type reaction. In the second isothermal step (A → B in the Figure), the partial pressure $p(O_2)$ is slowly increased after a short dwell time for homogenization.

Solidification from any Nd-rich composition starts at a $[p(O_2),T]$ couple placed on an equilibrium line between m_1 and p_2 or p'_1, respectively, depending on the starting mixture. Here, m_1, p_2, and p'_1 represent the corresponding univariant reactions [Eqs. (7.3) – (7.5)].

$$m_1: Ln211 \text{ (Nd422 for } Ln = \text{Nd)} + L(m_1) \rightleftarrows Ln123 \tag{7.3}$$

$$p_2: Ln211 \text{ (Nd422)} + L'(p_2) \rightleftarrows Ln123ss + 201 \tag{7.4}$$

$$p'_1: Ln\ 201 + L'(p'_1) \rightleftarrows 123ss + CuO \tag{7.5}$$

This clearly indicates the solidification of a solid solution $Ln123$ss over a wide range for y > 0 from the Nd-rich melt L'.

On the other hand, the stoichiometric 123 (or Ba-rich solid solution with only small deviation from y < 0) can crystallize over the entire parameter field between m_1 and p_3, thus indicating its preference for properties control.

The third step is cooling down as usual.

Post-Growth Annealing of Ln123 Bulk Materials in Controlled p(O$_2$)

Phase relationships with $Ln123$ phases imply that stoichiometry and properties can simply be tuned replacing OCMG by a high temperature *Post-growth Annealing in Controlled Oxygen atmosphere* under reduced $p(O_2)$ **(PACO)**. PACO is applied

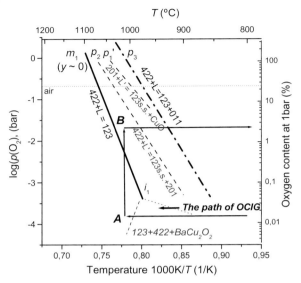

Figure 7.4 Part of the log $p(O_2)$ vs T^{-1} phase diagram and reaction path during "Oxygen-Controlled Isothermal Growth" (OCIG) of $Nd_{1+y}Ba_{2-y}Cu_3O_7$ bulk material.

to bulk samples which have been grown in ambient air atmosphere [7.17]. The advantage of a simple and reproducible growth process technology in air is faced with the complicated transformations under the annealing condition, which can affect the materials depending on composition, homogeneity, and microstructure of the as-grown sample. Two situations have to be distinguished with respect to the composition y_{m_1} at the peritectic temperature $T(m_1)$.

1. *123 phase with $y > y_{m_1}$:* The homogeneity range is restricted at lower $p(O_2)$, and the Nd (Ln)-rich phase tends to decompose partially, thus shifting the composition of the remaining 123 phase towards y_{m_1}. After annealing, the samples have to be cooled down and oxidized as usual at about 300 – 400 °C. Because of the extension of the solidus line toward large excess of Nd (Ln), the samples with the initial value $y_0 > y_{m_1}$ are less favored with respect to homogeneity, and the amount of secondary phases which are formed during annealing is above the optimum. Even partial re-melting has been observed during annealing under reduced $p(O_2)$ [7.18], depending on the local composition.

2. ***123 phase with*** $y \leq y_{m_1}$**:** The stoichiometric composition is stable in oxygen or air only in a very narrow range of temperature. The stability field is extended and shifted even to $y < 0$ in reduced $p(O_2)$. Formation of stoichiometric or Ba-rich Nd123 phase is therefore favored by low content of oxygen in the environmental gas.

Since it is accepted that structural or composition fluctuations affect the flux pinning, one has to consider conditions which favor processes resulting in restructuring or defect formation. Fluctuations in the Nd/Ba ratio as well as in the oxygen concentration have been proven by experiments [7.4]. Both can appear on passing a region of instability or phase separation during post-growth treatment of the as-grown bulk materials, which also was applied to Sm- (Nd, Sm)-based 123 materials, e.g., in Ref. [7.19].

7.2.4
Composition Control in Oxidizing Atmosphere for Growing (CCOG)

CCOG is an alternative control mechanism to be applied even in an air atmosphere, based on the controlling function of chemical potentials of metal oxide constituents in the melt on the stoichiometry of the crystallizing *Ln*123 solid solution.

This approach for stoichiometry control in solid-solution type bulk *Ln*123 materials consists of a proper choice of a precursor composition which is suitable to control chemical potentials of metal oxide components according to the phase diagram [7.21], [7.22].

The relationships are demonstrated in Figure 7.5 with gross compositions appearing after adding Nd_2O_3 and Nd_2BaO_4, respectively. The sketches of Figure 7.5 represent situations in the phase diagram which appear for the three precursor mixtures:

    I a mixture of (stoichiometric) Nd123 and Y_2O_3,
    II a mixture of Nd123 and Nd_2BaO_4 ("210"),
    III a mixture of Nd123 + $BaCuO_2$ ("011").

Figure 7.5a represents the projection of a part of the liquidus surface on to the ternary subsolidus section at temperatures $T' < T(m_1)$, indicating the compositions of the coexisting melt for the Nd-rich and Ba-rich boundary of Nd123ss, respectively, left and right from the apex on the liquidus surface at $T(m_1)$. Figures 7.5b and 7.5c are isothermal sections at $T < T(m_1)$.

Figure 7.5 b is focused on the situation that T passes the liquidus temperature for the precursor composition I, i.e. the thermodynamic criteria for the starting point of Nd-rich 123ss crystallization from the actual precursor mixture. Thereafter, 123ss grows with permanently increasing parameter y. The process is terminated by approaching the peritectic solidification temperature of "210" + "422" at $T(p_2)$. From Figure 7.1 it can be concluded that the composition with the maximum Nd content has not been achieved at this point. Therefore, samples resulting from precursor I are superconducting after appropriate oxidation between 300 and 380 °C. However, T_c does not exceed 70 K, in accordance with the Nd concentration in the solid solution. The behavior is in contrast with that of Y123, which always crystallizes with the fixed 1:2:3 cation ratio from any point of its primary crystallization field.

In contrast, the crystallization of nearly stoichiometric Nd123ss $y \to 0$ is initialized from a Ba-rich precursor mixture (e.g. Nd123 + Nd$_2$BaO$_4$, II in Figure 7.5c). With further cooling, y can decrease even more, as discussed before. Similar situations also appear with admixtures of Ba-rich Nd422 (Nd$_{4-x}$Ba$_{2+x}$Cu$_2$O$_{10}$) or BaO$_2$ [7.21]. It has to be noted here that, despite the wide homogeneity field of Nd422 in

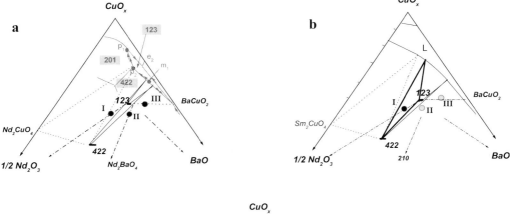

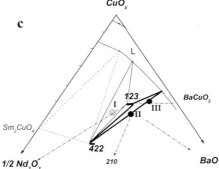

Figure 7.5 Phase relationships appearing during the processing of NdBaCuO with different ingot compositions: **a** The isothermal section in air at a temperature close to $T(m_1)$ indicating three compositions representing (I) Nd123 + Y$_2$O$_3$, (II) Nd123 + Nd$_2$BaO$_4$, and (III) Nd123 + BaCuO$_2$. Additionally, the projection of a part of the liquidus surface is represented. The marks indicate the primary crystallization fields for 201, 422 and 123, respectively, the latter limited by m_1, p_2, p_1 and e_2. Solid compositions are indicated by italics. **b** The situation at the liquidus temperature for precursor I. **c** The same situation for a Ba-rich precursor II

the subsolidus region, the Nd422 is always found with its stoichiometric composition nearly in equilibrium with the peritectic melt. Stoichiometric composition and homogeneity can obviously be proven by $T_c = 96$ K and the width of the transition $\Delta T_c = 1$ K [7.22], as is obtained from precursor compositions adjusted to mixtures of Nd123 and $BaCuO_2$. It becomes obvious from Figure 7.5c that the liquidus composition is equivalent, which results either from precursor mixtures according either to III or to II. The amount of 422 phase inclusions however is much less in bulk materials grown from precursor III. Composition control has been applied advantageously in combination with OCIG [7.16] or with oxygen-controlled single-crystal growth [7.14].

7.3
Alternative Seeding Techniques

Seeding becomes a serious problem for processing Nd, Sm and further *Ln*123 materials, since the liquidus and solidus temperatures are considerably higher than for YBCO. Reliable alternatives for *Ln*123 bulks have not yet been found. Seeding with MgO, which is sometimes used, does not proceed in a reliable manner. Recently, attempts to use *Ln*123 film coatings on MgO or MgO blended Nd123 seeds have been reported [7.23], [7.36].

Alternatively, the liquidus and solidus temperatures can be reduced by replacing *Ln*123 by mixed compounds. Therefore, growth and properties of solid solutions like $(Sm,Nd)Ba_2Cu_3O_7$, $(Y,Nd)Ba_2Cu_3O_7$, etc. have been investigated (Sections 7.4 and 7.5).

A more sophisticated technique is the "hot seeding" method, in which the seed is put onto the heated sample after it was cooled down from the overheated dwell temperature nearly to the equilibrium temperature (see, e.g., Ref. [7.24]), which allowed a reasonable sample quality in large Nd123 to be grown also in air [7.25].

Experimental complications appear in batch processing especially under controlled gas atmospheres. Nevertheless, hot seeding has been proven to be the most promising method so far with respect to properties of large bulk samples, and it is also applicable to Ag/*Ln*BaCuO composites [7.26]. Large-scale application, however, needs further technological development.

7.4
Further $LnBa_2Cu_3O_7$-Based Materials

Similarly to Nd123, $Ln_{1+y}Ba_{2-y}Cu_3O_{7-\delta}$ phases with Ln = Sm, Eu, Gd, form a solid solution-related homogeneity range, and the coexistence region with the melt is shifted to higher temperatures compared with $YBa_2Cu_3O_{7-\delta}$. In the case of high-quality Sm123-, Gd123-, and also Dy123-based materials, melt solutions containing 10–20 wt% of silver have been applied, which will be dealt with in Section 7.5.

Here, selected examples of melt-grown solid solutions with mixed Ln cations will be referred to.

(Sm,Nd)Ba$_2$Cu$_3$O$_7$ and (Sm, Y or Gd or Nd)Ba$_2$Cu$_3$O$_7$ are examples of applying concentration control in an oxygen atmosphere (CCOG) and oxygen potential control (OCMG) to improve pinning behavior in mixed systems [7.19], [7.27].

The recent development of bulk materials with the ternary solid solution phase Nd$_{0.33}$Eu$_{0.33+x}$Gd$_{0.33-x}$Ba$_2$Cu$_3$O$_{7-\delta}$ is worth emphasizing, since it permitted magnetic levitation in liquid oxygen (90 K) to be shown for the first time. This example demonstrated the remarkable levitation force only a few degrees below the critical temperature [7.28]. The precursor consisted of the 123ss and (Nd$_{0.33}$Eu$_x$Gd$_{0.66-x}$)$_2$BaCuO$_5$ solid solution and Gd$_2$BaCuO$_5$ which was intensively ball milled to an average particle size of 70 nm. Processing by the OCMG method resulted in further refinement, with a high content of 211 particles between 20 and 50 nm. Critical current densities of 50 kA cm^{-2} at 77 K in zero field and up to 50–70 kA cm^{-2} at 2.6–2.8 T (due to the expressed peak effect) have been reported, and the highest irreversibility field ever realized at 77K was reported to be 14 T for a composition ratio Nd:Eu:Gd = 33:38:28. Features obtained by TEM analysis were attributed to a secondary "nano-twin" structure and anti-site defects. A trapped field of only 0.7 T at 77 K indicates that single-grain size in the material has been restricted.

7.5
Ag/LnBaCuO Composites with Large Lanthanide Ions

As is known from YBaCuO, the mechanical strength can be improved in composite materials with Ag. Furthermore, solidus and liquidus temperatures of Ag/LnBaCuO composite materials are decreased compared with the pure materials. This is expected to diminish the complication of seeding (as explained in Section 7.3).

7.5.1
Fundamentals of Processing

The topology in the LnBa$_2$Cu$_3$O$_7$–Ag phase diagram is similar to the corresponding section 1/6 (YBa$_2$Cu$_3$O$_7$ + 0.24 Y$_2$BaCuO$_5$)–Ag (Figure 6.7), and the quasi-monotectic reaction m' controls decomposition and crystallization:

$$m' \quad (Ln)123 + L'' \rightleftarrows (Ln)211/422 + L' \tag{7.6}$$

In contrast to the peritectic-type decomposition m_1' of pure LnBa$_2$Cu$_3$O$_7$, the reaction m' is not an invariant reaction in the quaternary system. Therefore, the temperature $T(m')$ as well as the composition of the appearing melts depend on the initial ratio Ln123 : Ln211 (or Ln422).

7.5 Ag/LnBaCuO Composites with Large Lanthanide Ions

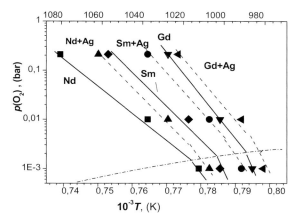

Figure 7.6 Relationship between oxygen partial pressure and temperature of quasi-monotectic decomposition of $Ln123/Ag$ composites (10 wt% Ag_2O) compared with peritectic temperatures in Ag-free systems. The dot and dash line represents the development of invariant points for Nd-, Sm-, (Gd)-, Y-123 compositions. Information on the behavior of corresponding Ag-containing systems is not available

Figure 7.6 represents the quasi-monotectic temperatures for $Ln123$–Ag mixtures (Ln = Nd, Sm, Gd) compared with the decomposition temperatures of the same components without silver. The concentration of Ag in the figure corresponds to ~9 at% related to metal oxides. The quasi-monotectic temperatures decrease by only 20 K from Nd to Gd in air, whereas another decrease by about 60 K results from reducing the oxygen partial pressure from 0.21 bar to the invariant point. Note that the spacing between different Ln metals becomes even less below 10^{-2} bar oxygen gas pressure and amounts to only 5 or 10 K from NdBaCuO to SmBaCuO or to GdBaCuO, respectively. The limitation for any melt processing is set near 10^{-3} bar oxygen pressure by the solid-state decomposition of $Ln123$.

7.5.2
Reactions Near the Seed–Melt Interface

Ag/YBaCuO and Ag/(Ln)BaCuO should behave similarly if processed at appropriate temperatures. In contrast to YBaCuO, however, one needs an independent mechanism to control the Ln:Ba stoichiometry in the superconducting $Ln123$ phase. Furthermore, even if Ag-free seeds are applied, the dissolution of seed material was found to proceed near the interface during the short dwell time for a $NdBa_2Cu_3O_7$ seed plus SmBaCuO/Ag bulk. Ag inclusions in this area prove that the seed was molten up to a depth of 0.5 mm [7.7]. At $T \geq 950\,°C$, an Ag-rich liquid L″ is formed in the sample. The presence of Ag at the seed–sample interface causes the decomposition temperature of the (originally Ag-free) Nd123 seed to decrease, and the interface propagates because of the rapid diffusion of the Ag into the appearing liquid phase. Unfortunately, this tendency is still more pronounced at reduced $p(O_2)$, since the gap between the temperatures of quasi-monotectic decomposition of Nd123 and Sm123 becomes smaller.

7.5.3
Growth and Properties of Ag/LnBaCuO Composites

Ln = **Sm or Nd composite materials** are the prototypes to demonstrate the performance of Ag/*Ln*BaCuO composite materials. Appropriate single-grain size and quality enable trapped fields up to 1.2 T to be achieved for *Ln =* Nd [7.29] and even up to 2.1 T at 77 K ($d = 36$ mm) and 8 T at 40 K for Ag/$Sm_{1+y}Ba_{2-y}Cu_3O_7$ composites [7.30]. Processing at low $p(O_2)$ and hot seeding are obligatory to achieve high quality, as a consequence of phase equilibria restrictions. Typically, single-domain Ag/*Ln*BaCuO samples are prepared from precursors of *Ln*123 and *Ln*211 additions in appropriate molar ratios, and 0.5 wt% Pt and 10–30 wt% Ag_2O are admixed and pressed. The thermal procedure is analogous to that for Ag/$YBa_2Cu_3O_7$, taking into account the higher monotectic temperature $T(m')$. Crystallization proceeds according to the quasi-monotectic reaction m' [Eq. (7.3)].

Ag/$Sm_{1+y}Ba_{2-y}Cu_3O_7$ processing in air and post-growth annealing have the same motivation as for Ag-free materials, and result in composition and properties control analogous to Ag-free Nd123-based materials [7.7]. Similar fluctuations of the Sm/Ba ratio were observed in the Sm123 phase as for Nd in the Ag-free material. Thus, the results should be explained by the same arguments as those for Ag-free materials This is in accordance with the low solubility of Ag in the melt-textured 123 phases represented by the formula $(RE)Ba_2(Cu_{0.984}Ag_{0.016})_3O_z$ for both $RE =$ Sm and $RE =$ Y.

Ag/$Gd_{1+y}Ba_{2-y}Cu_3O_7$ composites and **Ag/$Dy_{1+y}Ba_{2-y}Cu_3O_7$ composites** have been grown with diameters up to 50 mm in low $p(O_2)$ with hot seeding (OCMG), which revealed the extremely high trapped field 2.6 T at 77 K for a Gd-based sample and 1.7 T for Dy123. This is in accordance with high j_c, (for Gd123: 60–80 and 30–45 kA cm^{-2} at 0 T and 2 T, respectively), but also indicating a significant reduction in the current barriers which result from a mosaic structure and cracks parallel to the *c*-axis [7.31], [7.32], [7.33]. On the other hand, j_c and trapped field are less sensitive to stoichiometric control by OCMG or CCOG than in Nd- or Sm-based materials, in agreement with the much narrower solid-solution range for Gd123 and Dy 123 (Figure 2.1).

Composites of silver and $(Ln_A Ln_B Ln_C)Ba_2Cu_3$ have also been reported for (Sm, Gd) and (Nd, Sm, Gd) solid solutions with $B_0(77K) \cong 0.6$ and 1.2 T, respectively [7.34].

References

7.1 S. Yoo, I. Yoo, N. Sakai, H. Takaichi, H. Higuchi, N. Murakami, Appl. Phys. Lett. **65**, 633 (1994).

7.2 T. Wolf, H. Küpfer, A.C. Bornarel, *Applied Superconductivity 1997*, Inst. Phys. Ser. No. 158, ed. H. Rogalla et al. (IOP Publishing, Bristol, 1997) 925.

7.3 M. Murakami, N. Sakai, T. Higuchi, S. I. Yoo, Supercond. Sci. Technol. **9**, 1015 (1996).

7.4 T. Hirayama, Y. Ikuhara, M. Nakamura, Y. Yamada, Y. Shiohara, J. Mater. Res. **12**, 293 (1997).

7.5 E. Goodilin, M. Limonov, A. Panfilov, M. Khasanova, A. Oka, S. Tajima, Y. Shiohara, Physica C **300**, 250 (1998).

7.6 M. Nakamura, Y. Yamada, T. Hirayama, Y. Ikuhara, Y. Shiohara, S. Tanaka Physica C **259**, 295-303 (1996).

7.7 G. Krabbes, Th. Hopfinger, C. Wende, P. Diko, G. Fuchs, Supercond. Sci. Technol. **15**, 665 (2002).

7.8 H. Wu, M. J. Kramer, K. W. Dennis, R. W. McCallum, Physica C **290**, 252 (1997).

7.9 H. Schmalzried, Chemical Kinetics of Solids, Verlag Chemie Weinheim, 1995.

7.10 G. Krabbes, W. Bieger, P. Schätzle, G. Fuchs, J. Thomas, Adv. Solid-State Phys. **39**, 384 (1999).

7.11 Yu. Tretyakov, E. Goodilin, Russ. J. Inorganic Chem. **46**, 203 (2001).

7.12 V. V. Petrykin, E. A. Goodilin, J. Hester, E. A. Trofimenko, M. Kakihana, N. N. Oleynikov, S. Yu, D. Tretyakov, Physica C **340**, 16 (2000).

7.13 F. Giovanelli, M. Ferretti, J.-F. Bardeau, M. Hervieu, I. Monot-Laffez, Supercond. Sci. Technol. **17**, 8 (2004).

7.14 Y. Shiohara, E. A. Goodilin, Single-crystal growth for science and technology, in: Handbook on the Physics and Chemistry of Rare Earths Vol. 30 (2000), Chapter 189, Elsevier Science B. V.

7.15 N. H. Babu, M. Kambara, D. A. Cardwell, A. M. Campbell, Supercond. Sci. Technol. **15**, 702 (2002).

7.16 W. Bieger, G. Krabbes, P. Schätzle, A. Leistikow, J. Thomas, P. Verges, Mater. Sci. Eng. B **53**, 100 (1998).

7.17 A. M. Hu, Z. X. Zhao, M. Z. Wu, C. Wende, T. Strasser, B. Jung, G. Bruchlos, W. Gawalek, P. Görnert, Physica C **278**, 43 (1997).

7.18 A. Leistikow, Materialentwicklung im System NdBaCuO, (Thesis), Shaker, Aachen 2001 p.65 (ISBN 3-8265-9306-5).

7.19 A. Hu, G. Krabbes, P. Schaetzle, P. Verges, Inst. Phys. Conf. Ser. **167**, 131 (2000).

7.20 W. Bieger, G. Krabbes, P. Schätzle, L. Zelenina, U. Wiesner, P. Verges, J. Klosowski, Physica C **257**, 46 (1996).

7.21 A. Vecchione, M. Gombos, P. Tedesco, S. Pace, Intern. J. Modern Physics B **14**, 2670 (2000).

7.22 P. Schätzle, G. Ebbing, W. Bieger, G. Krabbes, Physica C **330**, 19 (2000).

7.23 C. Cai, K. Tachibana, H. Fujimoto, Supercond. Sci. Technol. **13**, 698 (2002).

7.24 H. Ikuta, A. Mase, Y. Yanagi, M. Yoshikawa, Y. Itoh, T. Oka, U. Mizutani, Supercond. Sci. Technol. **11**, 1345 (1998).

7.25 E. Guilmeau, F. Giovannelli, I. Monot-Laffez, S. Marinel, J. Provost, G. Desgardin, Eur. Phys J. Appl. Phys. **13**, 157 (2001).

7.26 Wai Lo, K. Salama, Supercond. Sci. Technol. **13**, 725 (2000).

7.27 S. Nariki, S.-J. Seo, N. Sakai, M. Murakami, in: Advances in Superconductivity XII, ed. T. Yamashita, K. Tanabe, Springer Tokyo, 1999, p. 485.

7.28 M. Muralidhar, N. Sakai, Y. Wu, M. Murakami, M. Jirsa, T. Nishizaki, N. Kobayashi, Supercond. Sci. Technol. **15**, 1357 (2002); **16**, L46 (2003).

7.29 H. Ikuta, T. Hosokawa, M. Yoshioka, U. Mizutani, Supercond. Sci. Technol. **13**, 1559 (2000).

7.30 H. Ikuta, A. Mase, U. Mizutani, Y. Yanagi, M. Yoshikawa, Y. Itoh, T. Oka IEEE Trans. Appl. Supercond. **9**, 2219 (1999).

7.31 S. Nariki, N. Sakai, M. Murakami, Supercond. Sci. Technol. **15**, 648 (2002).

7.32 S. Nariki, H. Hinai, N. Sakai, M. Murakami, M. Otsuka, Physica C **378-381**, 764 (2002).

7.33 S. Nariki, N. Sakai, M. Murakami, Physica C **357-360**, 814 (2001).

7.34 M. Muralidhar, M. Jirsa, N. Sakai, M. Murakami, Supercond. Sci. Technol. **16**, R1 (2003).

7.35 M. Yoshizumi, M. Kambara, Y. Shiohara, T. Umeda, Physica C **334**, 77 (2000).

7.36 Y. Shi, N. Hari Babu, A. Cardwell, Supercond. Sci. Technol. **18**, L13 (2005).

8
Peak Effect

In many cases, doped and undoped bulk HTSCs show a flux pinning anomaly which is known as the "peak" or "fishtail" effect. The peak effect denotes an anomalous increase in the critical current density with magnetic field (see Figure 8.1), while the term "fishtail" effect refers to a corresponding second peak in the magnetization curve. The peak effect, which is also well known in low-temperature superconductors [8.1], has been reviewed by Weber [8.2]. Present knowledge clearly shows that a number of different defect structures and crystal features can lead to a peak effect in HTSC. The effect can originate from the twin structure [8.3–8.5], from intrinsic pinning of vortices at CuO_2 planes [8.6], from second-phase particles [8.7, 8.8], or from clusters of oxygen vacancies [8.9].

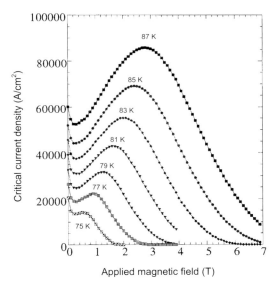

Figure 8.1 Field dependence of the critical current density of a bulk (Nd,Eu,Gd)-Ba-Cu-O sample measured for applied fields $H\|c$ at various temperatures. A pronounced peak effect develops at temperatures below 80 K. Taken from Ref. [8.10]

High Temperature Superconductor Bulk Materials.
Gernot Krabbes, Günter Fuchs, Wolf-Rüdiger Canders, Hardo May, and Ryszard Palka
Copyright © 2006 WILEY-VCH Verlag GmbH & Co. KGaA, Weinheim
ISBN: 3-527-40383-3

In undoped bulk YBCO, oxygen-deficient clusters and twin structures are mainly responsible for the peak effect. Clusters of oxygen vacancies are of particular importance in understanding the peak effect in bulk YBCO. In the following, several relevant results concerning the peak effect due to this defect structure reported for twin-free single crystals will be summarized.

8.1
Peak Effect (due to Cluster of Oxygen Vacancies) in Single Crystals

In the first model for the peak effect in HTSC proposed by Däumling et al. [8.11], clusters of oxygen vacancies were assumed to be responsible for the peak effect. These clusters, having a lower H_{c2} than the fully oxygenated matrix, represent weak pins at a low applied field, since the superconducting order parameter is only slightly suppressed. These pins become stronger with increasing field because the order parameter in these disturbed regions is much more strongly suppressed by the applied field than the order parameter of the matrix. Hence, there is a certain range of magnetic fields in which j_c increases with increasing applied field before it starts to decrease above the field B_p at which the j_c peak appears. This explanation of the peak effect in HTSC was later supported by investigations of extremely pure YBCO single crystals prepared in non-reactive $BaZrO_3$ crucibles [8.12, 8.13]. No peak effect has been observed for a statistical distribution of oxygen vacancies obtained by a high-temperature and high-pressure oxygenation. However, the peak effect was found to reappear after a subsequent low-pressure post-annealing, which only could alter the local distribution of the oxygen vacancies but not the overall oxygen content [8.14]. Thus, the authors concluded that the peak effect in the YBCO single crystals was correlated with clustering of oxygen vacancies. Clustering of oxygen vacancies in YBCO single crystals even occurs on extended annealing at room temperature [8.5]. This is shown in Figure 8.2, where j_c data for a twin-free $YBa_2Cu_3O_{7-\delta}$ single crystal are plotted against the applied magnetic field. After high-pressure oxygenation, the oxygen content was reduced to $\delta = 0.03$. The $j_c(B)$ dependence shows a slight j_c maximum at about 4 T immediately after this procedure. In contrast, a very well pronounced peak effect appeared after aging the crystal at room temperature for 5 weeks.

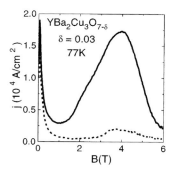

Figure 8.2 Field dependence of the critical current density of a twin-free $YBa_2Cu_3O_{6.97}$ single crystal at 77 K before (dashed line) and after (solid line) it was stored at room temperature for 5 weeks. Taken from Ref. [8.5]

It should be noted that statistically distributed oxygen vacancies can be obtained only in very pure single crystals. For less pure single crystals or in the case of bulk melt-textured YBCO, clustering of oxygen vacancies on impurities seems to be unavoidable even after high-temperature and high-pressure oxygenation, and therefore the possible generation of the peak effect is a common feature in YBCO.

The evolution of the j_c peak in twin-free YBa$_2$Cu$_3$O$_{7-\delta}$ single crystals was studied for varying oxygen deficiency [8.9]. In Figure 8.3, $B_p(T)$ data are shown for two values of the oxygen deficiency δ. For $\delta = 0.007$, $B_p(T)$ intersects the melting line $B_m(T)$ at the critical point, at which also the order-disorder transition field $B^*(T)$ and the irreversibility line $B_{irr}(T)$ of the Bragg glass meet each other (see Section 4.1.3 and Figure 4.7). In the vortex matter phase diagram shown in Figure 4.7, $B_p(T)$ would be located within the vortex glass phase above $B^*(T)$. On reducing the oxygen content, the critical point shifts to lower magnetic field along the melting line, as discussed in more detail in Section 4.1.3. At $\delta = 0.06$, the critical point vanishes and $B_p(T)$ starts to increase with decreasing temperature from T_c, as shown in Figure 8.3.

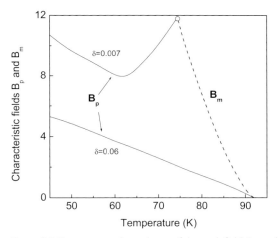

Figure 8.3 Temperature dependence of the peak field B_p at which the current passes its maximum for two twin-free YBa$_2$Cu$_3$O$_{7-\delta}$ single crystals with $\delta=0.007$ and $\delta=0.06$. For $\delta=0.007$, $B_p(T)$ intersects the melting line $B_m(T)$ at the critical point below 12 T which is marked by an open circle. For details see text. Data from Ref. [8.9]

Above the peak field B_p, the $j_c(H)$ data have been found to be almost independent of the strength of the interaction between pinning sites and vortices [8.9]. This is a strong indication of plastic deformations of the flux line lattice in the field region above B_p. For fields below B_{min} (with B_{min} as the field at which j_c has its minimum value) the j_c data were interpreted in the framework of the collective pinning model [8.9].

A model in which the peak effect is explained by a crossover from elastic to plastic creep has been proposed by Abulafia et al. [8.15]. Investigating YBCO single crystals by local magnetic relaxation measurements, a strong increase in the relaxation has been observed above the peak field B_p, which was explained by plastic creep of vortices mediated by dislocations within the vortex lattice [8.16]. Comparing the activation energies for elastic and plastic vortex creep, the relation

$$B_p \propto \left[1 - \left(\frac{T}{T_c}\right)^4\right]^2 \tag{8.1}$$

has been derived for the peak field B_p. This same temperature dependence of B_p as predicted by Eq. (8.1) has been found for the investigated YBCO single crystal [8.15]. The corresponding vortex-creep phase diagram for YBCO is shown in Figure 8.4. The plastic creep regime in this phase diagram covers the area between $B_p(T)$ and the melting line $B_m(T)$.

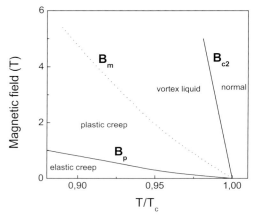

Figure 8.4 Vortex-creep phase diagram for a YBCO single crystal (taken from Ref. [8.15]). B_p – peak field of the maximum of $j_c(B)$, B_m – melting line, B_{c2} – upper critical field.

The peak effect in clean, twin-free YBCO single crystals can be attributed to clusters of oxygen vacancies. However, the mechanism which explains the anomalous rise of j_c with increasing applied field in the field range between B_{min} and B_p is a controversial topic and still a subject of debate. The simple explanation considering oxygen deficient clusters as field-induced pins as proposed originally [8.11] has been ruled out, taking into account the evolution of the peak effect with temperature at constant oxygen deficiency [8.9]. For increasing temperature, $B_p(T)$ has been observed to increase in a certain temperature range (see Figure 8.3 for $\delta=0.007$), in contrast to the increasing $B_{irr}(T)$. The resulting violation of the temperature scaling $j(B,T)/j(B_p,T)$ of the current rejects the model of field-induced pins. Instead, the anomalous rise of j_c with increasing field between B_{min} and B_p in single crystals has been interpreted by postulating a field-driven disorder transition within the vortex lattice from an almost perfect vortex lattice or *Bragg* glass to a highly-disordered vortex glass [8.17]. With increasing applied field, the ordered lattice in the *Bragg* glass was assumed to become unstable because of a proliferation of large dislocation loops favored by point disorder.

Other possible explanations for the increase of j_c with field are based on the collective interaction of the pins with the vortex lattice, which governs the flux pinning properties in these superconductors. In the static collective pinning theory [8.18], the increase of j_c with field is attributed to a variation of the size of the *Larkin-Ovchinnikov* correlation volume. In the dynamic model for the peak effect [8.19] based on the collective creep theory [8.20], which is an extension of this model by relaxation effects, the unrelaxed current is assumed to decrease monotonously with increasing field. The rise of j_c with field is related to field-dependent changes in the relaxation behavior. A large relaxation rate in the low-field single vortex state is connected with a low relaxed current density, whereas a low relaxation rate within the small flux bundle regime at higher fields leads to a higher current density.

8.2
Peak Effect in Bulk HTSC

In this section, the peak effect in bulk YBCO and Nd-based HTSC will be considered. The majority of models interpreting the increase of j_c with applied magnetic field are based on field-induced pins. These low-T_c clusters embedded in a high-T_c matrix are considered to be superconducting below B_{min} and reach their normal state for applied magnetic fields approaching B_p. Examples of field-induced pins are oxygen-deficient clusters or clusters forming in *RE*-123 compounds (with the light rare-earth elements *RE* = Nd,Sm,Eu,Gd and mixtures therefrom) by compositional fluctuations due to substitution of *RE* on Ba sites. These $RE(Ba,R)_2Cu_3O_{7-x}$ (*RE*-123) clusters having diameters in the range of 10–100 nm have been observed by TEM and STM.

A pronounced peak effect has also been reported for bulk $(Nd_{0.33}Eu_{0.38}Gd_{0.28})$-$Ba_2Cu_3O_y$ (NEG-123) compounds [8.25–8.27], which was attributed to an NEG-rich phase (*RE*-123 clusters) with weaker superconducting properties. By means of controlled addition of NEG-211 powders, the critical current density j_c(0,77 K) at self-field and 77 K was found to increase from about 13 kA cm^{-2} for pure NEG-123 to 100 kA cm^{-2} for 50 mol% NEG-211 addition. With increasing j_c(0,77 K), a systematic suppression of the peak effect was observed changing into a plateau for 40 mol% NEG-211 and being destroyed for 50 mol% NEG-211 [8.26]. The increase in j_c was related to submicron-sized insulating Gd211 particles which were formed together with the large NEG-211 inclusions during the melt processing. In order to describe the experimental $j_c(H)$ data, two different contributions to j_c were assumed: "background" pinning due to small Gd211 particles producing a monotonously decreasing $j_c(H)$ dependence and field-induced pinning due to *RE*-123 clusters. Without Gd211 particles and for low concentrations of the 211 particles, a pronounced peak effect appears due to field-induced pinning. With increasing concentration of 211 particles, the "background" pinning becomes stronger and the crossover between "background" and field-induced pinning shifts to higher applied fields, until the peak effect completely disappears if the crossover field exceeds the peak field B_p.

NEG-123 compounds with 5 mol% NEG-211 were found to have not only a pronounced peak effect, but also extremely high irreversibility fields up to $B_{irr} = 12$ T (measured by a SQUID) or 15 T (measured by a VSM) at 77 K for applied fields $H \| c$ [8.27]. The improved pinning in this material was attributed to a nanoscale laminar structure of aligned NEG-rich clusters.

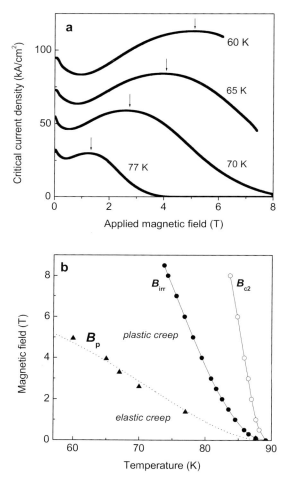

Figure 8.5 a Field dependence of the critical current density of an Li-doped bulk YBCO sample for fields $H \| c$ at various temperatures. The peak field B_p of the maximum of $j_c(B)$ is marked by arrows. b Vortex-creep phase diagram for the same YBCO sample. B_p – peak field (▲), B_{irr} – irreversibility field (●), B_{c2} – upper critical field (○). The dotted line is a fit by $B_p \propto [1-(T/T_c)^4]^2$. Data from Ref. [8.30]

In bulk Nd123, both kinds of field-induced pins, oxygen-deficient cluster [8.5] and *RE*-123 cluster [8.21–8.24] have been considered to be responsible for the peak effect. The disappearance of the peak effect has been reported in bulk Nd123 after neutron irradiation [8.28]. Characteristic features of the strong pinning centers produced by neutron irradiation are a huge enhancement of j_c at low applied fields and a monotonously decreasing $j_c(H)$ dependence which dominates the experimental $j_c(H)$ dependence of the irradiated sample. Similar data have been also reported for bulk Zn-doped YBCO after neutron irradiation (see Figure 6.6).

A scenario with two kinds of pinning sites has been proposed by Reissner et al. [8.29] for the interpretation of the peak effect in bulk melt-textured YBCO with a nearly optimum oxygen content. Oxygen vacancies were assumed to be responsible for the peak effect, whereas the low-field region of j_c ($\mu_0 H < B_{min}$) was related to "background" pinning. After analyzing relaxation data within the framework of the collective pinning theory, a crossover from 3D behavior at low fields for the "background" pinning to pinning of 2D planar defects in the region of the peak-effect was proposed [8.29].

The peak effect in bulk Nd123 [8.24] and in bulk Li-doped YBCO [8.30] (see Figure 8.5a) has been interpreted as crossover from elastic to plastic creep according to the model proposed by Abulafia et al. [8.15]. In both cases, relaxation data below and above B_p have been compared. It was found that data below the peak field $B_p(T)$ can be described within the collective creep model. The data above $B_p(T)$ showed strongly enhanced flux creep rates corresponding to very low values of the exponent μ [8.30] or even negative μ [8.24] which are inconsistent with the collective creep theory.

The vortex-creep phase diagram derived for bulk Li-doped YBCO [8.30] is plotted in Figure 8.5b. The peak field B_p was determined from the $j_c(H)$ data shown in Figure 8.5a, whereas B_{irr} and B_{c2} were derived from magnetization vs temperature data obtained in the field cooled and zero-field cooled mode as described in Section 4.1. The phase diagram for Li-doped YBCO is similar to that obtained for YBCO single crystals (Figure 8.4) taking into account that the melting line $B_m(T)$ in Figure 8.4 is in Figure 8.5b replaced by the irreversibility line of the bulk sample. The peak field $B_p(T)$ of the Li-doped YBCO can be described by Eq. (8.1), as shown in Figure 8.5b. According to this interpretation, plastic creep covers a remarkably large area between $B_p(T)$ and $B_{irr}(T)$.

References

8.1 A. M. Campbell, J. E. Evetts, *Critical Currents in Superconductors*, ed. by B. R. Coles et al. (Taylor & Francis Ltd, London, 1972).

8.2 H. W. Weber, Flux Pinning, in Handbook on the Physics and Chemistry of Rare Earths, edited by K. A. Gscheidner, Jr., L. Eyring, M. B. Maple (Elsevier, Amsterdam, 1999).

8.3 W. K. Kwok, J. A. Fendrich, C. J. van der Beek and G. W. Crabtree, Phys. Rev. Lett. **73**, 2614 (1994).

8.4 A. A. Zhukov, G. K. Perkins, J. V. Thomas, A. D. Caplin, H. Küpfer, T. Wolf, Phys. Rev. B **56**, 3481 (1997).

8.5 Th. Wolf, A.-C. Bornarel, H. Küpfer, R. Meier-Hirmer, B. Obst, Phys. Rev. B **56**, 6308 (1997).

8.6 M. Oussena, P. A. J. de Groot, R. Gagnon, L. Taillefer, Phys. Rev. Lett. **72**, 3606 (1994).

8.7 M. R. Koblischka, A. J. J. van Dalen, T. Higuchi, K. Sawada, S. I. Yoo, M. Murakami, Phys. Rev. B **54**, R6893 (1996).

8.8 G. Krabbes, G. Fuchs, P. Schätzle, S. Gruß, J. W. Park, F. Hardinghaus, G. Stöver, R. Hayn, S. – L. Drechsler, T. Fahr, Physica C **330** 181(2000).

8.9 H. Küpfer, Th. Wolf, C. Lessing, A. A. Zhukov, X. Lancon, R. Meier-Hirmer, W. Schauer, H. Wühl, Phys. Rev. B **58**, 2886 (1998).

8.10 M. Muralidhar, M. R. Koblischka, M. Murakami, Supercond. Sci. Technol. **13**, 693 (2000).

8.11 M. Däumling, J. M. Seuntjens, D. C. Larbalestier, Nature **346**, 332 (1990).

8.12 A. Erb, E. Walker, R. Flükiger, Physica C **245**, 245 (1995).

8.13 A. Erb, E. Walker, R. Flükiger, Physica C **258**, 9 (1996).

8.14 A. Erb, E. Walker, J. Y. Genoud, R. Flükiger, Physica C **282-287**, 2145 (1997).

8.15 Y. Abulafia, A. Shaulov, Y. Wolfus, R. Prozorov, L. Burlachkov, Y. Yeshurun, D. Majer, E. Zeldov, H. Wühl, V. B. Geshkenbein, V. M. Vinokur, Phys. Rev. Lett. **77**, 1596 (1996).

8.16 V. B. Geshkenbein, A. I. Larkin, M. V. Feigel'man, M. V. Vinokur, Physica C **162-164**, 239 (1989).

8.17 H. Küpfer, Th. Wolf, R. Meier-Hirmer, A. A. Zhukov, Physica C **332**, 80 (2000).

8.18 A. I. Larkin, Yu. N. Ovchinnikov, J. Low Temp. Phys. **34**, 409 (1979).

8.19 L. Krusin–Elbaum, L. Civale, V. M. Vinokur, F. Holtzberg, Phys. Rev. Lett. **69**, 2280 (1992).

8.20 G. Blatter, M. V. Feigel'man, V. B. Geshkenbein, A. I. Larkin, V. M. Vinokur, Rev. Mod. Phys. **66**, 1125 (1994).

8.21 N. Chikumoto, M. Murakami, in Proc 10th Anniversary HTS Workshop on Physics, Materials and Application, edited by B. Batlogg et al. (World Scientific, Singapore, 1996), p. 139.

8.22 M. Murakami, N. Sakai, T. Higuchi, S. I. Yoo, Supercond. Sci. Technol. **9**, 1015 (1996).

8.23 M. R. Koblischka, A. J. J. van Dalen, T. Higuchi, K. Sawada, S. I. Yoo, M. Murakami, Phys. Rev. B **58**, 2863 (1998).

8.24 T. Mochida, N. Chikumoto, M. Murakami, Phys. Rev. B **64**, 064518 (2001).

8.25 M. Muralidhar, M. R. Koblischka, T. Saitoh, M. Murakami, Supercond. Sci. Technol. **11**, 1349 (1998).

8.26 M. R. Koblischka, M. Muralidhar, M. Murakami, Physica C **337**, 31 (2000).

8.27 M. Muralidhar, N. Sakai, M. Nishiyama, M. Jirsa, T. Machi, M. Murakami, Appl. Phys. Lett. **82**, 943 (2003).

8.28 H. W. Weber, in Proc 10th Anniversary HTS Workshop on Physics, Materials and Application, edited by B. Batlogg et al. (World Scientific, Singapore, 1996), p. 163.

8.29 M. Reissner, J. Lorentz, Phys. Rev. B **56**, 6273 (1997).

8.30 L. Shlyk, G. Krabbes, G. Fuchs, K. Nenkov, B. Schüpp, J. Appl. Phys. **96**, 3371 (2004).

9
Very High Trapped Fields in YBCO Permanent Magnets

In recent years, considerable progress has been achieved in the development of bulk YBCO and other HTSCs with improved pinning properties for applications at and below 77 K. In general, the poor mechanical properties of bulk HTSCs constitute a serious problem with respect to their practical application. In particular, the trapped fields of HTSCs below 77 K become limited by the rather low tensile strength of the material, which is exposed to large magnetic tensile stresses.

Several strategies have been developed to overcome these limitations and achieve high trapped fields in bulk HTSC. Attempts to improve the mechanical properties by the addition of silver (a result mainly due to the reduction of number and size of microcracks) have already been discussed.

Another possible method of avoiding cracking at low applied fields is to reinforce the YBCO disks with steel tubes, which resist the magnetic tensile stress. Improved mechanical properties of bulk YBCO were also reported after epoxy resin impregnation. In the following sections, these two techniques are discussed, and the results obtained are presented.

9.1
Bulk YBCO in Steel Tubes

9.1.1
Magnetic Tensile Stress (in Reinforced YBCO Disks)

The idea to compensate the magnetic tensile stress in YBCO permanent magnets by coating the YBCO disks with a metal casing, generating a compressive stress in the superconductor after cooling from 300 K to 77 K, is very simple. However, several problems have to be solved in order to realize this idea. First of all, one needs a material which has not only a higher coefficient of thermal expansion than YBCO, but also a *Young*'s modulus large enough to withstand strong mechanical forces. Steel tubes have turned out to be a good solution. Austenitic steel has, over the temperature range 50 – 300 K, an average coefficient of thermal expansion of $a_{av} \approx 10^{-5}$ K^{-1} [9.1], which is comparable with the a value for YBCO

High Temperature Superconductor Bulk Materials.
Gernot Krabbes, Günter Fuchs, Wolf-Rüdiger Canders, Hardo May, and Ryszard Palka
Copyright © 2006 WILEY-VCH Verlag GmbH & Co. KGaA, Weinheim
ISBN: 3-527-40383-3

in the c direction, but significantly larger than its value in the ab plane (see Section 4.5.3). Furthermore, austenitic steel has a large Young's modulus of 194 GPa. In order to ensure the force transmission from the steel tube to the YBCO disk, the uneven space between the tube and disk was filled with Stycast epoxy using a vacuum impregnation technique.

The effect of the steel tube is illustrated in Figure 9.1. Shown is the maximum tensile stress $\sigma(0)$ within a bulk YBCO sample as a function of the maximum trapped field B_0. Without the steel tube, $\sigma(0) \propto B_0^2$, in accordance with Eq. (5.2). The critical magnetic tensile stress at which cracking occurs is assumed to be 30 MPa. If a steel tube is used, then a (negative) compressive stress acts on the encapsulated YBCO disk after the sample is cooled to the low temperature in the absence of a trapped field. With increasing trapped field, this negative stress decreases, changing its sign at a certain value of B_0, and reaching the tensile strength $\sigma(0)$ at a considerably higher trapped field than it does without the steel tube. Thus, much higher trapped fields can be achieved without cracking. This enhancement of the trapped field increases with the magnitude of the compressive stress.

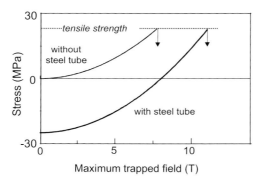

Figure 9.1 Relation between maximum tensile stress and maximum trapped field in a YBCO sample reinforced by a steel tube and without the steel tube. Because of the compressive (negative) prestress acting on the reinforced YBCO at zero-field, higher trapped fields are achievable without cracking

In order to estimate this enhancement, the distribution of the stresses acting on a pair of two bulk YBCO disks under the influence of an applied magnetic field and a steel tube was calculated by a finite-element procedure using a Quick-field software package [9.2]. Input parameters for these calculations are the coefficient of thermal expansion, Young's and shear modulus, Poisson's ratio, and the geometry of the superconductor. In Figure 9.2, the calculated stress distribution is shown for different cases. In the left panel of this figure, the field-induced *tensile* stress in an YBCO disk-pair without steel tube is plotted after applying a magnetic field of 14.5 T. The highest tensile stress of 100 MPa appears in the central part of the disk-pair in the gap region, as expected. The middle panel of Figure 9.2 shows

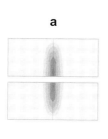 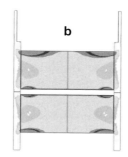 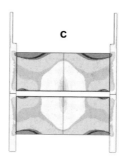

Figure 9.2 Calculated stress distribution in a YBCO mini-magnet. **a** Magnetic tensile stress acting on a YBCO mini-magnet after a magnetic field of 14.5 T has been applied. Dark regions show strong tensile stress. **b** Compressive stress developing in the mini-magnet encapsulated in a steel tube after cooling down from 300 K to 77 K. In this case, dark regions show strong compressive stress. **c** Stress distribution in the encapsulated mini-magnet which was magnetized in an applied field of 14.5 T in the field cooling mode. Compressive stress dominates the outer dark region. Tensile stress remains only in the small dark area in the center of the bright region, as described in the text

the distribution of the *compressive* stress generated in the encapsulated disk-pair by cooling from 300 K to 77 K. Finally, the combined influence of compressive and tensile stresses on the disk-pair is shown in the right panel of Figure 9.2. The disk-pair in the steel tube is assumed to be magnetized in the field-cooling mode, applying a magnetic field of 14.5 T at 300 K, cooling down the disks in the steel tube from 300 to 77 K, and reducing the applied magnetic field to zero. In a colored version of the stress distribution of Figure 9.2c, compressive and tensile stress are distinguished by different colors. In the grayscale version shown, dark areas show both compressive and tensile stress. Nevertheless, compressed and expanded regions, being spatially separated, can easily be distinguished. Whereas the outer dark part of the YBCO disks is completely dominated by compressive stress, the almost stress-free bright region in the middle part of the YBCO disks contains in its center a small dark region corresponding to tensile stress. The maximum tensile stress in this region is about 75 MPa, which has to be compared with a tensile stress of 100 MPa acting on the YBCO disks without a steel tube.

9.1.2
Trapped Field Measurements

For trapped field measurements, bulk YBCO disks were encapsulated in steel tubes (see Figure 9.3). The cylindrical YBCO pellets were ground to a diameter of 24 mm for YBCO(+Ag) and YBCO(+Zn) and to a diameter of 22 mm for YBCO(+Ag/Zn) and embedded in austenitic Cr-Ni steel tubes with a wall thickness of 2 mm and 4 mm, respectively. The larger wall thickness of the steel tubes for the YBCO(+Ag/Zn) disks and mini-magnets was used in order to generate a larger compressive stress on the superconductor after cooling from 300 K to the measuring temperature below 77 K. The small gap of a few μm between the steel tube and the YBCO disks was filled with epoxy resin using a vacuum impregnation technique. For the investigation of single disks, an axial Hall sensor was positioned on top of the disks in the center, whereas a transverse Hall sensor was placed in the gap between the two disks of the investigated mini-magnets. The size of the gap was 1.7 mm [for YBCO(+Ag) and YBCO(+Zn) mini-magnets] and 2.6 mm [for YBCO(+Ag/Zn) mini-magnets]. Additionally, a temperature sensor was fixed on top of one sample in each case.

The temperature dependence of the maximum trapped field $B_o(T)$ of bulk YBCO disks was investigated in the variable temperature cryostat of a superconducting 18 T magnet [9.3]. Both magnetic field and sample temperature were continuously recorded at intervals of 1s during the measurements. After applying an external field at 100 K, the temperature was reduced at 2.5 K min^{-1} until the measuring temperature was reached. When thermal equilibrium was reached, the external field was ramped down at a rate of 0.1 T min^{-1} to minimize heating of the samples by dissipative flux motion. Nevertheless, an increase in sample temperature between 0.2 K at 60 K and 1.8 K at 25 K was observed. The trapped fields B_o presented in Section 9.3 correspond to the field value recorded immediately after reaching an external field of zero. Afterwards, the disks were heated to 100 K in order to remove the remanent field. To avoid thermally induced flux jumps the heating rate was limited to 0.5 K min^{-1} below 24 K, to 1 K min^{-1} between 24 K and 30 K and to 5 K min^{-1} above 30 K.

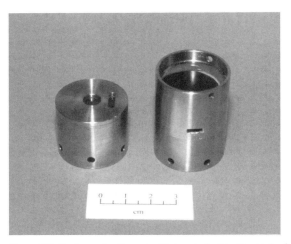

Figure 9.3 Photograph of two steel tubes. Left side: Steel tube for a single YBCO disk. Right side: Steel tube for two YBCO disks

Usually, the YBCO disks broke during the activation procedure, i.e. cracking was initiated by the increasing tensile stress exceeding the tensile strength of the reinforced material. However, especially in the temperature range between 20 and 30 K, in some cases the YBCO disks were damaged after the activating procedure and even after the trapped field measurement. In order to remove the trapped field, the YBCO disks were warmed up at a heating rate of 0.5 K min^{-1}. Despite this low heating rate, flux jumps due to thermomagnetic instabilities could not be avoided completely [9.4]. Examples of the time dependence of the trapped field and the corresponding temperature rise observed in YBCO disks during flux jumps are shown in Figure 9.4.

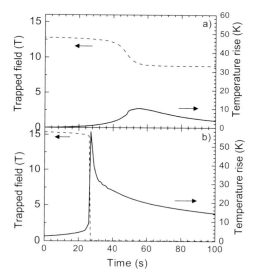

Figure 9.4 Trapped field (dotted lines) and temperature rise (solid lines) versus time during flux jumps observed in YBCO mini-magnets encapsulated in steel tubes. **a** Self-terminating flux jump accompanied by a temperature rise of about 10 K and a reduction of the trapped field from 13 T to about 8 T. **b** Complete flux jump rapidly releasing the stored magnetic energy (at $\mu_o B_o \approx 15$ T) into thermal energy

The features of a small self-terminating flux jump are shown in the upper panel of this figure. This partial flux jump released only a small part of the magnetic energy stored in the YBCO disk. The corresponding temperature rise of about 10 K did not damage the YBCO disk. On the other hand, the complete flux jump shown in the lower panel of Figure 9.4 caused a rapid release of the entire magnetic energy into thermal energy, returning the superconductor to the normal state. This flux jump led to irreversible damage of this YBCO disk.

9.2
Resin-Impregnated YBCO

As an alternative, the mechanical properties of bulk YBCO have been improved by resin impregnation [9.5]. Molten resin was found to penetrate into the bulk and into voids within the superconductor through microcracks having openings on the surface. In this way, an enhancement of the tensile strength from 18.4 to 77.4 MPa was achieved. In order to overcome the problem of the large difference in the thermal expansion coefficient between YBCO and resin causing damage of the surface resin layer during thermal cycles, the YBCO disks were wrapped with carbon fiber fabric before resin impregnation. Carbon fiber has a much lower thermal expansion coefficient (about $a = 2 \times 10^{-5}$ K^{-1}) than resin with $a = (3-4) \times 10^{-5}$ K^{-1}. Therefore, the strong thermal contraction of the resin is less pronounced in bulk YBCO wrapped with carbon fiber fabric.

Trapped field measurements for YBCO disk-pairs wrapped with carbon fiber fabric and impregnated with resin have been performed in an 18 T superconducting magnet. After applying an external field of 17.9 T at 100 K, the sample was cooled to 29 K and the applied field was swept down at a low sweep rate of 0.1 T min^{-1} to minimize sample heating. However, flux jumps have been observed to result in damage of the YBCO disks.

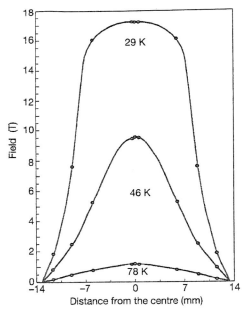

Figure 9.5 Distribution of the trapped field in a bulk resin-impregnated YBCO disk measured at several temperatures. Taken from Ref. [9.5]

In the next step, the heat transfer along the c axis of the YBCO disks was improved by including a Bi-Pb-Sn-Cd alloy and an Al wire in a hole at the center of the sample which was drilled mechanically into the sample. The Al wire was inserted in the hole and the YBCO was impregnated with the Bi-Pb-Sn-Cd alloy. This alloy and the Al wire improve the thermal conductivity along the c axis. In this way, flux jumps could be avoided, as was confirmed by trapped field measurements [9.5].

In Figure 9.5, the distribution of the trapped field is shown for several temperatures. Whereas the trapped field data at $T = 78$ K and 46 K show saturation, the field is far from saturated at $T = 29$ K. The field distribution at this temperature has a plateau-like shape with a trapped field magnitude of 17.2 T [9.5].

9.3
Trapped Field Data of Steel-reinforced YBCO

In order to achieve high trapped fields in bulk YBCO, the combined influence of steel tubes and 10 wt% Ag additions was studied. Internal reinforcement (improving strength and fracture toughness) by Ag addition and the application of compressive stress by a bandage of stainless steel are completely *independent* and, therefore, additive effects. Further, the effect of 0.1 wt% Zn doping, a combination of 0.1 wt% Zn and 10 wt% Ag doping (denoted in the following as Ag/Zn), and neutron irradiation on the trapped field $B_o(T)$ of single YBCO disks and of YBCO mini-magnets have been investigated [9.6, 9.7]. The temperature dependence of the trapped field B_o found for YBCO single disks and mini-magnets is plotted in Figures 9.6 and 9.7, respectively.

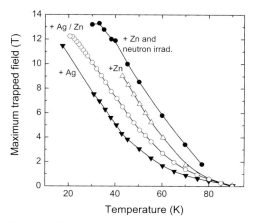

Figure 9.6 Temperature dependence of the maximum trapped field B_o for steel-reinforced YBCO single disks

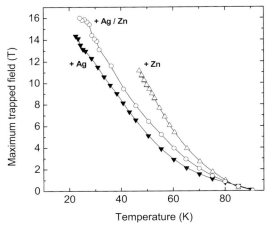

Figure 9.7 Temperature dependence of the maximum trapped field B_o for steel-reinforced YBCO mini-magnets

The $B_o(T)$ curves for YBCO(+Zn) show a strong increase of the trapped field with decreasing temperature. The improved pinning properties of this material, compared, for instance, with YBCO(+Ag), lead to a considerable shift of the $B_o(T)$ curves to higher temperatures by about 18 K (single disks) and 15 K (mini-magnets). Maximum B_o values of 9 T (at 43 K) and 11.2 T (at 47 K) were obtained in YBCO(+Zn) single disks and mini-magnets, respectively, before cracking occurred. The addition of Ag to YBCO, resulting in a reduced density of microcracks in the material [9.8], improves the mechanical stability of the sample and the achievable trapped field. Indeed, the YBCO disks containing silver broke in most cases at a significantly higher trapped field than the YBCO disks without silver. Furthermore, different cracking behavior was observed for the YBCO/Ag composite material. This is demonstrated in Figure 9.8, in which field profiles of two damaged YBCO disks are compared. The typical damage for YBCO without Ag is that one crack divides the YBCO disk into two superconducting domains (Figure 9.8a). For the YBCO/Ag composite, the existence of a crack is indicated by a small disturbed region in the field distribution (Figure 9.8b). This crack, which is much shorter than the diameter of the YBCO disk, is located in the outer region of the superconductor, thus disturbing the symmetry of the distribution of the trapped field, but not destroying its single-domain character.

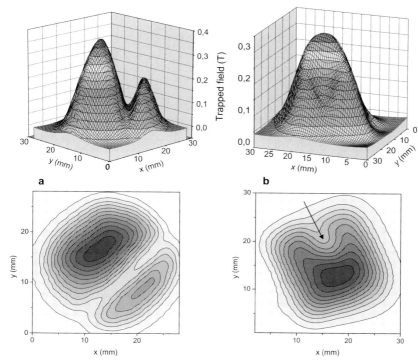

Figure 9.8 Field profiles of two damaged YBCO disks as 3D plots (top panels) and as top views (bottom panels). **a** Typical field profile for YBCO without silver, and **b** field profile of a YBCO disk with 10 wt% Ag addition

In Figure 9.6, trapped field data of an irradiated, Zn-doped YBCO disk are included. The almost parallel shift of the $B_0(T)$ curves to higher temperatures corresponds to a stepwise improvement of the pinning properties. In a first step the pinning effect in YBCO(+Ag) is improved by Zn doping, and a further improvement is obtained by removing the silver from the Zn-doped YBCO. The most effective improvement is achieved by an additional neutron irradiation of the Zn-doped YBCO. This strong enhancement of the trapped field observed for neutron-irradiated YBCO is in agreement with the high j_c values of this material (see Section 6.1.3). Note that all single YBCO disks, except for the unirradiated one without silver, trap almost the same maximum field of around 13 T. This obviously represents the mechanical limit for the steel-reinforced samples and is not related to the superconducting properties. Because of the improved pinning in the neutron irradiated, Zn-doped YBCO disk, this highest achievable field is shifted to higher temperatures. Thus, 13.3 T could be reached at 33 K, i.e. around 10 K higher than for the unirradiated YBCO(+Ag/Zn) disk.

Obviously, considerably higher fields than those measured on the surface of a single YBCO disk are measured in the gap between two equivalent YBCO disks (see Figure 9.7). The highest trapped field was obtained in mini-magnets consisting of YBCO(+Ag/Zn) disks showing improved mechanical and pinning properties. In particular, trapped fields of up to 16.0 T (at 24 K) were achieved in YBCO(+Ag/Zn) mini-magnets [9.4, 9.6, 9.7]. The corresponding $B_o(T)$ curve lies between those for YBCO(+Zn) and YBCO(+Ag), as expected. For the YBCO(+Ag/Zn) disks, cracking was observed at higher B_o values than for the YBCO(+Ag) disks. This can be explained by the improved reinforcement of the YBCO(+Ag/Zn) disks, for which a steel tube with a wall thickness of 4 mm was used, whereas the other steel tubes had a wall thickness of 2 mm. As a result of this improved reinforcement, the YBCO(+Ag/Zn) mini-magnet was found to withstand the trapped field of 16.0 T at T = 24 K without cracking.

9.4
Comparison of Trapped Field Data

In order to summarize this chapter, trapped field data for YBCO mini-magnets are collected in Figure 9.9. In this figure, data obtained by different techniques (mechanical reinforcement, resin impregnation) are compared with trapped field data of a YBCO mini-magnet composed of four proton-irradiated YBCO tiles 20 mm in diameter. For this mini-magnet, which was used without "internal" or external reinforcement of the superconductor against magnetic tensile stress, a maximum trapped field of 10.1 T was reported at 42 K as early as 1996 [9.9, 9.10]. The highest trapped field, 17.24 T, was achieved in resin-impregnated YBCO with carbon fiber wrapping [9.5]. This mini-magnet consisted of two YBCO disks 26.5 mm in diameter. Slightly lower trapped fields, up to 16 T, have been obtained in a steel-reinforced mini-magnet consisting of two YBCO(+Ag/Zn) disks 21 mm in diameter [9.4].

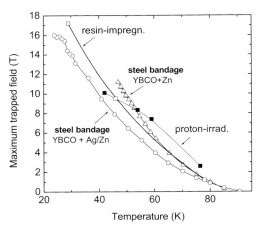

Figure 9.9 Temperature dependence of the maximum trapped field B_o for YBCO mini-magnets (see text)

The data shown in Figure 9.7 and Figure 9.9 demonstrate the extremely high trapped fields which can be achieved in bulk YBCO magnets. However, it should be borne in mind that the relevant parameter for applications of bulk superconductors is the trapped field on the surface of the bulk YBCO, which is shown in Figure 9.6.

References

9.1 F. Pobell, *Matter and Methods at Low Temperatures* (Springer-Verlag, Berlin, Heidelberg, 1992).

9.2 Tera Analysis.Quickfield 4.2. Tera Analysis Ltd., Knasterhovvej 21, DK-5700 Svendborg, 1999.

9.3 G. Fuchs, P. Schätzle, G. Krabbes, S. Gruss, P. Verges, K.-H. Müller, J. Fink, L. Schultz, Appl. Phys. Lett. **76**, 2107 (2000).

9.4 S. Gruss, G. Fuchs, G. Krabbes, P. Verges, G. Stöver, K.-H. Müller, J. Fink, L. Schultz, Appl. Phys. Lett. **79**, 3131 (2001).

9.5 M. Tomita, M. Murakami, Nature **421**, 517 (2003).

9.6 R. Gonzalez-Arrabal, M. Eisterer, H.W. Weber, G. Fuchs, P. Verges, G. Krabbes, Appl. Phys. Lett. **81**, 868 (2002).

9.7 G. Fuchs, G. Krabbes, K.-H. Mueller, P. Verges, L. Schultz, R. Gonzalez-Arrabal, M. Eisterer, H.W. Weber, J. Low Temp. Phys. **133**, 159 (2003).

9.8 P. Diko, G. Fuchs, G. Krabbes, Physica C **363**, 60 (2001).

9.9 R. Weinstein, J. Liu, Y. Ren, R.-P. Sawh, D. Parks, C. Foster, V. Obot, in Proc. 10th Anniversary HTS Workshop on Physics, Materials and Applications, edited by W. K. Chu, D. Gubser, and K. A. Müller (World Scientific Press, Singapore, 1996) p. 625.

9.10 J. Liu, R. Weinstein, Y. Ren, R.-P. Sawh, C. Foster, V. Obot, in Proceedings of the 1995 International Workshop on Superconductivity, Maui (ISTEC and MRS, 1995) p. 353.

10
Engineering Aspects: Field Distribution in Bulk HTSC

Phenomena with relevance for engineers can be derived from the magnetic phase diagram of the oxide HTSC, which is shown in Figure 10.1. The various parts of this figure can be explained as follows. As long as the external flux density does not exceed B_{c1}, superconductors are in a diamagnetic *Meissner* state, where the magnetic flux is completely expelled from the interior by surface currents. In the area between $B_{c1}(T)$ and $B_{c2}(T)$, magnetic flux penetrates the superconductor in the form of flux lines (*Shubnikov* or mixed state). Their motion has to be prevented by flux pinning, i.e. an interaction of the flux lines with imperfections in the material. As long as the flux lines are pinned, a maximum supercurrent of density j_c can flow without any loss. The critical current density j_c depends on the applied magnetic field and on the temperature. If the current density exceeds j_c, then the flux line lattice (or parts of it) starts to move, which gives rise to losses, as described in more detail in Chapter 1. Because of thermal fluctuations, the mobility of the flux line lattice strongly increases with applied magnetic field and temperature. Above the irreversibility field B_{irr}, currents cannot flow without losses, i.e. the critical current density j_c disappears, although the superconductor is not yet in the normal state. This means that applications of HTSC are restricted to the field range below the irreversibility field $B_{irr}(T)$.

Figure 10.1 is intended to illustrate different vortex phases of bulk HTSC in the H–T plane. This figure does not claim to show realistic relations between the characteristic fields of HTSC. Experimental data for $B_{irr}(T)$ and $B_{c2}(T)$ of bulk YBCO are presented in Figure 4.6. Typical values of B_{irr} and B_{c2} at 77 K, for fields applied along the c axis, are 7 T and 22 T, respectively (see Figure 4.6), whereas B_{c1} (77K) is, at approximately 0.02 T [10.2], considerably smaller.

High Temperature Superconductor Bulk Materials.
Gernot Krabbes, Günter Fuchs, Wolf-Rüdiger Canders, Hardo May, and Ryszard Palka
Copyright © 2006 WILEY-VCH Verlag GmbH & Co. KGaA, Weinheim
ISBN: 3-527-40383-3

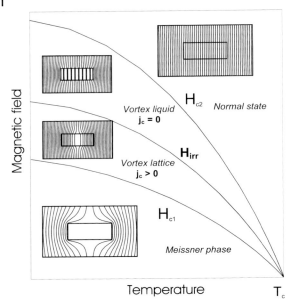

Figure 10.1 Schematic view of the magnetic phase diagram of high-temperature superconductors [10.1], including the lower and upper critical field $B_{c1}(T)$ and $B_{c2}(T)$, respectively, and the irreversibility line $B_{irr}(T)$

10.1
Field Distribution in the Meissner Phase

10.1.1
Field Cooling

The typical response of HTSC bulks to external fields in the *Meissner* state is illustrated in Figure 10.2. While the superconductor is cooled below the critical temperature T_c in the presence of a magnetic field $H < H_{c1}$, surface currents completely expel the magnetic flux from the interior of the superconductor. This perfect diamagnetism is observed in type I superconductors (for $H < H_c$, where H_c is the thermodynamic critical field) and type II superconductors (for $H < H_{c1}$). This effect, which was discovered by *Meissner* and *Ochsenfeld* [10.3] in 1933, is an inherent property of superconductors, as well as the disappearance of the electric resistivity.

The strongly inhomogeneous field distribution of the ring magnet in Figure 10.2 results in forces which are strong enough to support the weight of the HTSC-bulk. These repulsive forces lead to contact-free levitation, which is a visible proof of the *Meissner* effect.

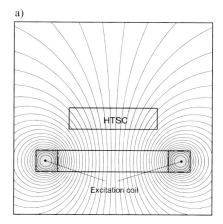 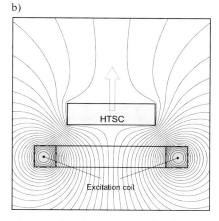

Figure 10.2 Field distributions when an HTSC bulk is cooled to below T_c in the presence of an external field. **a** Temperature of the HTSC-bulk above T_c, **b** Temperature of the HTSC bulk below T_c. Surface currents within a submicron thin sheet [10.5] expel the flux from the interior of the SC. The strongly inhomogeneous field of the magnet generates a repulsive levitation force

10.1.2
Zero-Field Cooling

If the HTSC bulk is cooled to below the critical temperature before the magnetic field is applied, then surface currents in the HTSC bulk prevent the penetration of this magnetic field (Figure 10.3). This diamagnetism leads to a strong field concentration between the excitation coils and the HTSC sample. The vertical and lateral forces [10.5] are associated with this deformation of the magnetic field and are explained in later sections.

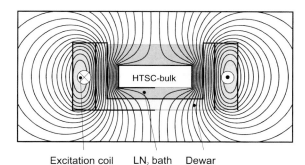

Figure 10.3 The field distribution when the external field is excited after the cooling of the HTSC

10.2
Field Distribution in the Mixed State

Type II superconductors are characterized by the existence of the mixed state between B_{c1} and B_{c2}, i.e. the internal magnetic field becomes non-zero because of the penetration of flux lines.

10.2.1
Field Cooling

Typical field profiles after cooling the superconductor in the presence of an external magnetic field are shown in Figure 10.4. The superconductor is assumed to be in the critical state according to the *Bean* model [10.6], i.e. field gradients within the superconductor are constant and determined by the critical current density j_c of the superconductor. In this approximation, the field dependence of j_c is neglected. Starting from a constant internal field, variation of field distributions can be realized by changing the external magnetic field. In Figure 10.4a, the applied field H_1 is smaller than the penetration field H_0. Under this condition, the internal field H_1 remains unchanged in the central part of the superconductor after switching off the applied field. The resulting field distribution has a trapeze-like shape (in 2D view). In Figure 10.4b, the applied field H_1 is larger than the penetration field H_0. In this case, supercurrents flow in the whole superconduc-

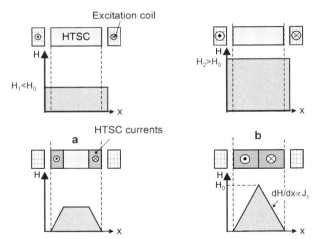

Figure 10.4 Field profiles in an HTSC bulk after field cooling. The superconductor is cooled in the presence of a magnetic field generated by an excitation coil. Shown are field profiles, the current direction in both the coil and the superconductor, and the activated part of the superconductor. **a** A magnetic field $H_1 < H_0$ is applied before the superconductor is cooled (upper panel). After switching off the applied field, the remaining field distribution has (in 2D view) a trapeze-like shape (lower panel). **b** As in **a** except that the applied magnetic field H_1 is larger than the penetration field H_0. Then, a triangle-shaped field distribution (in 2D view) remains in the superconductor

tor, and therefore the typical triangle-shaped field distribution in a 2 dimensional projection is observed after switching off the applied field. A maximum field H_o is trapped in the center of the superconductor. The gradient of the trapped field is proportional to the critical current density j_c.

It should be noted that measured field profiles of bulk HTSC have, in contrast to the field profile in Figure 10.4b, a more or less curved shape (see, for instance, Figure 5.3). This curved shape reflects the field dependence of the critical current density. In Chapter 12, experimental data for $j_c(H)$ (or at least model functions [10.7, 10.8] for $j_c(H)$ describing experimental data) will be used to calculate levitation forces between superconductors and permanent magnets.

10.2.2
Zero-Field Cooling

Examples of field profiles obtained after zero-field cooling are shown in Figure 10.5. The superconductor is cooled below T_c in the absence of an external magnetic field. This is applied afterwards. The field starts to penetrate the superconductor, which is again assumed to be in the critical state according to the *Bean* model. This means that the field decreases with a constant field gradient (corresponding to the critical current density) from its value outside of the superconductor to zero within the superconductor. The increase of the applied magnetic field is stopped before the penetrated field reaches the center of the sample (Figure 10.5b). The applied field is then reduced to zero. The remaining field distribution within the superconductor is triangle-shaped (in 2D view) in the outer region, whereas the internal magnetic field remains zero in the central part of the sample, as shown in Figure 10.5c. Afterwards, the applied field was increased to $H_1 = 2H_o$, where H_o is the penetration field at which the flux front within the superconductor reaches the center of the superconductor. The gradient of the resulting field profile (see Figure 10.5d) is the same as for the other field profiles. Finally, the applied magnetic field is reduced and switched off. The resulting field profile with its triangle shape (Figure 10.5e) is typical for a superconductor in which supercurrents of constant critical current density flow in the entire volume. The maximum trapped field in the center corresponds exactly to the penetration field H_o.

At first glance, the field profiles obtained after Zero-field Cooling (ZFC) and Field Cooling (FC) for switched-off external magnetic field look similar. However, there is an important difference. Field profiles after ZFC always have (in 2D view) a triangle-based shape, whereas after FC a triangle-based or trapezoid shape can be obtained. The possibility to realize a trapezoid-shaped field profile is important for the use of HTSCs as superconducting permanent magnets (SPM) in energy converters (see Chapter 13), as they use the total flux of the excitation system.

Another significant difference between the two activation methods, which has already been mentioned in Section 5.2, concerns the magnitude of the activating field. For the same maximum trapped field H_o, the magnetizing field should be at least twice as large as H_o in the case of ZFC, whereas only a magnetizing field of about H_o is necessary in the case of FC.

The 2-dimensional field distributions in and around an HTSC cuboid (32 × 32 × 10 mm³) activated after ZFC are shown in Figure 10.6. A high excitation field $H_1 = 2H_0$ has to be applied to achieve the field distribution shown in Figure 10.6a.

The vertical component of the calculated trapped field distribution shown in Figure 10.6b was scanned at 4.6 mm above the superconductor. This field profile is shown in Figure 10.7a. The magnetic forces between this HTSC magnet and ferromagnetic bodies are demonstrated in Figure 10.7b.

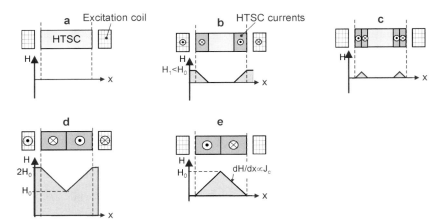

Figur 10.5 Field profiles in an HTSC bulk after zero-field cooling. The magnetic field applied after the superconductor is cooled is generated by an excitation coil. Shown are field profiles and current direction in both the coil and the superconductor. The activated part of the superconductor in which supercurrents flow is also marked. **a** The HTSC-bulk is cooled below the critical temperature T_c in the absence of an external field. **b** The current of the field excitation coil is switched on. The applied field is smaller than the penetration field H_0. **c** The current is switched off. The resulting field profile has a triangle shape in the outer region. **d** The applied field is enhanced to $H_1 = 2H_0$, where H_0 is the penetration field. **e** After switching off the current, a typical triangle-shaped field profile is obtained

10.2 Field Distribution in the Mixed State

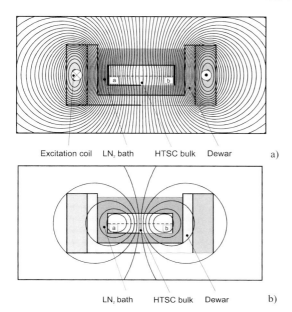

Figure 10.6 Field distribution in the cross-sectional center plane of the magnetization unit. The dimensions of the HTSC cuboid are $32 \times 32 \times 10$ mm^3. **a** The superposition of the fields excited by the currents of the magnetization coil and the internal shielding currents of the HTSC. The magnetization state is indicated in Figure 10.5d. The flux between adjacent field lines amounts 1×10^{-4} Vs per meter depth. **b** Trapped field distribution due to the internal currents after the excitation current is switched off. The magnetization state is indicated in Figure 10.5e. For better resolution, the flux between adjacent field lines is reduced to 5×10^{-5} Vs per meter depth

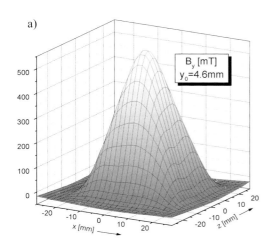
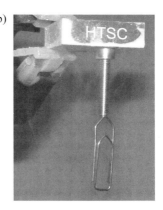

Figure 10.7 a Physical proof of the magnetization state of the HTSC bulk based on the magnetic field distribution. Scanned normal component at 4.6 mm above the surface of the HTSC bulk with the dimensions indicated in Figure 10.6. **b** Technical proof of the magnetization state of the HTSC bulk based on the evidence of magnetic forces on iron parts (a screw and a paper clip hanging on the SPM)

References

10.1 M. Murakami, Melt Processed High-Temperature Superconductors, World Scientific Publishing Co. Pte. Ltd. (1992).

10.2 L. Krusin-Elbaum, A. P. Malozemoff, Y. Yeshurun, D. C. Cronemeyer, F. Holtzberg, Phys. Rev. B **39**, 2936 (1989).

10.3 W. Meissner, R. Ochsenfeld, Naturwissenschaften **21**, 787 (1933).

10.4 J. D. Doss, Engineer's Guide to "High-Temperature Superconductivity", A Wiley-Interscience Publication. John Wiley & Sons (1989).

10.5 R. P. Feynman, The Feynman Lectures on Physics, R. Oldenbourg Verlag, München – Wien Volume II, Part 1; (Bilingua) (1974).

10.6 C. P. Bean, Rev. Mod. Phys. **36**, 31 (1964).

10.7 Y. B. Kim, C. F. Hempsteead, A. R. Strnad, Phys. Rev. **129**, 528 (1963).

10.8 W. A. Fietz, M. R. Beasley, J. Silcox, W. W. Webb, Phys. Rev. A **136**, 335 (1964).

10.9 P. J. Lee, Engineering superconductivity, John Wiley & Sons, Inc., ISBN 0-471-41116-7 (2001).

11
Inherently Stable Superconducting Magnetic Bearings

Superconducting magnetic bearings (SMBs) constructed with bulk HTSC materials offer various advantages over conventional magnetic bearings. Their most important advantage is that they operate in an inherently stable fashion, i.e. without the need for sensors and control electronics, thus avoiding the problems associated with the use of conventional actively controlled magnetic bearings. Their use in high-speed machines is of outstanding interest for turbo machinery, flywheels, and transportation systems operating under ambient, clean-room, or even vacuum environments. Therefore, SMBs and integrating systems are expected to become a priority field of application of bulk HTSC. Principles of design and operation are introduced in the present chapter. The generation and handling of forces in SMBs are key issues. Results of numerical calculations are important both for a deeper understanding and for the optimization of designs. Basic principles of numerical methods will therefore be discussed to complete this chapter.

11.1
Principles of Superconducting Bearings

11.1.1
Introduction to Magnetic Levitation

The first reference to the stability conditions for a body in a magnetic field was made as early as 1842 by Earnshaw [11.1]. He proved that under no circumstances could any particle reach stable equilibrium in any static force field (such as a magnetic field) which follows an inverse second order characteristic law [11.1]. The reason is that the inverse square force law governed by the *Laplace* equation allows only saddle-type equilibria, and this always results in unstable levitation. Braunbeck [11.2] was able to show how to deal with this unpromising situation by including diamagnetic materials in his considerations, and he concluded that a static,stable, free suspension of one system in the magnetic field of a second system may become possible if diamagnetic material is present in one of the two systems. It should be mentioned that analogous conclusions have been drawn for

High Temperature Superconductor Bulk Materials.
Gernot Krabbes, Günter Fuchs, Wolf-Rüdiger Canders, Hardo May, and Ryszard Palka
Copyright © 2006 WILEY-VCH Verlag GmbH & Co. KGaA, Weinheim
ISBN: 3-527-40383-3

electrostatic or gravitational fields, diamagnetic materials being replaced by materials with equivalent properties appropriate to the type of field under consideration.

Thus, diamagnetic superconductors, which repel a magnetic flux, can be used for magnetic levitation. Stable levitation of a permanent magnet above a superconducting saucer-shaped surface was demonstrated for the first time by Arkadiev [11.6] in 1947. Later, the suspension of a low-T_c superconducting sphere within a superconducting coil was reported [11.7]. Small superconducting bearings have also been developed using low-T_c superconductors [11.8–11.10]. However, the necessary use of liquid helium for the cooling of the superconductors discouraged further work in superconducting bearings. Interest in superconducting magnetic bearings then increased significantly following the discovery of HTSCs, which can be cooled with liquid nitrogen, a much easier operation.

There are two other circumstances under which non-superconducting magnetic systems can avoid the consequences of Earnshaw's theorem, namely the presence of active feedback or time-varying fields, e.g., in the case of eddy currents. Both have been utilized since the late 1960s to make magnetic levitation possible in transportation systems. One is the superconductor-based *electrodynamic levitation* system used for the Japanese MAGLEV [11.3], and the other is the feedback-controlled *electromagnetic levitation* system used for the German TRANSRAPID [11.4]. In the electro-dynamic levitation system, eddy currents induced in the ground coils of the track interact with the alternating field of the superconducting coils on the moving train, whereas in the electromagnetic levitation system set-up, the field in the gap between the electromagnet (of the train) and the ferromagnetic reaction rail must be actively controlled by means of a power converter [11.5].

11.1.2
Attributes of Superconducting Magnetic Bearings with Bulk HTSC

The discovery of HTSC in 1986 was soon followed by a table-top demonstration of the new superconducting material. The photographs in Figure 11.1 show a permanent magnet levitating above a bulk HTSC (cooled by liquid N_2) and a bulk HTSC suspended beneath the permanent magnet, respectively. This contact-free levitation or suspension is completely stable in free space, offering a variety of technical applications [11.11].

Superconducting magnetic bearings (SMBs) with bulk HTSC materials differ significantly from conventional magnetic bearings, and have a number of beneficial features. They permit contact-free frictionless operation up to the highest speeds, and they are inherently stable because of the elastic response of the pinned flux line lattice. Therefore, they can operate without the need for any electronic feedback control [11.12] and offer the highest electromagnetic compatibility (EMC), preventing problems associated with conventional active magnetic bearings. Furthermore, SMBs are free from contaminating lubricants. Therefore, they can easily be incorporated in physical and chemical processes.

Figure 11.1 Demonstration of inherently stable forces between a permanent magnet (PM) and bulk HTSC. The HTSC was cooled under the influence of the PM. **a** PM suspended above an HTSC by repelling forces; **b** PM hanging beneath the HTSC by attractive forces

11.2 Forces in Superconducting Bearings

The most important feature of superconducting bearings is the levitation force between the superconductor and the exciting magnet arrangement, this in most cases consisting of permanent magnets. Other relevant properties of a superconducting bearing are its stiffness and dynamic performance. Extended details on levitation forces and their application in superconducting bearings can also be found in reviews [11.11], [11.13].

11.2.1 Forces in the Meissner and the Mixed State

In order to explain the generation of the force between superconductor and magnet forming an SMB, one has to decide whether the magnetic field of the magnet penetrates the superconductor or not.

For applied magnetic fields for which $H < H_{c1}$ (H_{c1} being the lower critical field), surface currents completely expel the magnetic flux from the interior of the superconductor. The levitation in this case is characterized by large vertical repulsive forces and very small horizontal forces. However, as the latter are indispensable in order to stabilize the levitation, this situation represents levitation with very restricted stability, which is not suitable for technical applications.

In Figure 11.2, the repulsive levitation force between a YBCO disk and a permanent magnet is shown as a function of the distance between magnet and superconductor. The YBCO disk was cooled at a large distance from an Sm-Co permanent magnet ("*zero-field cooling*"). By reducing the distance between permanent magnet and superconductor, a repulsive force appears due to the diamagnetic response of the superconductor, which remains in the *Meissner* phase as long as the field of the permanent magnet on the surface of the superconductor does not exceed H_{c1}.

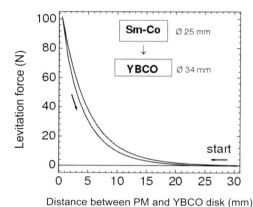

Figure 11.2 Dependence of the repulsive levitation force on the distance between an Sm-Co magnet and a YBCO disk after cooling outside the field (ZFC) at $T=77$ K. The remanence of the Sm-Co PM is 1 T.

This repulsive force increases as the magnet approaches the superconductor and reaches a maximum value of 100 N (corresponding to a *levitational pressure* of 20.4 N cm^{-2} with respect to the face surface of the PM) for the smallest distance between superconductor and magnet. Hysteretic behavior is observed when the magnet is subsequently removed from the superconductor. This is a clear indication that magnetic flux has begun to partly penetrate the superconductor at a certain distance during the approach. Therefore the force appearing at small distances (i.e. higher fields) is influenced by pinning of flux lines. Consequently, the force measured for increasing distances is influenced by a small attractive component due to flux pinning.

Zero-field cooling produces the largest vertical force, but is impractical, because it requires the cooling of the superconductor before the bearing is assembled. For practical reasons the components of SMBs – superconductor and permanent magnet – are usually assembled at room temperature. The superconductor has to be cooled *subsequently in the presence of the magnetic field* of the permanent magnet by an appropriate *field cooling (FC)* procedure. This procedure has an important influence on the force characteristics of the bearing, since it leaves the superconductor in the mixed state, i.e. partially penetrated by magnetic flux in discrete vortex-like structures (cf. Chapters 4 and 10).

An example of the vertical force-vs-distance relationship in an SMB after FC is shown in Figure 11.3. The force, which is zero after field cooling, becomes attractive (negative) as the permanent magnet is moved away from the superconductor. This attractive force can be used to achieve stable levitation of a permanent magnet which is located below the superconductor (Figure 11.1b). Strong hysteresis is observed in the force vs distance relationship as the PM is brought back toward the superconductor. Numerical calculations of the currents and forces associated with such a hysteretic loop after FC are presented in Section 11.8.

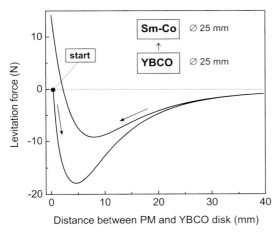

Figure 11.3 Levitation force (between an Sm-Co magnet and a YBCO disk) vs distance at $T=77$ K after cooling in the field of the PM (FC: field cooling)

The appearance of forces after FC can easily be explained by flux pinning. The inhomogeneously distributed field within a superconductor generated by a permanent magnet persists below T_c as a distribution of curved flux lines which are pinned at the pinning centers of the material. This inhomogeneous field distribution "frozen" in the superconductor by flux pinning is stabilized against displacements both in the vertical *and* the horizontal direction by loss-free supercurrents of maximum density $j_c(H,T)$. A *vertical displacement* changes the field, i.e. the number of flux lines in the superconductor. Additional supercurrents are induced by the moving flux lines pulling the superconductor back to its initial position. In the case of *small lateral displacements*, the flux lines remain pinned and respond by stronger bending. The resulting elastic deformations within the flux line lattice produce a restoring force similar to that of a mechanical spring attached to the superconductor and the permanent magnet. For *large lateral displacements*, the flux lines move from their original pinning centers to new ones. The supercurrents induced by the moving flux lines produce a restoring force and tend to pull the superconductor back to its initial position. The corresponding field distributions are illustrated in Figure 11.4.

Concluding this section, it should be emphasized that pinning-assisted levitation after FC has the advantage of producing significantly greater *lateral* forces than those obtained after ZFC. Furthermore, ZFC which produces larger *vertical* forces is impractical for SMBs, because it requires cooling of the superconductor before the bearing is assembled.

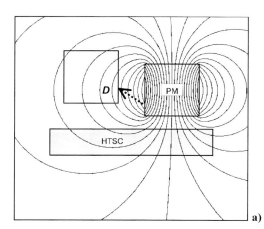
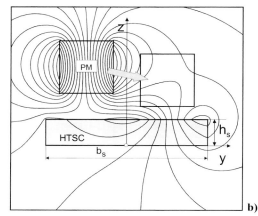

Figure 11.4 Field distributions in an SMB before (**a**) and after (**b**) a simultaneous lateral and vertical displacement of the permanent magnet PM. The field-cooled HTSC is assumed to have infinitely large current density. Then, according to the Perfectly Trapped Flux Model (see Section 11.8.1), the field distribution within the HTSC remains completely unchanged in the displaced position. The arrow in **b** indicates the restoring force after the displacement of the PM to the position D.

11.2.2
Maximum Levitational Pressure in Superconducting Bearings

The levitation force between a permanent magnet and a superconductor cooled in the field of the magnet (field cooling) and containing a trapped magnetic flux can be written as

$$F \propto j_c \, R \, dB/dz \, , \tag{11.1}$$

where R is the size of the current loops in the superconductor and dB/dz is the field gradient in the axial direction produced by the permanent magnet within the superconductor. The levitation force of a superconducting bearing increases with the magnetization of the superconductor, which is proportional to the critical current density j_c and the radius R of a single-domain YBCO sample. For precise calculation of the levitation force according to Eq. (11.1) one needs to know the field distribution within the superconductor for integrating the force over the volume of the superconductor.

The maximum levitation force which can be achieved in superconducting bearings, however, can easily be estimated by considering the repulsive force in the *Meissner* state. The estimation of maximum force values for two types of pairs, PM – Bulk HTSC and HTSC permanent magnet – Bulk HTSC will be considered, the latter as a candidate for a novel design of SMBs in which the permanent magnet is replaced by a bulk superconducting YBCO magnet.

Superconducting Bearings using Conventional Permanent Magnets

The upper limit of the levitation force, F_{max}, achievable for a given permanent magnet, would be obtained for an ideal superconductor of infinite size and infinitely large critical current density that would perfectly shield the magnetic field of the permanent magnet. The resulting levitation force is equivalent to the interaction between the permanent magnet and its mirror image. The maximum magnetic pressure between a pair of magnets which are in direct contact is given by

$$F_{max}/A = B^2/(2\mu_0), \tag{11.2}$$

where A is the magnet pole face and B is the magnetic flux density. Taking into account the fact that the magnetic flux density B is twice the surface field B_s on the magnet face one gets from Eq. (11.2)

$$F_{max}/A = 2\,B_s^2/\mu_0. \tag{11.3}$$

For a maximum field density of $B_s = 0.4$ T, which is typical for rare earth magnets (SmCo, remanence $B_r \approx 1.0$ T), the maximum achievable magnetic pressure is $F_{max}/A \approx 25.5$ N cm^{-2}. The experimental value of $F = 100$ N obtained for the levitation force between a permanent magnet (diameter 25 mm) and a YBCO sample after zero-field cooling (see Figure 11.2) achieves about 80% of this upper limit.

All Superconducting Bearing using a Bulk YBCO Magnet

At temperatures below 77 K, much higher magnetic fields can be realized in superconducting permanent magnets (SPM) made from bulk YBCO than in conventional permanent magnets. Therefore, the properties of superconducting bearings in which a YBCO SPM is used instead of a conventional magnet have been investigated in detail. In Figure 11.5, the levitation forces between a YBCO levitator (ZFC, 77K) and an SPM (also YBCO but magnetized at 50K with a maximum trapped field of $B_0 = 4$ T) is compared with the forces generated by a conventional NdFeB permanent magnet after zero field cooling of the YBCO levitator to $T = 77$ K (ZFC mode).

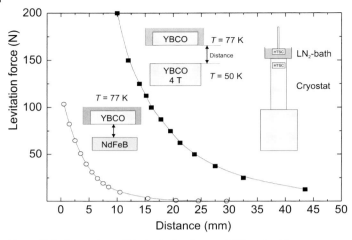

Figure 11.5 Comparison of the levitation forces of different cold excitation systems for the pairs PM (NdFeB, warm)–YBCO, left, and SPM (YBCO)–YBCO, right

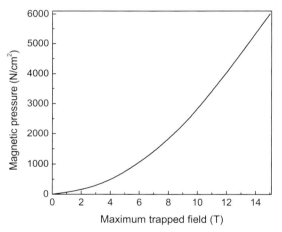

Figure 11.6 Levitational pressure vs maximum trapped field in the YBCO–SMB arrangement as displayed in Figure 11.5 (right) for a distance of 10 mm

In both cases, the excitation arrangement interacts with a YBCO sample operating at $T = 77$ K after a ZFC activation. The levitation force using the YBCO SPM achieves a value of 200 N (or a levitational pressure of 40 N cm^{-2}) at a distance of 10 mm between YBCO-SPM and YBCO-bulk, which is larger by a factor of 20 than the levitation force of the NdFeB magnet.

These investigations have proven that bearings with remarkably higher levitational pressure can be designed using bulk superconducting permanent magnets instead of conventional permanent magnets. FEM calculations of the levitational pressure based on materials properties, as shown in Figure 11.5, predict that a magnetic pressure of up to 4000 N cm^{-2} can be achieved for a YBCO magnet with a maximum trapped field of 12 T (see Figure 11.6). Such trapped fields can be generated in bulk YBCO at temperatures around 30 K [11.14], [11.15]. However, to utilize these novel types of superconducting bearings for practical applications, serious technical problems have to be solved, including the separate cooling of the two YBCO samples to the low temperature and the magnetizing procedure of the YBCO magnet. Pulsed field magnetization could be a promising alternative (Section 5.5).

11.3
Force Activation Modes and Magnet Systems in Superconducting Bearings

11.3.1
Cooling Modes

A superconductor can be held in a stable position with respect to mechanical displacement if external and internal forces are equilibrated. Under the influence of gravity it can be *levitated* inherently stably above a *permanent magnet*, which generates a repulsive force, or it can be suspended by "hanging" it below the magnet, which generates an attractive force (Figure 11.1b). The generation of the appropriate forces has to be realized by an optimal functional design and the use of an appropriate mode of force activation during cooling of the superconductor and during operation of the SMB. The following modes of field cooling will be considered in more detail in the sections below:
- Operational Field Cooling (OFC): $g_{op} = g_{act}$, i.e. the gap remains unchanged
- Operational Field Cooling with a vertical offset (OFCo): $g_{op} < g_{act}$
- Maximum Field Cooling (MFC): $g_{op} > g_{act}$

The positions of the superconductor during cooling with respect to its operational position (and the corresponding distances g_{act} and g_{op} between superconductor and exciting magnet) are illustrated in Figure 11.7a.

Examples of calculated (vertical) force vs displacement characteristics in different FC modes are shown in Figure 11.7b. These forces are normalized to the force obtained for the ZFC mode. The position of the bulk HTSC during activation (by cooling) is marked by large dots on the graphs. By reducing the gap between HTSC and permanent magnet, a *repelling normal force F* is generated, which is indicated by a *positive sign* in Figure 11.7b. *Attractive normal forces* due to enhancing the gap after cooling the HTSC are indicated by a *negative sign* in Figure 11.7b.

Another essential parameter is the stiffness in the *operational position*, which is defined as dF/dg. In Figure 11.8, calculated values for the vertical and lateral stiffness are compared for the considered FC modes of activation. Again, the stiffnesses in Figure 11.8 are normalized to the vertical stiffness in the case of ZFC.

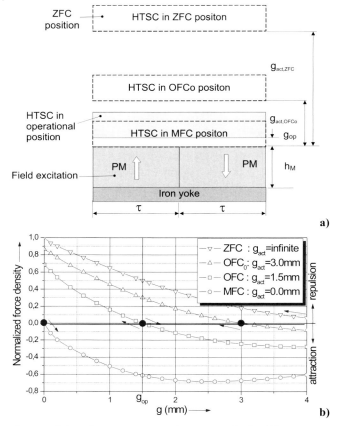

Figure 11.7 SMB force activation modes: **a** Positions of the superconductor during cooling in the ZFC (zero field), MFC (maximum field), and offset to operational field (OFCo) mode in relation to the operational position (filled); **b** Development of force densities for the different activation modes as a function of the gap g. An operation gap g_{op} of 1.5 mm is assumed for the SMB; the cooling positions g_{act} are marked by black dots and given in the legend

Note that despite the large vertical stiffness in the ZFC mode, its lateral stiffness is negligible, thus disqualifying this mode for most levitation applications.

Large restoring forces are generated in an SMB after FC due to flux pinning in the bulk HTSC if the superconductor or the permanent magnet is displaced from its initial position. The normal and lateral force components can easily be calculated by numerical methods for the considered activation modes. These numerical methods will be presented in Section 11.8.

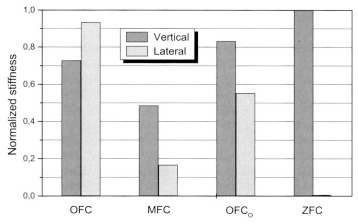

Figure 11.8 Vertical and lateral stiffness calculated for the operation gap of 1.5 mm for the different activation modes of Figure 11.7

11.3.2
Operational Field Cooling with an Offset

Vertical Displacement after Operational Field Cooling

A *repelling force* is required to levitate a bulk superconductor *above* a PM because its weight must be compensated. This repelling force can easily be achieved by reducing the gap between PM and superconductor after its cooling. Thus, the superconductor is cooled in a position with a certain offset from the operational position (Operational Field Cooling with an Offset, OFCo) choosing $g_{act} > g_{op}$. During cooling at the distance g_{act} from the PM, the superconductor is supported to compensate its weight and left free for operation at a smaller distance g_{op}. The operational gap depends on the weight of the superconductor and the force-vs-displacement characteristics of the SMB. It becomes smaller if an additional external force (load) acts on the cooled superconductor of a superconducting transportation system.

Lateral Displacement after Operational Field Cooling

The generation of a lateral force in the case of the OFCo mode is illustrated in Figure 11.9. As long as the pole pitch is in the range of its optimal values (Section 11.4), the sinusoidal development of the guidance force in Figure 11.9b is typical for all periodic excitation systems and does not depend on the arrangement or type of excitation. The figure illustrates that, in addition to the restoring lateral guidance force, a repelling force in the normal direction is also generated because of the lateral displacement. This kind of interaction between normal force (or stiffness) and lateral displacement is known as cross stiffness [11.16], [11.17]. The increasing repelling force due the lateral displacement has been explained by the attractive interaction between the displaced permanent magnet and the trapped flux within the superconductor, which remains in the initial location. This attractive interaction produces not only a lateral restoring force, but also a slight increase in the vertical force [11.13].

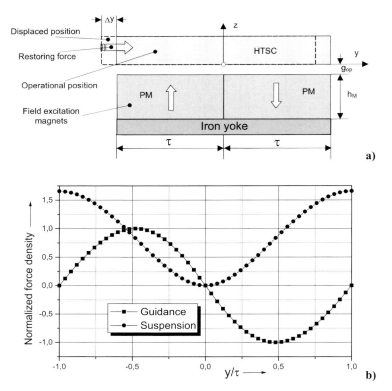

Figure 11.9 a Periodic flat magnet arrangement; b Guidance and suspension force densities as a function of the lateral displacement for $g_{act} = g_{op} = 2.5$ mm. The HTSC has been cooled in the symmetrical position $\Delta y = 0$

This effect can be used to improve the quality of axially loaded cylindrical radial bearings (see Sections 11.3.4 and 11.6.2). Because of its weight, the shaft of such an SMB is axially displaced after the superconducting hollow cylinder is cooled. The main component of the resulting restoring force points in the axial direction, but there is additionally a radial force component, which improves the radial stiffness.

It should be emphasized that the cross link between a normal force component in an SMB and a lateral displacement of the PM can be simply explained by the linear superposition of the force components of the attractive force vector.

11.3.3
Maximum Field Cooling Mode

For a suspended superconductor hanging *under* a PM, an *attracting force* is required to compensate its weight (Figure 11.1b). The largest attractive force is obtained by cooling the superconductor in a position close to the magnet (Maximum Field Cooling, MFC). After the cooled superconductor is left free, the gap between the PM and the superconductor increases. Again, the operational position is self-adjusting according to the load and the field profile.

The attractive force-vs-distance characteristics in the case of MFC is shown in Figure 11.7b. The stiffness decreases progressively with increasing gap. The absolute value of the attractive force achieves its maximum at g_{max}, where the stiffness becomes zero. Therefore, g_{max} is the largest gap for any stable suspension. Also in the MFC mode, force components and stiffness in the vertical direction are generated by a lateral displacement of the superconductor.

11.3.4
Magnet Systems for Field Excitation in Superconducting Bearings

In an optimally designed excitation system, a set of permanent magnets or current-carrying coils of an electromagnet should be arranged to yield a maximum force density and stiffness if interacting with the superconductor counterpart of the SMB.

Levitation Stabilized in Two Dimensions
Most SMBs are used for contact- and friction-less motion in one direction, e.g., for linear transport or rotating magnetic bearings. An appropriate excitation arrangement must provide field gradients in *two dimensions* and a constant field in the third dimension. Two examples of such SMBs are presented schematically in Figures 11.10 and 11.11 together with the associated field distributions.

The arrangement shown in Figure 11.10 consists of a number of alternating magnetized poles fixed on an iron yoke. In practice, each of the three poles is realized by stacking a number of cuboidic permanent magnets of high performance, because such magnets (NdFeB or SmCo alloys) can be manufactured only of limited size. They are magnetized normal to both the iron back yoke and the desired direction of motion. The field which is generated by this extended linear flat excitation system is shown in its projection on the cross-section (Figure 11.10b) and along the direction of motion (Figure 11.10c).

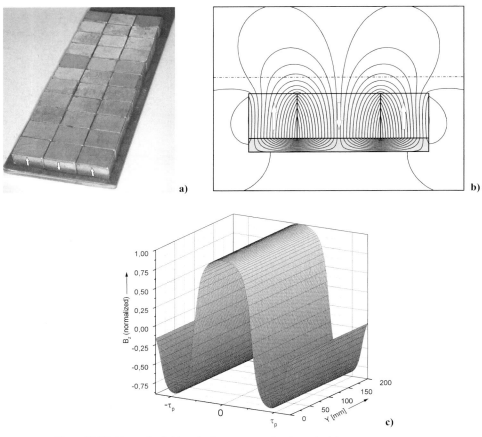

Figure 11.10 Linear flat field excitation system: **a** Design principle, each pole consisting of a stack of individual PMs; **b** Field distribution in the cross-section of the excitation system; **c** normal component of the flux density along the dash-dotted line in Figure 11.10b and the direction of motion

In a similar way, the cylindrical excitation system shown in Figure 11.11 is constructed by stacking hollow cylinders from permanent magnets which can be manufactured in pieces up to 200 mm in diameter.

Figure 11.11 Cylindrically shaped excitation system derived from a circular bending of the linear system from Figure 11.10

Levitation Stabilized in Three Dimensions

Stabilization in three dimensions is necessary to fix equipment precisely in a certain position in relation to its surrounding without direct mechanical contact. This condition is consistent with vanishing degree of freedom for any motion. It can be achieved by activating the bulk HTSC in the presence of an array of magnets with field gradients in three dimensions. The planar excitation system shown in Figure 11.12 consists of cuboidic permanent magnets which are magnetized in alternating directions, forming a chessboard-like pattern of magnetization. These

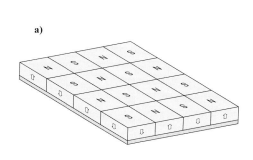
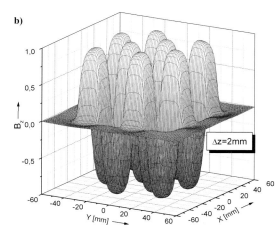

Figure 11.12 Planar field excitation systems composed of a chessboard arrangement of alternating vertically magnetized PM exhibiting field gradients in all three dimensions. The arrows indicate the directions of magnetization of the permanent magnets: **a** Design; **b** Distribution of the normal field component of the excitation system measured in a plane 2 mm above the top surface

magnets are mechanically fixed on an iron plate. The field generated by such an array has steep gradients in the two directions of the plane parallel to the surface of the magnets. An example of this type of SMB is demonstrated in Section 11.6.1.

Arrays with Flux Concentration
Manifold variations of shape and magnitude of the magnetic field in the excitation arrangement can be achieved by incorporating tiles of soft iron. Because of its outstanding magnetic conductivity, they are used as a "flux collector". This is demonstrated in Figure 11.13, where the flux in the permanent magnet has been turned through 90° with respect to the flat arrangement of permanent magnets in Figure 11.10.

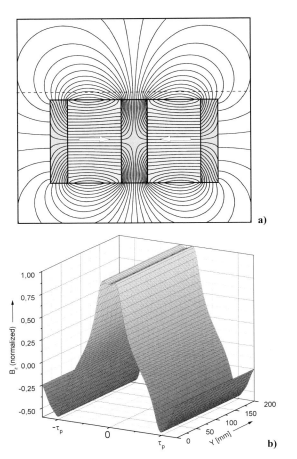

Figure 11.13 Linear field excitation systems composed of horizontally magnetized magnets placed between iron yokes (flux concentration arrangement): **a** Field distribution of the excitation system; **b** Normal component of the flux density along the dashed line (in Figure 11.13a) and the direction of motion

11.3 Force Activation Modes and Magnet Systems in Superconducting Bearings

In comparison to the flat arrangements shown in Figures 11.10 and 11.11, a significant increase of the flux density is achieved using this flux-collecting assembly. The performance of flux-collecting magnet systems can be further improved by the reduction of stray fields (see Section 11.4.1).

Flux concentration arrangements are often used in cylindrically shaped excitation systems (see Figure 11.14a). A realized system composed of ring-shaped permanent magnets and iron poles is shown in Figure 11.14b. This option has been applied beneficially in rotating cylindrical bearings, which are introduced in Section 11.6.2. In the radial SMB presented in Figures 11.37 and 11.39, an HTSC tube surrounds the cylindrical flux-collecting system, which is integrated in the rotating shaft of these machines.

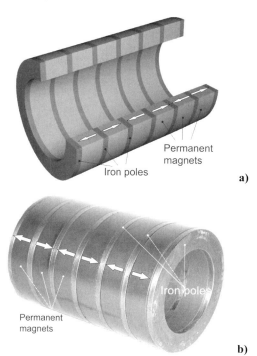

Figure 11.14 Cylindrically shaped flux-collecting system. This typical configuration is composed of a stack of ring-shaped permanent magnets separated by iron shims. Neighboring PMs are oppositely magnetized similarly to the linear configuration shown in Figure 11.13: **a** Schematic view; **b** Photograph of a flux-collecting system with 7 iron poles.

Electromagnetic Excitation Systems

Alternatively to the excitation of the SMB by permanent magnets, an electrically fed excitation system can be applied to realize flux densities up to 1.8 T in the pole region. All activation modes demonstrated for permanent excited systems can also be realized with electrically fed systems, which however require considerable wattage.

A small example of an electrically operated excitation system is shown in Figure 11.15. The electromagnetic field for levitation is generated by a stationary horizontal coil. The long stator of the linear drive is assembled in the central part of the track. Bulk YBCO superconductors arranged in the moving carrier are cooled by liquid nitrogen.

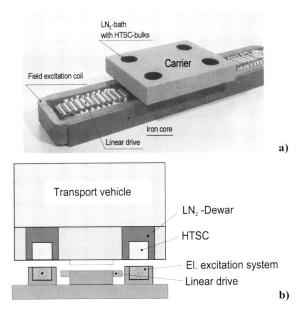

Figure 11.15 Demonstration of linear transport with bulk HTSC levitated in the magnetic field of a stationary horizontal coil; **a** Experimental assembly; **b** Schematic view

11.3.5
Force Characteristics

Analysis of force characteristics by numerical modeling is both a helpful guide through experimental results and an important tool for designing bearing details. Despite this, consideration is focused on the properties of the HTSC and permanent magnetic material (PM) which are used in the excitation systems. In this context, a value of 1.0 T has been presumed for the remanence of PM in the presented calculations. Two activation modes, namely Zero-Field Cooling and Maximum-Field Cooling will be considered. Details of the Operational Field Cooling Mode will be discussed in the subsequent sections. Only results are considered here, while the numerical methods can be found in Section 11.8.

Zero-Field Cooling

Although the lateral stiffness resulting from activation by Zero-Field Cooling (ZFC) is too small for applications on the technical scale, it is worth consideration in order to give a semi-quantitative description of the main parameters which influence the properties of SMBs. To introduce the features of force characteristics of SMBs more generally, we will consider the periodic magnetic systems shown in Figure 11.16.

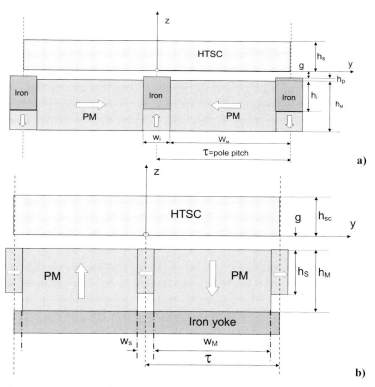

Figure 11.16 Layout and notations of periodic excitation systems. Flux concentration system (**a**) and flat-magnet system (**b**) with stray field compensation magnets in each case (see Section 11.4.2)

ZFC for a periodic flat excitation system is illustrated in Figure 11.17. Supercurrents are induced in the superconductor while it approaches from infinity (a) to a position near the magnets (b). The flux is concentrated in the gap between the HTSC and the permanent magnet, assuming an infinitely large critical current density. The characteristic relationship between *force* and *gap* for a flat excitation system has already been shown in Figure 11.7b. The force in a flat excitation system significantly depends on the pole pitch (see Figure 11.16b). The relation between *force density* and *pole pitch* is shown in Figure 11.18 for several values of the operational air gap. For a gap of 2 mm, the optimal pole pitch is found to be

30 mm, shifting with increasing gap to higher values. The repulsive force density decreases monotonously with increasing gap as shown in Figure 11.19. The slope of this dependence indicates that for small gap values $g < 2$ mm, the vertical stiffness can be significantly enhanced by a design with a smaller pole pitch.

The corresponding analysis has also been applied to excitation systems with flux concentration (design in Figure 11.16a). A maximal *repulsive* force density of 17 N cm^{-2} was obtained, which is about half as large as the value of 35 N cm^{-2} found for the flat excitation system (see Figure 11.19) when the same volume of magnet material is assumed. Otherwise, the thickness of the permanent magnet has to be increased in the considered example from 16 to 20 mm to achieve the same force density for the flux-collecting excitation system. However, the outstanding advantage of flux concentration systems is their significant increase of lateral stiffness.

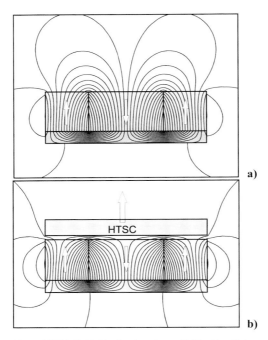

Figure 11.17 Field distributions in an SMB with a flat magnet excitation system in the case of ZFC: **a** The HTSC-bulk is placed remote from the magnets for cooling, **b** The HTSC-bulk has approached from infinity to a position near to the surface of the magnets. The arrow marks the repelling force

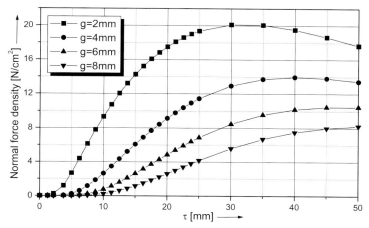

Figure 11.18 Repulsive force density vs pole pitch for a periodic flat magnet arrangement after a ZFC activation. Basic data (see Figure 11.16b): $h_M = 16$ mm, $w_s = 0$, $B_r = 1.0$ T

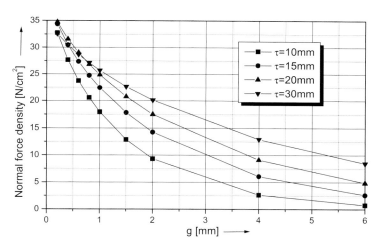

Figure 11.19 Repulsive force density vs air gap for a periodic flat magnet arrangement after a ZFC-activation. Basic data (see Figure 11.16b): $h_M = 16$ mm, $w_s = 0$, $B_r = 1.0$ T

Maximum Field Cooling Activation

Maximum field cooling (MFC) is the choice for *hanging suspension*. In this case, the magnetic flux of the magnet completely penetrates the superconductor by field cooling. *Attractive forces* are generated while the superconductor is shifted to the position of its operation (Section 11.4). The result is that the field distribution within the HTSC remains almost constant and is compensated by surface currents, whereas a part of the flux is partially redistributed from the space of the excitation system to the opening gap. In contrast to the ZFC case, the *vertical attractive forces* in the periodic flux concentration design are *larger* than they are for the periodic flat arrangement. The *optimal pole pitch* for a given gap is much smaller in the MFC case than in the ZFC case.

The *vertical stiffness* is determined by the slope of the *attractive* normal force density as a function of the gap, which is illustrated in Figure 11.20. The figure indicates a rather high stiffness for small pole pitches (< 10 mm in the considered configuration). On the other hand, it is evident that stable suspension is strictly limited by that gap at which the attractive force achieves its maximum.

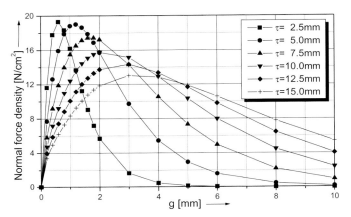

Figure 11.20 Attractive force density vs air gap of an SMB for a flux concentration excitation system for MFC-activation. Basic data (see Figure 11.16a): $h_M = 20$ mm, $w_M/\tau = 0.8$, $h_i/h_M = 1$, $h_p = 0$, $B_r = 1.0$ T

11.4
Optimized Flux Concentration Systems for Operational-Field Cooling (OFCo)

Operational-Field Cooling with an offset (OFCo) is the most effective mode by which to activate the superconductor of an SMB. This procedure has a number of interacting parameters, which have to be optimized for an individual setup. Specified constructions have to account for static and dynamic aspects of the machine under consideration. Numerical model calculations used to optimize a superconducting magnetic bearing with respect to its specific dimensions and the field shaping can be performed by methods given in Section 11.8. In this section, some guidelines with respect to proper dimensions of components and field shaping are discussed for a *periodic SMB with flux concentration and OFCo activation*. The basic design of the considered flux-collecting system is shown in Figure 11.16a. Included in this design are special features to reduce the stray field. Stray fields are the first topic of this chapter before the optimization of the parameters of the considered SMB are discussed.

11.4.1
Stray Field Compensation

Stray fields are an inherent accompaniment of a flux collecting arrangement, and thus conduct a significant part of magnetic flux into the back space. This is illustrated in Figure 11.21a for the case where the iron collectors have the same height as the permanent magnets ($h_i/h_M = 1$). The stray field into the back space can be reduced if the iron collectors are shortened at their lower sides, as shown in Figure 11.21b. A further reduction of the stray field can be achieved if a permanent magnet is introduced into the field path of the iron pole as shown in Figure 11.21c. This arrangement with permanent magnets compensating the stray field is included in the basic design of the periodic excitation systems shown in Figure 11.16a.

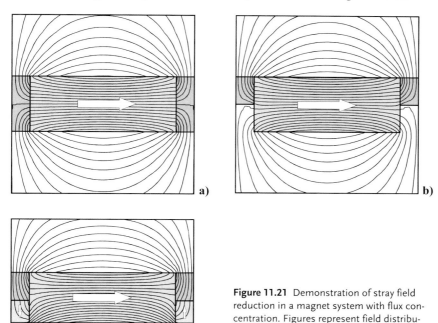

Figure 11.21 Demonstration of stray field reduction in a magnet system with flux concentration. Figures represent field distributions for **a** conventional flux concentration arrangement ($h_i = h_M$), **b** reduced height of the iron poles ($h_i < h_M$), and **c** additional introduction of permanent magnets into the stray field path of the iron pole.

11.4.2
Dimensional Optimization of System Components

The starting parameter for the parameter optimization of the considered SMB is the *gap* between the PM (excitation system) and the HTSC. A minimal gap clearance is desirable. The limits for activating and operational gaps are set by the wall thickness of the cryostat (~2 mm) and other parameters of operation. Assuming a range of displacement of + 0.5 to − 0.5 mm and an offset for OFCo activation of 1 mm, one obtains as smallest value g_{op} = 2.0 mm for g_{act} = 3.5 mm. The field distribution calculated for the considered system is represented in Figure 11.22.

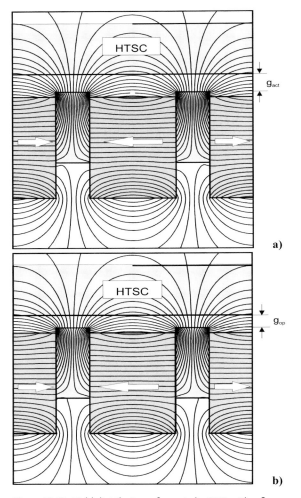

Figure 11.22 Field distribution of a periodic SMB with a flux concentration excitation arrangement: **a** HTSC in the cooling position at a distance of g_{act} = 3.5 mm; **b** The HTSC is approached to the excitation system. Operational air gap: g_{op}=2.5 mm

Typical parameters which influence the performance of an SMB with flux concentration (see Figure 11.16a) are the *pole pitch* τ and the *width w_i of the iron collectors*. These parameters are considered later with respect to an optimal layout of a periodic SMB for use under OFCo activation mode.

11.4 Optimized Flux Concentration Systems for Operational-Field Cooling (OFCo)

Pole pitch τ

The influence of the pole pitch on the lateral and normal stiffness density is shown in Figure 11.23 for the considered SMB. The optimal values for the pole pitch of about 11.5 mm (normal stiffness) and 13 mm (lateral stiffness) do not coincide. It seems to be reasonable to operate close to the maximum lateral stiffness (because its values are lower than the normal stiffness) and to use, for example, a pole pitch of 12 mm.

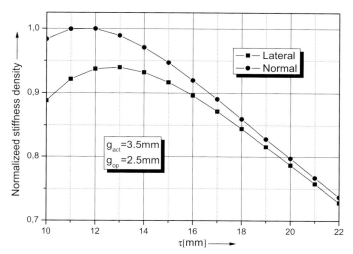

Figure 11.23 Normal and lateral stiffness as a function of the pole pitch τ for a periodic SMB with flux concentration arrangement. Design parameters: $h_M = 20$ mm, $h_p = 1.0$ mm, $w_i/\tau = 0.28$, $h_i/h_M = 0.65$

Width of iron collectors

In Figure 11.24, the stiffness densities are plotted against the iron collector width for a pole pitch of 12 mm. Both curves have a maximum at the same optimum iron width of $w_i = 1.2$ mm. For such a small width of the iron collector, saturation effects in the iron parts can only be excluded if permanent magnets of reduced remanence flux density are used. For permanent magnets of high remanence, wider iron collectors are recommended to avoid saturation effects. An iron width of $w_i = 2.4$ mm has been used for the further optimization.

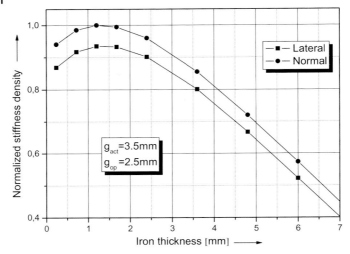

Figure 11.24 Normal and lateral stiffness as a function of the iron width w_i for a periodic SMB with flux concentration arrangement. Parameters: $h_M = 20$ mm, $h_p = 1.0$ mm, $w_i/\tau = 0.28$, $h_i/h_M = 0.65$, and $\tau = 12$ mm as the design parameter so far optimized

Height of iron collectors
The influence of the height h_i of the iron collectors on normal and lateral stiffness has been evaluated in Figure 11.25 for $w_i = 2.4$ mm using as x coordinate the relative iron height h_i/h_M where h_M is the height of the permanent magnets. Maxima of both lateral and normal stiffness were found to occur at an optimum iron height of $h_i/h_M = 0.75$. The stiffness increases by more than 10% if the height of the iron collectors is reduced from $h_i = h_M$ to $h_i = 0.75 h_M$. Here, the iron collectors are shortened at the lower side of the SMB as shown in Figure 11.16a. This improvement is achieved by a reduction of the stray field, as mentioned in Chapter 11.4.1.

In order to compensate for the imperfection of technical permanent magnets, the iron collectors should be slightly shifted into the gap space as shown in Figure 11.16a. The *iron pole* height h_p is that part of the total height of the iron collectors which freely extends into the gap space. Both lateral and normal stiffness decrease linearly (by 8% per mm) with increasing h_p. Typical values of h_p used in SMBs are in the range < 1 mm.

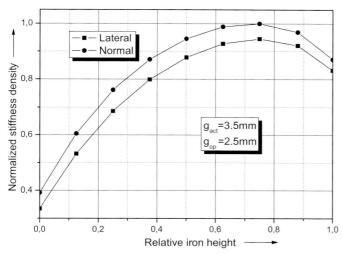

Figure 11.25 Normal and lateral stiffness of a periodic SMB with flux concentration arrangement as a function of the relative height h_i/h_M of the iron collector, where h_M is the height of the permanent magnet. Parameters: $h_M = 20$ mm, $h_p = 1.0$ mm, and the optimized parameters so far: $\tau = 12$ mm and $w_i/\tau = 0.2$

11.5
Parameters Influencing the Forces of Superconducting Bearings

11.5.1
Critical Current Density

In the previous sections, levitation forces and the stiffness of SMBs have been calculated for ideal superconductors exhibiting infinitely large critical current densities j_c. Under this condition, any field variation within the HTSC is suppressed completely by currents induced in a surface layer of the HTSC. An example has been shown in Figure 11.4, where the field distribution within a Field-Cooled HTSC (Figure 11.4a) remained unchanged after the permanent magnet above the HTSC had been vertically and laterally displaced (Figure 11.4b).

In real HTSCs, however, j_c has a finite value, and levitation forces become lower than those for ideal HTSCs with infinitely large j_c. This is demonstrated in Figure 11.26. In Figure 11.26 a and b, the field penetration in an HTSC cylinder after Zero-Field Cooling is compared for $j_c = 100$ A/mm² and infinitely large j_c, respectively. The levitation force normalized to its maximum value F_{max} for infinitely large j_c is shown in Figure 11.26c as a function of the critical current density. The levitation force strongly decreases with j_c and becomes smaller by a factor of two for $j_c = 25$ kA cm^{-2} compared to the force F_{max} achievable with an ideal superconductor.

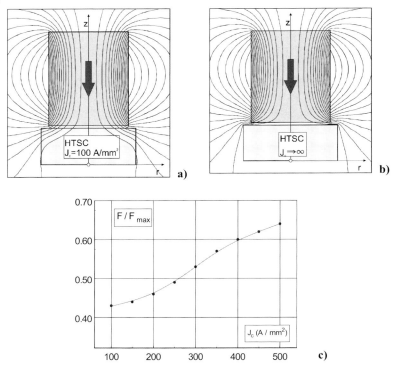

Figure 11.26 Influence of the critical current density j_c in bulk HTSC on field distribution and levitation force in an SMB: **a** Field distribution after ZFC for $j_c = 100$ A mm^{-2}; **b** Field distribution after ZFC assuming infinitely large j_c; **c** Normalized levitation force vs critical current density. The force is normalized to its value F_{max} for an HTSC with infinitely large j_c

11.5.2
Temperature

Because critical currents of HTSCs strongly depend on temperature, levitation forces would also be expected to depend on temperature. The temperature dependence of the levitation force of an SMB has been investigated using an excitation system as shown in Figure 11.27a. It consists of two 200-mm long permanent bar magnets between iron collectors with a pole pitch of 17.5 mm. A cluster of 4 HTSC blocks, each with lateral dimensions of 33 mm × 33 mm and a height of 10 mm, fixed in the cryostat was cooled to the measuring temperature before the permanent magnets were caused to approach the superconductors. The vertical component of the repelling force was measured at a distance of 3 mm.

It is obvious from Figure 11.27b that the levitation force increases considerably at operation temperatures below 77 K (cf. Section 11.5.1). The levitation force was found to increase by 90% on cooling from 80 to 50 K or by 20% by reducing the temperature by only 3 K from 80 to 77 K.

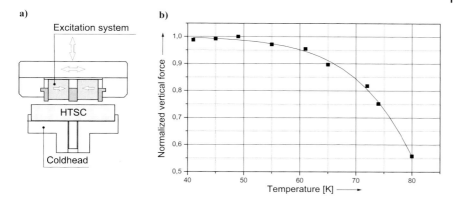

Figure 11.27 Influence of temperature on normal levitation forces: **a** Sketch of the equipment for 3D force measurements. The excitation system can be moved by stepping motors. The HTSC cluster is placed on a coldhead inside a Dewar. The coldhead is in thermal contact with the cryocooler via Cu blocks. **b** Influence of the temperature of the bulk HTSC on the repelling forces after ZFC activation (g_{op} = 3 mm)

11.5.3
Flux Creep

Magnetic relaxation in HTSC due to thermally activated flux motion causes an approximately logarithmic decay of the magnetization M with time t. The relaxation rate $S = -(1/M_o)dM/d(\ln t)$ of YBCO is, with $S \approx 0.08$ at 77 K, relatively large (see Section 1.1.6). Flux creep reduces the trapped field in the HTSC and obviously also causes a temporal degradation of the levitation force in SMBs. The levitation force F, which is proportional to the magnetization of the HTSC, would be expected to decay at the same rate as the magnetization, i.e. $S = -(1/F_o)dF/d(\ln t)$.

The relaxation of the levitation force in bulk YBCO has been investigated at different temperatures [11.33–11.36]. Experimental data for the levitation force-vs-time relationship between a YBCO disk and an Sm-Co ring magnet [11.35] are shown in Figure 11.28. After field cooling the YBCO disk at different initial distances below the PM, the distance was enhanced to a common operational distance of 5 mm. Then, the time dependence of the levitation force was measured. The data shown in Figure 11.28a are normalized to the levitation force F_o measured after 10 min. The time dependence of the levitation force follows the expected logarithmic decay. The decreasing relaxation rate observed for increasing cooling distance reflects different penetration of magnetic flux in the superconductor. The characteristic penetration length δ becomes smaller as the initial applied magnetic field decreases due to the large cooling distance, as illustrated in Figure 11.28b. Obviously, the highest relaxation rate would be obtained in the case of full penetration.

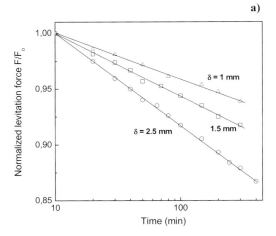

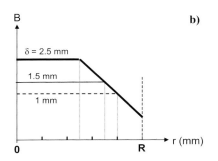

Figure 28 a Normalized attractive force vs time between a YBCO disk and an Sm-Co ring magnet measured at 77 K for different penetration length δ as indicated in the figure. Initial force F_o = 260 mN (circle), 205 mN (square) and 150 mN (triangle) (taken from Ref. [11.35]); b Sketch of the field profile within the YBCO disk (radius R = 5 mm) after field cooling of the superconductor at a distance g_{act} = 0 (thick line), 4 mm (thin line) and 4.5 mm (dashed line) from the ring magnet and positioning it at an operational distance of g_{op} = 5 mm; c Field profile with two opposite field gradients (thick line). The near-surface layer with reverse field gradient disappears after a certain time because of flux creep (dotted line).

Flux creep in superconductors can be strongly reduced by lowering the temperature after the critical state in the superconductor has been established. This procedure and experimental data have been described in Section 4.3.2. Alternatively, flux creep can be suppressed if opposite field gradients are established within the superconductor by reversing the applied magnetic field [11.35]. An example of such a field profile is shown in Figure 11.28c. For a field distribution with a near-surface layer having an opposite field gradient, the vortices diffuse to the region with a smaller vortex density, i.e. vortices flow from the volume to the near-surface layer rather than to the surface of the superconductor. Therefore, flux creep is strongly suppressed as long as the near-surface layer preserves its opposite field gradient. The near-surface layer tends to shrink because of flux creep, and finally the opposite field gradient completely disappears, as illustrated in Figure 11.28c (dotted line). Then, the magnetic relaxation proceeds as normal across the surface of the superconductor.

It has been demonstrated that the relaxation of the levitation force in a YBCO disk can be almost completely suppressed during a certain period of time [11.35]. In the experiment described by Smolyak et al. [11.35], this time window had a duration of up to about 4 h. It has also been shown that this time window of suppressed force relaxation increases exponentially with the width of the near-surface layer having an opposite field gradient.

11.5.4
HTSC Bulk Elements Composed of Multiple Isolated Grains

So far, the bulk HTSC of the SMB has been considered to consist of one single domain. The behavior of composed bulk HTSCs was of interest in the pioneering stage of this material because of the very restricted grain size in the inferior samples available at this time. Now, interest in arrays of bulk HSTCs is again rising since power machines of larger dimensions are under development. The subsequent force analysis in SMBs refers to arrays of bulk HTSCs consisting of a certain number of isolated grains, which are assumed to be aligned along their c axes. The influence of size and critical current density on the levitation force in the case of Zero-Field Cooling and Field Cooling will be discussed.

Zero-Field Cooling

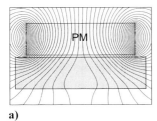
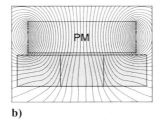

a) b)

Figure 11.29 Field distribution (based on 3D field calculation) of an SMB (consisting of PM and bulk HTSC) after ZFC at an operational gap of 1 mm: **a** Single-domain HTSC cuboid; **b** Array of 3 × 3 HTSC grains

The field distribution in an SMB after Zero-Field Cooling is shown for a single-domain HTSC (Figure 11.29a) and an array of 3 × 3 isolated HTSC grains covering the same area (Figure 11.29b). In both cases, the same critical current density j_c was assumed. The j_c value corresponding to full flux penetration in the single domain HTSC was used, i.e. $j_c \propto H/R$, where H is the applied magnetic field and R the radius of the single-domain HTSC. The critical current $I_c = j_c \times$ (active cross-section) depends on the grain size and significantly decreases in the small grains exhibiting the same j_c as the single domain HTSC. Therefore, the local currents in the outer regions of the small grains are strongly reduced, and flux penetrates the center of the HTSC as shown in Figure 11.29b. Consequently, a drastic drop in the levitation force with decreasing grain size is expected.

The dependence of the repulsive force on the number $n \times n$ ($n = 1,2,3,4$) of isolated grains is shown in Figure 11.30 for two data sets. The data for low j_c (defined by the full flux penetration in the single-domain grain) are normalized to the levitation force of the single-domain grain. Roughly, the levitation force decays hyperbolically with $1/n$.

The magnetic field of the permanent magnet will penetrate the single-domain HTSC only partially if its critical current density j_c is enhanced. This case is considered in Figure 11.30 for fourfold enhanced j_c. The calculated levitation force of the single-domain grain is more than twice as large as that for the lower j_c. Moreover, the data set for high j_c (partial flux penetration) shows a much weaker force decay with increasing number of grains than the data set for low j_c (full penetration).

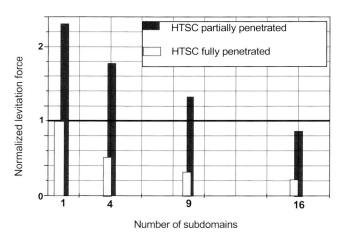

Figure 11.30 Normalized repulsive force of an SMB vs number n of isolated subdomains in an $n \times n$ HTSC array after ZFC activation. Two data sets are shown: low j_c (fully penetrated single-domain HTSC) and fourfold higher j_c (partially penetrated single-domain HTSC).

Field Cooling

The influence of the grain size of HTSC subdomains on the restoring force has been analyzed for a cylindrical SMB with flux-collecting excitation system [11.18]. Its geometry is shown in Figure 11.31. A stack of permanent magnets with flux-collecting iron shims between the magnets is fixed to the rotor shaft. The pole pitch p is given by the distance of neighboring permanent magnets. The rotor is surrounded by an HTSC hollow cylinder which is composed of single-domain rings (height h, width h) with $h < p$. These rings are separated from each other by small air gaps to prevent intergrain currents. The HTSC hollow cylinder is divided up to five times per pole pitch, which means that for $p/h = 5$ the HTSC hollow cylinder is fragmented into $5 \times 5 = 25$ rings per pole pitch.

For the calculation, it is assumed that the rotor shaft was lifted to its "freezing" position shown in Figure 11.31 before cooling the HTSC cylinder to 77 K. The restoring force has been calculated for a radial displacement of the rotor shaft from its initial symmetrical position using a finite-element method.

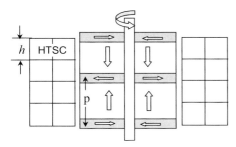

Figure 11.31 Geometry of the SMB for a ratio $p/h = 2$. The HTSC hollow cylinder is divided in 2×2 rings per pole pitch p

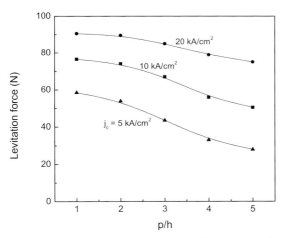

Figure 11.32 Calculated restoring force at ± 0.5 mm radial displacement for the SMB in Figure 11.31 as a function of the ratio between pole pitch p and grain size length h for three values of the critical current density of the bulk HTSC. Data taken from Ref. [11.18]

In Figure 11.32, the calculated restoring force at ± 0.5 mm radial displacement is shown as a function of the ratio p/h for several values of the critical current density of the bulk HTSC. As expected, the restoring force decreases with decreasing size of the single grains (expressed by growing pole pitch ratio p/h). The influence of fragmentation of the bulk HTSC on the restoring force becomes stronger for a lower critical current density. Thus, whereas for $j_c = 20$ kA cm^{-2} the restoring force of the single-domain HTSC decreases for the smallest grain size by only about 20 %, this reduction increases to 50 % for $j_c = 5$ kA cm^{-2}. This tendency is reasonable because supercurrents induced by a displacement of the rotor shaft penetrate the HTSC up to a depth $\Delta H/j_c$ from the surface of the HTSC. Therefore, for low j_c and grains with a small diameter d, the supercurrents are flowing in a larger volume fraction. Consequently, the magnetization of the superconducting grains (which is proportional to $j_c \times d$) is lowered and the resulting restoring force becomes smaller. For the same reason, the restoring force in Figure 11.32 decreases for decreasing critical current density j_c at a given ratio of p/h.

In calculating the restoring force, it was assumed that the position of the grain boundary of the multi-domain HTSC is in line with the iron shim as illustrated in Figure 11.31. Slightly enhanced levitation forces have been obtained for $p/h \leq 2$ in the case where the grain boundary of the multi-domain HTSC is shifted to a symmetrical position *between* two neighboring iron shims [11.18].

11.5.5
Number of Poles of the Excitation System

The influence of the number of poles on the normal force density was studied for an SMB consisting of a single-domain bulk HTSC and a flux-collecting excitation system with a variable number of poles n. In Figure 11.13, the considered SMB is shown for an excitation system with three poles. The forces were calculated after an OFCo activation at a distance of $g_{act} = 3.5$ mm followed by an approach to $g_{op} = 2.5$ mm. Assuming an infinitely large critical current density of the bulk HTSC, the repulsive stiffness was found to increase monotonously with the number of poles approaching a maximum value. The periodic arrangement of Figure 11.22 corresponds to $n \to \infty$.

For an HTSC with a finite critical current density (j_c), too large field gradients within the HTSC resulting from the large number of poles becomes unfavorable because supercurrents induced by a relative displacement of the excitation system against the HTSC are limited by j_c. It is to be expected that the levitation force and stiffness become maximal for an optimum number of poles generating field gradients within the HTSC, which should be roughly equivalent to j_c. To determine the optimum number of poles, one has to take into account the gap, because the resulting field gradients strongly depend on the distance between the excitating system and the bulk HTSC. For a large air gap in the range 4–5 mm, which may be necessary by the thermal insulation of the superconductor, the optimum number of poles would be expected to be less than that for a small air gap [11.19].

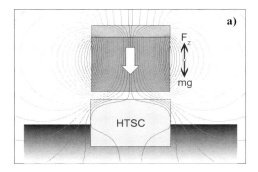
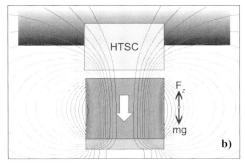

Figure 11.33 Flux distribution for arrangements with inherently stable force interactions between PM and HTSC bulks. The excitation system is composed of a single PM with an iron back yoke, with **a** the PM suspended above an HTSC by repelling normal forces (ZFC or OFCo activation) and **b** the PM hanging beneath the bulk HTSC by attractive forces (MFC or OFCo activation)

11.6
Applications of Superconducting Bearings

An SMB should have sufficient stiffness to ensure that external static and dynamic forces will be counterbalanced to hold the displacement within the specified limits. Two principles of construction have to be distinguished: a load can be levitated in an inherently stable fashion by repulsive forces (Figure 11.33a) or suspended in a hanging position by attractive forces (Figure 11.33b). Furthermore, the SMB has to counterbalance the weight of the mobile body. This is advantageously realized using the OFCo mode, i.e. by a displacement from the activation position (see previous sections).

The mobile part of an SMB has in most cases one degree of freedom. Examples are SMBs for rotary motion and linear transportation. Stationary SMBs are designed to levitate equipment in a fixed position.

The moving part of an SMB consists of either the field excitation system (usually high-performance permanent magnets) or the HTSC (arranged in a Dewar). Exceptionally, both parts can rotate in contactless magnetic clutchs (see, e.g., Ref. [11.63]). In this case the effective motion is represented by the appearent slip. Passive SMBs have also been constructed for possible use in equipment on satellites under microgravity [11.54], [11.55]. Of course, load compensation is not necessary in this environment.

Superconducting magnetic bearings for different tasks, including appropriate designs of excitation systems and the bearing characteristics will be discussed in the next Sections.

11.6.1
Bearings for Stationary Levitation

The suspension of equipment or containers with sensitive contents in a fixed position is a characteristic task to be solved by a stationary contactless bearing. This has been demonstrated in the case of a cryotank for liquid hydrogen (LH_2) for application in automobiles etc., which was levitated by an SMB.

Stationary SMB for Cryocontainer

The hydrogen tank shown in Figure 11.34a consists of an exterior vessel and a cold internal tank which is thermally insulated from the exterior tank by a vacuum in the interspace and additional radiation shields. The SMB replaces the suspension struts made of glass fiber-reinforced plastics which are used in a conventional design to fix the inner vessel. These suspension struts cause most of the undesired heat input to the inner tank and consequently the evaporation losses.

In the design shown in Figure 11.34, the HTSC blocks are arranged in the inner tank facing strong permanent magnets which are mechanically fixed inside the outer container [11.20]. The superconductors are cooled by thermal contact with the cold inner tank containing the liquid hydrogen. In this particular case, the heavy vessel has to be stabilized in all 6 degrees of freedom and the bearings have

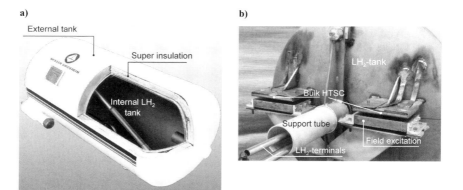

Figure 11.34 Mobile LH_2 tank suspended by a planar SMB. General view (**a**) and details of the planar bearing suspending the inner tank (**b**) (taken from Ref. [11.20])

to be arranged at well-defined fixing points. This has been realized by all together four individual planar SMBs, as shown in Figure 11.34.

A drastic reduction in the heat transfer from the surroundings to the tank was achieved using this SMB. Unfortunately, this planar SMB design is space consuming. Furthermore, to activate the SMB, the excitation system has to be moved down in the warm state and lifted up into the working position after cooling. An improved solution is considered in the next Section.

Cylindrically Shaped SMB for Cryocontainer

The use of cylindrically shaped SMBs affords a better alternative for supporting the cryocontainer. Two such SMBs were positioned inside a central thin-walled tube, as shown in Figure 11.35 [11.20]. The excitation system was cylindrical, and contained a central shaft fixed to the outer container. The hollow YBCO cylinders of these SMBs were glued directly onto the inner wall of the central bore of the tank (Figure 11.35).

The force activation is initiated by a memory metal actuator in the completely sealed cryocontainer during the filling process [11.22]. This actuator is extended in the warm state, thus lifting the inner tank. This brings the excitation system in contact with the lower part of the HTSC cylinder. Otherwise, when the tank is filled with LH_2 the memory metal spring is contracted and allows the inner tank to be suspended by the SMB in a stabilized position, where it is thermally insulated. This design is much more compact than that of Figure 11.34 and gives fully automatic activation.

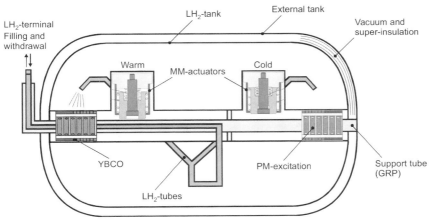

Figure 11.35 Cross-sectional view of an LH_2 container in a double-wall design with two radial SMBs and temperature-controlled force activation actuators [11.20]

11.6.2
Bearings for Rotary Motion

Basic Construction

Cylindrical SMBs for rotary motion can be used for high-speed rotating axes [11.37]. In the designs sketched in Figure 11.36, the descriptions planar, cylindrical, and conical are applied analogously to mechanical ball bearings. Forces and stiffness are generated in SMBs by displacements from the activation position. *Normal forces* in *radial* ("journal") *bearings* are caused by *radial displacements* of the axis. They are caused in the *axial* ("thrust") *bearings* by *axial displacements* of the axis [11.39]. On the other hand, *tangential forces* caused by a *lateral displacement* of the shaft are directed axially in radial bearings and radially in axial bearings.

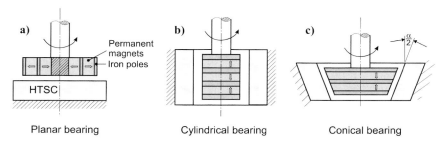

Planar bearing   Cylindrical bearing   Conical bearing

Figure 11.36 Principal types of SMB with a flux-collecting excitation system for rotary motion

The choice of the appropriate design depends primarily on the orientation of the body to be suspended, the requirements concerning the lift and guidance forces, and the corresponding stiffness. Variation of the apex angle for the conical SMB design in Figure 11.36 [11.21] enables the bearing performances to be adjusted to suit different force and stiffness demands [11.38].

It is important to recognize that for all SMB designs the axial stiffness is superior to the radial stiffness. Therefore, the use of vertical axes should be favored.

SMBs for Rotating Machines

Figure 11.37 shows an SMB designed for a test turbo machine for operation at 12 000 rpm [11.23]. It consists of a cylindrical field excitation system of stacked PM cylinders, which encloses the rotating shaft within the warm bore of a Dewar. A number of YBCO rings are assembled to form the hollow HTSC cylinder arranged on the inner wall of a copper tube, which is thermally connected to the coldhead of a cryocooler. The copper tube and the internal GRP tube in Figure 11.37 form a hermetically sealed enclosure, while a superinsulation foil ensures the thermal insulation of the bore of the Dewar.

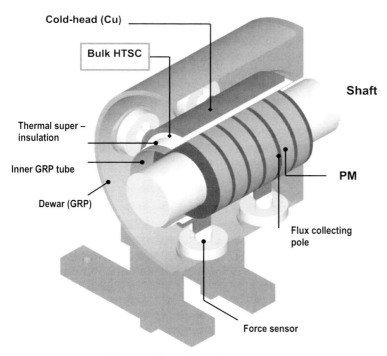

Figure 11.37 Design of the SMB for a turbo machine [11.23]. The bearing is integrated into a Dewar. The axis with the excitation system is placed in the warm bore

The application of a bulk HTSC in a contactless magnetic clutch is shown in Figure 11.38. The driver unit consists of an electric motor and the bulk HTSC in a cryostat. Integrated in the secondary shaft is a freely rotating plastic impeller which includes the permanent magnets (Figure 11.38a). The apparatus enables a liquid to be intensively stirred in a completely clean or sterile environment. The container materials usually considered are designed for single use, and stirring is with a hermetically sealed impeller with a built-in PM. Figure 11.38b shows a small mobile driving unit which can easily be connected to the equipment, joining each of the single-use containers without any mechanical contact and thus circulating the liquid inside.

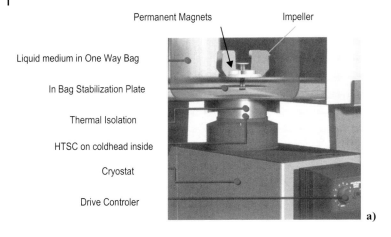

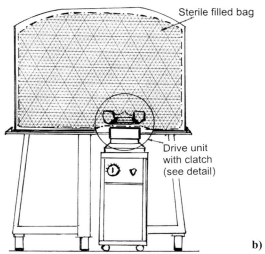

Figure 11.38 SMB for a contactless clutch connecting a separate passive impeller with the driver unit [11.63]. The detail **a** shows the electric drive with the bulk HTSC in a cryostat, while the complete system ready for use is shown in **b**

SMB Designed for Superconducting Motors

The construction of superconducting motors with a power of several 100 kW ([11.43], [12.13]) has stimulated the development of appropriate SMBs for such motors [11.30], [11.41], [11.44]. Radial bearing forces of several kN and a stiffness in the order of kN mm^{-1} are required to stabilize their rotors.

The main attribute of an SMB designed for a *400 kW model motor* with a superconducting winding is a YBCO hollow cylinder 203 mm in diameter consisting of a number of segments made from melt-textured material [11.30], [11.41], [11.42], [11.43]. The total length of the SMB is 250 mm. The superconducting cylinder is cooled indirectly by a copper tube which is flushed by liquid nitrogen. The application of reduced pressure enables temperatures even below 77K to be achieved. The rotor, consisting of 26 Nd-Fe-B ring magnets (outer diameter 200 mm, inner diameter 150 mm) and iron shims between them, had a weight of 500 kg. It is interesting to note that each of these large ring magnets was manufactured in one piece. Measurements of the force vs displacement at superconductor temperatures between 66 and 86 K revealed a significant influence of the temperature on the forces. A radial bearing force of 2700 N was measured at a stator temperature of 72 K for a rotation frequency of 25 Hz [11.30].

An SMB operating in a bath of liquid nitrogen at 77 K was constructed to complete a *superconducting 200 kW reluctance motor* ([12.13], Section 12.4). It consisted of a stainless steel stator housing the cylindrically ground bore of a YBCO hollow cylinder (Figure 11.39a) formed from 20 multi-seeded YBCO plates (90 mm × 35 mm) [11.44]. The gap between the YBCO hollow cylinder (diameter of inner bore 145 mm, length 90 mm) and the rotor was 1.5 mm. A flux-collecting excitation system with 7 poles was mounted on the shaft of the SMB. The Nd-Fe-B ring magnets were assembled from arc segments and protected against centrifugal forces by pre-stressed stainless steel rings. These ring magnets were separated by iron shims (Figure 11.39b).

This SMB, directly cooled by a bath of LN_2, is characterized by excellent stiffness data. The specific axial stiffness per unit area of the inner bore of the YBCO hollow cylinder is in the region of 5 N mm^{-1} cm^{-2}). The corresponding axial stiffness is about 2000 N mm^{-1}. Approximately 50% of this value was obtained for the radial stiffness [11.44]. Apart from the high-quality HTSC, this bearing benefits additionally from the high remanence of the Nd-Fe-B ring magnets at low temper-

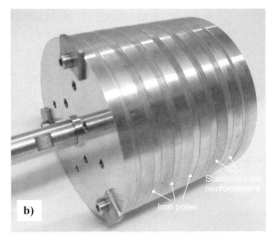

Figure 11.39 SMB operating in a bath of LN_2 (partially disassembled). **a** Stainless steel stator housing a YBCO hollow cylinder (inner diameter 145 mm) assembled from 20 multi-seeded ground YBCO plates (90 mm × 35 mm each); **b** Excitation system in a flux-collecting arrangement with 7 poles

atures. Because of the negative temperature coefficient of the remanence of Nd-Fe-B, the flux density at the poles of the SMB increases from 0.85 T to 1.06 T if the operating temperature is reduced from 300 K to 77 K. Taking into account this flux density and the measured force characteristics, an average engineering critical current density of $j_c \approx 25$kA cm^{-2} has been estimated for the bulk YBCO at the applied field [11.29].

SMB for High-Speed Rotating Flywheels

A flywheel used as an inertial energy storage system is one of the most efficient and thus promising devices that can be used for, e.g., the smoothing of energy transfer of wind farms or solar plants in the event of lack of the respective primary energy – wind or sunshine. As it is essential to keep energy dissipation as low as possible, the application of contactless operating magnetic bearings (preferably SMBs) is highly promising.

Essential parameters of different flywheel constructions with SMBs are listed in Table 11.1. As an example, the SMB developed for the German flywheel project "Dynastore" [11.25] is considered here in more detail. It is designed for use as a decentralized and mobile energy storage system with a rated energy storage of 2 MW for 20 s at a maximum speed of 12 000 rpm. The rotating ring of this flywheel is supported by a radial SMB, which is integrated into the spinning flywheel. The cross-section of the SMB is shown in Figure 11.40.

Table 11.1 Flywheel energy storage systems with SMB

Stored energy (kWh)	Diameter (mm)	Weight (kg)	Angular speed (rpm)	Bearings	Reference
0.08	394	9.3	20 400	SMB	[11.50]
0.3	294	10.3	15 000	SMB	[11.52]
0.3	296	28	20 000	SMB (horizontal axis)	[11.59]
0.48	400	37	30 000	SMB + AMB[1]	[11.47]
1.4	600	76	20 000	SMB + AMB[1]	[11.48]
10	1000	334	17 200	SMB + AMB[1]	[11.49], [11.56]
2		140	20 000	SMB+ HSMB[2]	[11.57]
10			10 000	SMB+ HSMB[2]	[11.51], [11.58]
10		500	10 000	SMB	[11.25], [11.53]

[1] AMB: active magnetic bearing
[2] HSMB: hybrid superconducting magnetic bearing

A flux-collecting arrangement is connected to the rotor. The YBCO hollow cylinder (diameter 540 mm) of the SMB is composed from trapezoidal bulk YBCO samples glued onto a copper coldhead, which is supported by a thin-walled insulating structure. This construction reduces the heat transfer from the surroundings to the coldhead and thus to the bulk YBCO samples. A radial stiffness of 2086 N mm^{-1} and an axial stiffness of 3020 N mm^{-1} have been obtained from measurements on a model arrangement of the bearing [11.53]. This is sufficient to support and stabilize the rotor, which has a mass of 500 kg.

Two general SMB concepts have been used in flywheels (see Table 11.1): the cylindrical SMB of Figure 11.36b, which was also applied in the example above and the planar thrust SMB as sketched in Figure 11.36a. The application of the latter is limited to small facilities or demonstrations because of its relatively low stiffness. Otherwise, the low specific stiffness for a given active area must be compensated for by a correspondingly large diameter of the rotating disk consisting of the permanent magnets. In this case, enormous tensile forces would act on the ring magnet during its high-speed rotation, exceeding the tensile strength of all known magnetic materials.

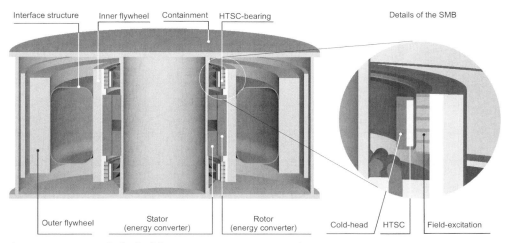

Figure 11.40 Design of a flywheel for an energy storage system with contact-free SMB in a radial construction. The flywheel is embedded within a vacuum and safety vessel. Taken from Ref. [11.25]

11.6.3
Bearings for Linear Motion

Superconducting magnetic bearings designed for linear motion require a magnetic field distribution with field gradients in two directions. Excitation systems for this particular field characteristic have been compiled in Section 11.3.

Fast and silent gliding vehicles attract attention whenever they are shown at exhibitions. Suspended toy trains – once they have been accelerated manually or by a short stator linear motor – keep on running over tens of turns without further propulsion [11.67], [11.68]. The freely gliding train in Figure 11.41 demonstrates the fascinating reality of levitation in free space thanks to bulk HTSC.

Figure 11.41 Contactless superconducting levitated model train gliding at the side of the track [11.68]

Meanwhile, more powerful personnel-carrying transportation systems have been tested on linear tracks up to 15.5 m long [11.60], [11.61]. The vehicle shown in Figure 11.42 is levitated by a total of 40 IFW-SupraBlock® pieces (380 g each) and carries a load of more than 450 kg (additional to its weight of 150 kg). It is driven by a conventional linear motor and the SMB is excited by NdFeB permanent magnets (see Figure 11.42b).

Permanent magnet excited field systems are expensive. Therefore, electromagnetic rail systems are under development. A functional design has been shown in Figure 11.15. Although controlled flux densities of up to 1.8 T can be achieved by electrically fed excitation systems, at least for a short time in the pole region, the high power consumption demands more sophisticated solutions. On the other hand, electromagnetic rail systems would enable switches in complex lines to be designed and would also enable the levitation position to be actively controlled independently of the load.

The vertical lift shown in Figure 11.43 was developed to demonstrate the application of linear SMBs for transport systems in clean rooms. The lift incorporates contactless linear drives and linear SMBs. This design can also be realized in the form of a high-speed and comfortable ropeless elevator, e.g., in high buildings or deep mines.

11.6 Applications of Superconducting Bearings

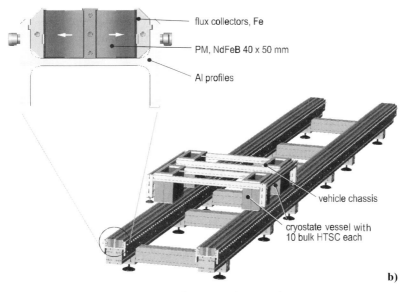

Figure 11.42 a Demonstration model of a personnel-carrying linear transportation system with PM-excited SMB and integrated linear drive. **b** Schematic view showing the track consisting of NdFeB permanent magnets and Fe poles as well the levitating vehicle supported by four SMBs. Taken from Ref. [11.61]

244 | 11 Inherently Stable Superconducting Magnetic Bearings

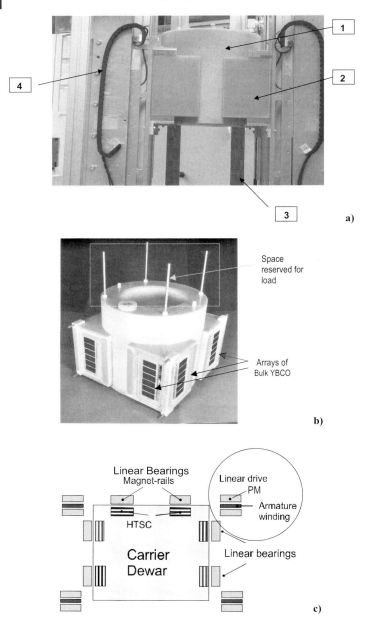

Figure 11.43 Ropeless lift with vertically acting linear drives and linear bearings for clean room applications [11.62]. **a** Set-up of the lift (1: cylindrical Dewar in the center of the carrier; 2: housing of the YBCO arrays; 3: magnetic rails of the linear bearing; 4: movable feed cable for the linear drive). **b** Photograph of the Dewar (partially opened) showing the HTSC blocks of the linear bearing which are mounted at the back on a copper plate, the latter in contact with the coolant. **c** Schematic drawing (top view) of the active parts with the linear superconducting bearing and the conventional linear drive

11.7
Specific Operation Conditions

11.7.1
Precise Positioning of Horizontal Rotating Axis

An indispensable precondition for high-speed rotating applications is to fix the machine axis precisely in the central position. Precise fixing of the horizontal axis in an SMB is much more complicated than it is for conventional magnetic or mechanical bearings. The solution proposed in Ref. [11.26] consists in dividing the superconducting cylindrical shell of the SMB into two movable half-shells. During cooling, the rotor is mechanically supported in the working position, whereas the two half-shells are cooled with an appropriate offset. They are then shifted toward the central axis after cooling (Figure 11.44). This radial approach causes an additional increase in radial stiffness. To keep the axis in its initial (i.e. the operational) position, the lower half-shell has to be cooled in a distance which – after cooling – compensates for the weight of the rotor (see Ref. [11.27]).

Each half-shell consists of a number of trapezoidal bulk YBCO blocks glued on a copper coldhead with a polygonal inner surface. Copper shims placed on both sides of and – in the axial direction – between the YBCO blocks promote the heat transfer. The precise position of the half shells is controlled by linear actuators.

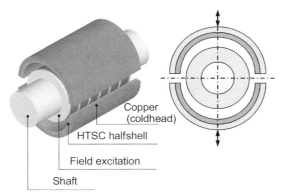

Figure 11.44 Design of an HTSC half-shell construction. Bulk HTSC blocks are glued on a copper coldhead; copper shims are located on both sides of the HTSC rings

11.7.2
Bulk HTSCs Cooled Below 77 K

The separation of the HTSC hollow cylinder from the field excitation system acting in the warm bore of a cylindrically shaped SMB is a favorable option for integrating the SMB into machines on an industrial scale. This is typically realized by using specially designed Dewar vessels and cryocoolers (Figure 11.45). The HTSC hollow cylinder in Figure 11.45 is mounted on a cylindrically shaped copper coldhead. The field excitation system is located in the warm bore of the Dewar vessel. The gap between HTSC cylinder and field excitation system which limits load and stiffness of the SMB is determined by the amplitude of the shaft oscillations and by the thickness of the Dewar wall and the insulation. The insulation should be at least 1 mm thick even if a super-insulation material (e.g., aluminum-coated Mylar) is used. Inferior insulation would cause an increase in the temperature at the surface of the HTSC, resulting in a dramatic decrease in the stiffness (see Figure 11.27b). The different thermal expansion coefficients of the materials constitute a serious problem which has to be solved. The Dewar design illustrated in Figure 11.45 was developed for the bearing of the high-speed turbo machine shown in Figure 11.37.

Here it should be emphasized that the temperature range between 66 K and 86 K has also been realized by connections to a cryocooler via massive copper rods. It is worth noting that the bearing force has been found to increase nearly by a factor 2 if the temperature is reduced from 77 K to 66 K [11.30].

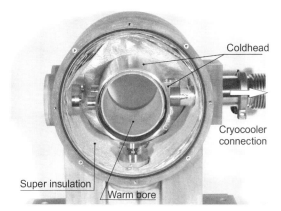

Figure 11.45 Partially opened Dewar for horizontal shaft bearings with the thermal and mechanical connections to coldhead and housing, respectively. A multi-layered super-insulation foil is used for thermal insulation

11.7.3
Cooling the Excitation System along with the Superconductor

Designs in which the excitation system is cooled along with the HTSC avoid the elaborate construction of a cryostat with super-insulation, which otherwise is necessary to reduce the heat transfer to the HTSC. Such simpler design allows the operation with reduced gaps and thus exhibits improved performances with respect to the force and stiffness densities [11.28].

A further advantage is associated with the negative temperature coefficient (ca. $-0.1\%/\,°C$ in the temperature range 20–100 °C) of the high-energy Nd-Fe-B magnets which are usually used in the excitation system. For a flux-collecting arrangement (Figure 11.39b) with Nd-Fe-B ring magnets, the stiffness was found to increase by 20% on reducing the operating temperature from 300 K to 77 K [11.29] in spite of the unfavorable 30° misalignment of the magnetization vector in Nd-Fe-B below 130 K due to a phase transformation.

Examples of SMBs using a cold excitation system are given in Refs. [11.30] and [11.44] (see also Figure 11.39).

11.7.4
Dynamics of Rotating Superconducting Bearings

The dynamic behavior of rotating machinery is well known and may be referenced to the existing standard literature. Although the levitation force, the bearing stiffness and damping of superconducting bearings are of different physical nature principally the same behavior as with conventionally suspended rotors will be observed.

Of interest are the losses in the magnetic bearing which occur despite the contact- and frictionless operation. The so-called coefficient of friction is defined by the rotational magnetic drag force normalized to the levitation force (or the weight of the levitated rotor). These drag forces are due to hysteric losses in the superconducting bulks and eddy current losses in the permanent magnets and the pole shoes or other conductive parts in the vicinity of the field excitation system. Experimental data for the coefficient of friction of rotating superconducting bearings are in the region of 10^{-7} [11.13], whereas typical values for active magnetic bearings are approximately 10^{-4}. The measurement of these low drag forces in superconducting bearings requires vacuum conditions in order to avoid air friction.

Assuming a disk shaped rigid rotor only two resonances will occur corresponding to the radial and axial degree of freedom. The amplitude of radial vibrations, which are most important for superconducting bearings, strongly increases due to the low damping of the bearings if the rotor is operated near *resonance frequency*. Above the resonance frequency the rotor vibration will become smaller and the rotor will approach rotation around its axis of polar inertia. Due to the low specific stiffness of superconducting bearings low resonance frequencies of several Hz only were observed for thrust bearings consisting of a disk shaped PM rotating above an HTSC [11.13], [11.14]. In a radial superconducting bearing with flux collecting excitation system and a YBCO hollow cylinder (inner diameter 51 mm, length 50 mm), the radial resonance has been measured at 34 Hz, which is a comparatively high value indicating a rotor with extremely low mass [11.65]. This is shown in Figure 11.46, where the amplitude of the vibration in horizontal and vertical direction is plotted against the frequency of rotation.

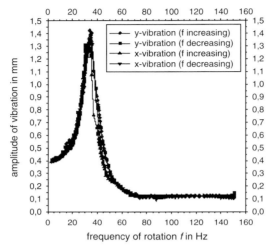

Figure 11.46 Amplitude of rotor vibration in the horizontal (x) and vertical (y) direction versus frequency of rotation. After Ref. [11.65]

Hysteretic losses will always occur if the flux distribution is oscillating with the rotation. This oscillation will be caused by flux inhomogeneities due to a not perfectly constructed excitation arrangement even if the rotor runs ideally centered. Additionally imbalance will cause the center of the rotor to run on a circular or elliptic orbit, which again is connected with an oscillation of the flux distribution. Below the resonance the rotor orbit will be of larger diameter than above the resonance when the rotor is increasingly centered by the inertial term of the equation of motion. Even a rotor with absolutely homogenous flux distribution will then show hysteric losses which are somewhat higher below the resonance than above. It is easy to understand that the hysteric losses will depend on the amplitude of the oscillations and on the hysteresis loop of the superconductor which is directly related to the quality of the bulks.

The *damping behavior* of translational vibrations in superconducting bearings of different sizes has been investigated over a wide range of frequencies and amplitudes using special apparatus equipped with shakers. The damping behavior measured as energy loss per cycle for constant amplitude was found to be almost independent of the vibration frequency over a frequency range of about four orders of magnitude. Slight damping was observed for low-amplitude vibrations, but it strongly increases with the vibrational amplitude. This increase in the energy loss per cycle was found to follow a quadratic law [11.30], [11.41], [11.65].

The *energy loss due to rotation* has been measured as a function of the rotation frequency for a large SMB with a YBCO hollow cylinder of inner diameter 203 mm and length 250 mm [11.30], [11.41]. This SMB was designed for a 400-kW motor (see Section 11.6.2). The *rotational losses* were found to contain contributions depending linearly and quadratically on the rotational frequency. Assuming a linear relationship between hysteresis losses (in the superconducting bulk material) and frequency and a quadratic law caused by eddy currents, roughly 35% of the experimental losses have been attributed to hysteresis losses in the bulk HTSC and 65% to eddy currents [11.30]. Eddy currents are expected to occur mainly in the NdFeB permanent magnets [11.30–11.32].

In principle, the rotational losses can be reduced by using less inhomogeneous PM rings. NdFeB high-performance PM rings with outer diameters of 160 mm can be manufactured having magnetic flux density inhomogeneities of less than 1% [11.66], whereas the larger PM rings (outer diameters 200 mm) used for the large SMB were found to exhibit flux density inhomogeneities of up to 10% [11.30].

11.8
Numerical Methods

The efficient use of bulk HTSCs in inherently stable magnetic bearings and electrical machines depends on the intensity of the force interaction between the magnetic field excitation system (e.g., permanent magnets) and the HTSC. Different calculation methods have been developed to describe this interaction within the wide range of possible applications [11.34], [11.70], [11.75], [11.76]. They are applicable to the design and optimization of any HTSC system.

Most mathematical models of superconductivity refer to the critical state model [11.77] explaining the hysteretic magnetic behavior of irreversible type-II superconductors. This model is based on the concept that an HTSC can carry only a limited macroscopic current density j_c. In most applications, the magnetic field distribution changes slowly and the HTSC can be considered to be in a quasi-equilibrium state. Under this condition, the critical state model can be used to describe the magnetic behavior accurately.

Depending on the specific HTSC-PM configuration, the basic model has to be modified in order to obtain a more accurate description. The most important conditions which have to be defined are:

1. Zero field cooling (ZFC) or field cooling (FC)
2. Small or large displacements between bulk HTSC and field excitation system
3. Hysteretic behavior or not
4. Static or dynamic interaction between the HTSC and the field excitation system.

According to the above conditions, many different calculation models for the force interaction have been developed. In the following, two of these models will be presented which have been used throughout this book.

11.8.1
Perfectly Trapped Flux Model (2D)

All numerical models for the determination of the force between superconductors and permanent magnets are based on *Maxwell's* laws. The classical differential equation for the vector potential A describes the magnetic field distribution in the region under investigation:

$$\nabla^2 A = -\mu_0(J - \mathrm{rot}M) \tag{11.4}$$

where J denotes the current density vector and M the magnetization vector. The boundary conditions are of the *Dirichlet* type. To solve Eq. (11.4), standard two- and three-dimensional finite element methods (FEM) have been applied. It should be noted here that the numerical implementation of any superconductivity model described below requires a very fine-tuned (FEM discretization) grid of both the HTSC and the air gap regions.

While cooling down an HTSC below its critical temperature T_C in the presence of a magnetic field generated by a permanent magnet (PM), the flux distribution within the sample is assumed to remain perfectly unaltered. The magnetic field penetrates the HTSC in the form of flux lines, which are locally pinned because of impurities and microstructural defects. In this state, a relative displacement of the PM raises a magnetic tension on the flux lines within the HTSC, and a subsequent restoring force can be observed. To evaluate the forces due to arbitrary displacements of 2D configurations, an approach described in [11.33] has been applied. The same approach has been used for the 3D configuration in Section 11.5.4.

The basic assumption of the Perfectly Trapped Flux Model is illustrated for a simple HTSC-PM configuration in Figure 11.4. In Figure 11.47, the same is demonstrated for an SMB with a flux-collecting excitation system. The magnetic field distribution within the HTSC bulk remains unchanged on displacing the HTSC from its initial position.

The initial field distribution within the HTSC is trapped by induced supercurrents. The current sheet on the HTSC surface (see Figure 11.48) has been calculated as the difference between the appropriate magnetic field strengths on the HTCS boundary.

By using the Perfectly Trapped Flux Model, maximal force values can be specified. In this way, a preliminary design of vertical or lateral action in HTSC-PM configurations for suspension and/or guidance applications can be prepared [11.39], [11.46].

11.8 *Numerical Methods* | 251

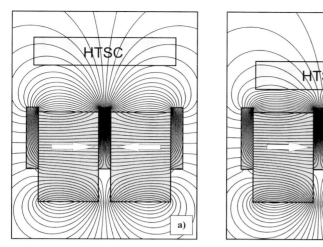

Figure 11.47 Field distributions in an SMB with flux-collecting excitation system. The initial field distribution inside the HTSC in the cooling position (**a**) remains due to surface current sheets completely unchanged after the HTSC is displaced to a new position (**b**).

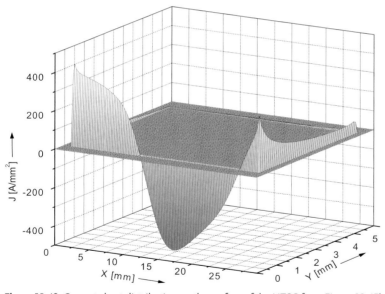

Figure 11.48 Current sheet distribution on the surface of the HTSC from Figure 11.47b

11.8.2
Perfectly Trapped Flux Model (3D)

The 2D algorithm described above can be extended to 3D field calculation. For this case, the scalar potential formulation has been used. By means of this method, three-dimensional magnetic field distributions in any HTSC-PM-iron configuration and, ultimately, forces acting between the HTSC bulk and the field excitation system can be calculated.

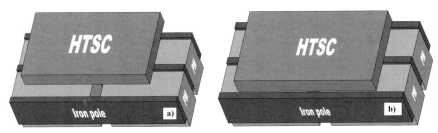

Figure 11.49 SMB with flux-collecting PM-iron arrangement: **a** Initial position of the HTSC bulk; **b** Final position of the HTSC bulk

This approach is demonstrated for the HTSC-PM-iron arrangement shown in Figure 11.49. The field distribution within the HTSC bulk has to be moved from the initial position (Figure 11.49a) to the new position (Figure 11.49b).

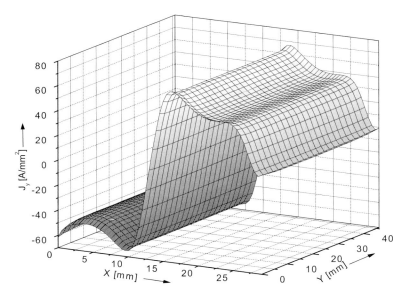

Figure 11.50 Current sheet distribution on the lower surface of the HTSC bulk of the SMB shown in Figure 11.49b in the final position

In Figure 11.50, the distribution of the current sheet on the lower surface of the HTSC in its final position is plotted. This current distribution has been calculated as the difference between the magnetic field strengths in the two positions.

11.8.3
Vector-Controlled Model (2D)

The so-called Vector-Controlled Model (VC model) takes into account both components of the varying magnetic field within the HTSC to determine the current density distribution and the corresponding forces. The VC model is based on the model of the critical state [11.77]: supercurrents (of current density j_c) are induced in the sample to suppress the variation of trapped magnetic flux within the superconductor. Two states of the volumetric current flow are simultaneously observable: in some regions, currents with a current density j_c flow perpendicularly to the field vector to maintain the field distribution as unchanged as possible, whereas other regions where the field is completely repelled are current-free.

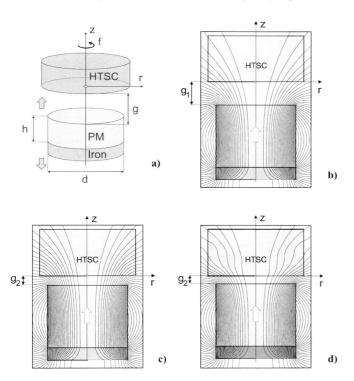

Figure 11.51 Field distributions for a typical SMB: **a** Cylindrical HTSC-PM arrangement (with axially magnetized PM) and iron yoke; **b** Field plot with HTSC in the initial (cooling) position ($g = g_1$); **c** Field plot after approaching the HTSC from g_1 (cooling position) to g_2; **d** Field plot after approaching from g_1 to g_{min} (minimal distance) followed by an increase of the gap from g_{min} to g_2

Figure 11.51 shows a typical HTSC-PM arrangement, which is used for the examination of this model. In this axially symmetric case, the superconducting currents can circulate in the r-ϕ planes. The induced supercurrents flow perpendicularly to the applied magnetic field. When the radial component B_r of the applied magnetic field is negligible compared to its z component B_z, very accurate results can be achieved by considering only B_z. However, if the HTSC is placed in an inhomogeneous field, as shown in Figure 11.51, much better results for the determination of the forces will be obtained by applying the VC Model.

During vertical displacements (of PM and/or HTSC), the direction of the critical currents is determined both by the signs of B_z and B_r and by the local history of the value of the applied magnetic field (increased or reduced). The influence of the history of the applied field on the field distribution is demonstrated in Figure 11.51c and d. In both cases, the distance between HTSC and PM was reduced to $g = g_2$ starting from $g = g_1$. In Figure 11.51c, the smaller distance g_2 was reached in one step, and in Figure 11.51d, the distance was first reduced to its minimum value $g_{min} < g_2$ and then increased to g_2. The resulting distribution of the current density causes a hysteretic force-vs-distance relationship which is typical for the levitation force of SMB after Field Cooling (see Figure 11.3).

In Figure 11.52, an experimental force-vs-distance cycle is plotted after a Maximum Field Cooling process, i.e. the HTSC is cooled in a position close to the PM. For comparison, results of numerical calculations are shown. A satisfactory agreement between experimental data and calculations is obtained.

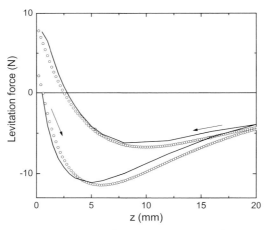

Figure 11.52 Levitation force-vs-distance loops after an MFC activation process starting from $z = 0$. Comparison of experimental data (circles) with calculated levitation forces (solid line)

Using the VC model, attractive and repulsive forces can be calculated. In order to reproduce experimental data, it is necessary to take into account the field dependence of the critical current density $j_c(B)$. Several approximations (see Refs. [11.71–11.73]) are available to describe $j_c(B)$ analytically. A sixth-order polynomial approach – the so-called "fish-tail" approximation – has turned out to be most suitable to describe experimental $j_c(B)$ data for bulk YBCO. An example of this approach is shown in Figure 11.53.

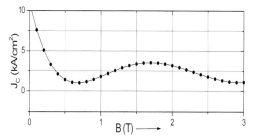

Figure 11.53 $j_c(B)$ function with fish-tail shape

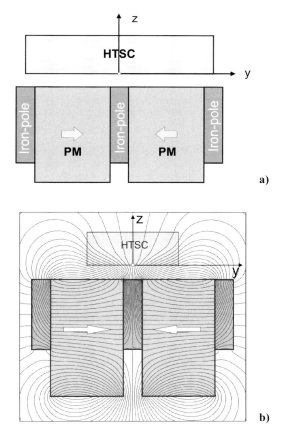

Figure 11.54 Magnetic flux distribution in an SMB with optimized flux-collecting excitation system (see Figure 11.16a) after the HTSC approaches from the cooling position at g_{max} to g_{min}

The VC model has also been used to estimate the force between an HTSC and flux-collecting PM arrangements with ferromagnetic poles. In Figure 11.54, the calculated field distribution for an optimized configuration (minimized HTSL- and PM-mass per force unit) is shown at a minimum distance g_{min} between HTSC and PM.

As the iron poles suppress any flux fluctuations in the direction of movement and, thus, parasitic losses in the HTSC, such flux concentration arrangements (linear and cylindrical) are especially favorable for high-speed bearings. Furthermore, the height of the flux-collecting iron poles has been shortened in order to reduce the stray field and, consequently, to increase the magnetic flux in the gap, which results in an enhanced levitation force.

References

11.1 S. Earnshaw, *On the Nature of the Molecular Forces which regulate the Constitution of the Luminiferous Ether*, Trans. Cambridge Phil. Soc. **7**, 97 (1842)

11.2 W. Braunbeck, Z. Phys. **112**, 753 (1938)

11.3 E. Masada, Development of Maglev and Linear Drive Technology for Transportation in Japan, in: *Proceedings of MAGLEV'95*, Nov. 26–29, 1995, Bremen (Germany), VDE-Verlag, 1995, ISBN 3-8007-2155-4

11.4 E. Eitlhuber, Eisenbahntechnische Rundschau **29**, 409 (1980)

11.5 H. Kemper, *Schwebebahn mit räderlosen Fahrzeugen, die an einer Fahrschiene mittels magnetischer Felder schwebend entlang geführt werden*, Deutsches Reichspatent 643 316, (1937)

11.6 V. Arkadiev, Nature **160**, 330 (1947)

11.7 I. Simon, J. Appl. Phys. **24**, 19 (1953)

11.8 T. A. Buchhold, Sci. Am. **202**, 74 (1960)

11.9 M. K. Bevir, T. C. Randle, M. N. Wilson, Electrical Engineer (Melbourne), **54**, 26 (1977)

11.10 G. J. Homer, T. C. Randle, C. R. Walters, M. N. Wilson, M. K. Bevir, J. Phys. D: Appl. Phys. **10**, 879 (1977)

11.11 F. C. Moon, *Superconducting Levitation* (John Wiley & Sons, Inc., 1994)

11.12 M. Marinescu, N. Marinescu. J. Tenbrink, H. Krauth, IEEE Trans. Magn. **25**, 3233 (1989)

11.13 J. R. Hull, Supercon. Sci Technol. **13**, R1 (2000)

11.14 S. Gruss, G. Fuchs, G. Krabbes, P. Verges, G. Stöver, K.-H. Müller, J. Fink, L. Schultz, Appl. Phys. Lett. **79**, 3131 (2001)

11.15 R. Gonzalez-Arrabal, M. Eisterer, H. W. Weber, G. Fuchs, P. Verges, G. Krabbes, Appl. Phys. Lett. **81**, 868 (2002)

11.16 T. Siugura, K. Matsunaga, Y Uematsu, T. Aoyagi, M. Yoshizawa, Adv. Supercond. **10**, 1349 (1997)

11.17 J. R. Hull, A. Cansiz, J. Appl. Phys. **86**, 6306 (1999)

11.18 C. Hofmann, G. Ries, Supercond. Sci. Technol. **14**, 34 (2001)

11.19 F. N. Werfel, U. Floegel-Delor, T. Riedel, R. Rothfeld, D. Wippich, B. Goebel, P. Kummeth, H.-W. Neumueller, W. Nick, Trans. Appl. Supercond. **13**, 2173 (2003)

11.20 Joint devolupment by Messer Griesheim GmbH, Hoechst AG, ZFW Göttingen, IEM TU Braunschweig

11.21 W.-R. Canders, H. May, *Magnetische Lagerung eines Rotors in einem Stator*, Patent DE 197 27 550 C 2

11.22 W.-R. Canders, H. May, *Device for fixing a body on superconducting bearings during its cooling phase*, Patent EP 0 936 367 B1

11.23 S. O. Siems, H. May, W. R. Canders, M. Leonhard, in *Supraleitung und Tieftemperaturtechnik*, Tagungsband zum

8. Statusseminar, ed. by VDI Technologiezentrum Düsseldorf Abt. Physikalische Technologien, (VDI Technologiezentrum, Düsseldorf 2003, ISBN 3-93-138444-6) abstract p.70, and full text CD attachment

11.24 Joint development by Atlas Copco Energas GmbH and IEM TU Braunschweig

11.25 Joint development by Pillar GmbH, IPHT Jena, IEM TU Braunschweig, ZFW Göttingen, Nexans GmbH

11.26 W.-R. Canders, H. May, *Magnetische Lagerung*, Patent DE 100 34 922 C 2

11.27 W.-R. Canders, J. Hoffmann, H. May, R. Palka, SMB Design based on Advanced Calculation Methods Validated by Practical Experience, in: *Proc. 9th Int. Symp. on Magnetic Bearings*, Lexington KY, USA, Aug. 3-6, 2004

11.28 F. Steinmeyer, *Magnetic bearing for suspending a rotating shaft using high Tc superconducting material*, Patent DE 100 42 962.9

11.29 Unpublished results of the authors

11.30 P. Kummeth, G. Ries, W. Nick, H. W. Neumüller, Supercond. Sci. Technol. **17**, S259 (2004)

11.31 A. N. Terentiev, H. J. Lee, C.-J. Kim, G. W. Hong, Physica C **290**, 291 (1997)

11.32 J. R. Hull, T. M. Mulcahy, J. F. Labataille, Appl. Phys. Lett. **70**, 655 (1997)

11.33 H. May, R. Palka, E. Portabella, W.-R. Canders, in: COMPEL (Int. J. Comput. Math., in Electr. Electron. Eng.) **23**, 286 (2004)

11.34 Y. Yoshida, M. Uesaka. K. Miya, IEEE Trans. on Magnetics **30**, 3503 (1994)

11.35 B. M. Smolyak, G. N. Perelshtein, G. V. Ermakov, Cryogenics **42**, 635 (2002)

11.36 Y. S. Tseng, C. H. Chiang, W. C. Chan, Physica C **411**, 32 (2004)

11.37 W.-R. Canders, High Speed Machines on Magnetic Bearings – Design Concepts and Power Limits, in: *Proc. Int. Conf. on Electrical Machines (ICEM'98)*, 1998, Istanbul, ISBN 975 429 126 8

11.38 W.-R. Canders, H. May, R. Palka, High speed rotating HTcSC magnetic bearing of new topology, in: *Proc. Conf. Non-Linear Electromagnetic Systems: Advanced Techniques and Mathematical Methods* ISBN 90 5199 381 1, (ISEM 1997, Braunschweig, Germany)

11.39 W.-R. Canders, H. May, R. Palka, COMPEL (Int. J. Comput. Math., in Electr. Electron. Eng.) **17**, 628 (1998)

11.40 S. O. Siems, W.-R. Canders, Int. J. Appl. Electromagn. Mech. **19**, 199 (2004)

11.41 F. N. Werfel, U. Floegel-Delor, T. Riedel, R. Rothfeld, D. Wippich, P. Kummeth, H. W. Neumüller, W. Nick, IEEE Trans. Appl. Supercond. **13**, 2173 (2003)

11.42 F. N. Werfel, U. Floegel-Delor, R. Rothfeld, D. Wippich, T. Riedel, Supercond. Sci. Technol. **18**, S19 (2005)

11.43 W. Nick, G. Nerowski, H. W. Neumüller, M. Frank, P. van Hasselt, J. Frauenhofer, F. Steinmeyer, Physica C **372**, 1506 (2002)

11.44 G. Fuchs, P. Verges, G. Krabbes, L. Schultz, in *DKV-Tagungsbericht 2003* (Benz Drucke, Stuttgart 2003, ISBN: 3-932 715-35-7) p. 163

11.45 S. O. Siems, W.-R. Canders, Supercond. Sci. Technol. **18**, S86 (2005)

11.46 E. Portabella, R. Palka, H. May, W.-R. Canders, in *Applied Superconductivity 1999*, Inst. Phys. Ser. No. 167, ed. by X. Obradors et al. (IOP Publishing, Bristol, 2000) p. 1047

11.47 Y. Miyagawa, H. Kameno, R. Takahata, H. Ueyama, IEEE Trans. Appl. Supercond. **9**, 996 (1999)

11.48 M. Minami, S. Nagaya, H. Kawashima, T. Sato, T. Kurimura, *Advances in Superconductivity X*, ed. by Osamura et al. (Springer-Verlag Tokyo 1998) p. 1305

11.49 K. Matsunaga, M. Tomita, N. Yamachi, K. Iida, J. Yoshioka, M. Murakami, Supercond. Sci. Technol. **15**, 842 (2002)

11.50 T. M. Mulcahy, J. R. Hull, K. L. Uherka, R. C. Nieman, R. G. Abhoud, J. P. Juna, J. A. Lockwood, IEEE Trans. Appl. Supercond. **9**, 297 (1999)

11.51 A. C. Day, M. Strasik, K. E. McCrary, P. E. Johnson, J. W. Gabrys, J. R. Schindler, R. A. Hawkins, D. L. Carlson, M. D. Higgins, H. R. Hull, Supercond. Sci. Technol. **15**, 838 (2002)

11.52 R. Koch, R. Wagner, M. Sander, H. J. Gutt, Development and Test of a

11.52 300Wh/10kW Flywheel Energy Storage System, in: *Applied Superconductivity 1999, Inst. Phys. Ser. No. 167*, ed. by X. Obradors et al. (IOP Publishing, Bristol, 2000) p. 1055

11.53 S. O. Siems, W.-R. Canders, H. Walter, J. Bock, Supercond. Sci. Technol. **17**, S229 (2004)

11.54 K. B. Ma, Y. V. Postrekhin, W. K. Chu, Rev. Sci. Instrum. **74**, 4989 (2003)

11.55 E. Lee, B. Kim, J. Ko, C. Y. Song, S. J. Kim, S. Jeong, S. S. Lee, IEEE Trans. Appl. Supercond. **15**, 2324 (2005)

11.56 NEDO (New Energy and Industrial Technology Development Organization), Firm prospect

11.57 A. Bitterman, Supercond. Cryoelectron. **12**, 25 (2000)

11.58 Boeing Flywheel Program Overview and Status (February 2003), Firm prospect

11.59 T.-H. Sung, J.-S. Lee, Y. H. Han, S.-C. Han , S.-K. Choi, S.-J. Kim, Physica C **372-376**, 1451 (2002)

11.60 J. Wang, S. Wang, Y. Zeng, Physica C **378-381**, 809 (2002)

11.61 L. Schultz, O. de Haas, P. Verges, C. Beyer, S. Röhling, H. Olsen, L. Kühn, D. Berger, U. Noteboom, U. Funk, presented at the 2004 Applied Superconductivity Conference, October 3-8, 2004, Jacksonville, to be published in IEEE Trans. Appl. Supercond. (2005)

11.62 G. Krabbes, G. Fuchs, W. Pfeiffer, P. K. Budig, B. Schumann, S. Pollack, *Vorrichtung zum vertikalen Transportieren von Produkten*, Patent DE 100 13 8 33.

11.63 A. N. Terentiev, *Pumping or mixing system using a levitating magnetic element*, Patent US 6 416 215 B1, and http://www.levtech.net

11.64 H. Teshima, M. Tanaka, K. Miyamoto, K. Nohguchi, K. Hinata, Physica C **256**, 142 (1996)

11.65 P. Kummeth, W. Nick, G. Ries, H.-W. Neumüller, Physica C **372-376**, 1470 (2002)

11.66 H. Yamamoto, H. Nasu, H. Kameno, Physica C **392-396**, 759 (2003)

11.67 J. R. Wang, M. Z. Wu, H. May, P. X. Zhang, L. Zhou, W. Gawalek, H. Weh, Rare Met. Mater. Eng. **27**, 240 (1998)

11.68 L. Schultz, G. Krabbes, G. Fuchs, W. Pfeiffer, K.-H. Müller, Z. Metallkd. **93**, 1057 (2002)

11.69 J. R. Hull, S. Hanany, T. Matsumura, B. Johnson, T. Jones, Supercond. Sci. Technol. **18**, S1 (2005)

11.70 D. Camacho, J. Mora, J. Fontcuberta, X Obradors, J. Appl. Phys. **82**, 1461 (1997)

11.71 D.-X. Chen, R. B. Goldfarb, J. Appl. Phys. **66**, 2489 (1989)

11.72 W. A. Fietz, M. R. Beasley, J. Silcox, W. W. Webb, Phys. Rev. A **136**, (1964) 335

11.73 Y. B. Kim, C. F. Hempstead, A. R. Strand, Phys. Rev. **139**, 1163 (1965)

11.74 M. Tsuchimoto, T. Kojima, H. Takeuchi, T. Honma, IEEE Trans. Magn. **29**, 3577 (1993)

11.75 M. Tsuchimoto, H. Takeuchi, T. Honma, Trans. IEE Jpn., **114**, D 741 (1994)

11.76 M. Tsuchimoto, T. Kojima, T. Honma, Cryogenics **34**, 821 (1994)

11.77 C. P. Bean, Rev. Mod. Phys. **36**, 31 (1964)

12
Applications of Bulk HTSCs in Electromagnetic Energy Converters

Almost 70% of the primary energy in Germany is converted into mechanical energy by the use of electrical drives, actuators for automobiles and planes, or machine tool drives. This situation is typical for all highly industrialized countries. Therefore, maximization of the efficiency of the energy conversion is extremely important from both the economical and the environmental point of view. Remarkable progress has been achieved during the past 20 years by the construction of new direct-current (DC) and synchronous machine (SM) designs [12.1].

Many expectations of further increases in the power density and efficiency of electrical machines have been raised with the discovery of the HTSCs. The application of *bulk HTSCs in machines* is associated with their ability to trap and shield a permeating magnetic field. Here, the bulk material is used as either a superconducting permanent magnet (SPM) or a shield. The two principles cover different performance and power areas for various fields of application. The availability of frequency-controlled power sources has encouraged these developments. Principles of the application of bulk materials and characteristics of the corresponding machines are presented here.

12.1
Design Principles

Energy conversion from electromagnetic to mechanical energy in rotating or linear machines is based on the generalized laws of *Faraday* and *Ampere*, which describe the generation of a magnetic field in the air gap from a voltage source. The Lorentz force appearing because of the interaction between the current and the flux density in the air gap is the basis for the conversion from electromagnetic to mechanical energy in a *motor*. In contrast, a *generator* converts mechanical to electrical energy, which is available as a voltage at the terminal of the machine, if an electrical conductor is moved by external forces inside the air gap. In this case the Lorentz force acts upon the charge carriers within the conductor. Ignoring aspects which are important for an optimal layout, any electrical machine can operate as either a motor or a generator – it is in principle reversible.

High Temperature Superconductor Bulk Materials.
Gernot Krabbes, Günter Fuchs, Wolf-Rüdiger Canders, Hardo May, and Ryszard Palka
Copyright © 2006 WILEY-VCH Verlag GmbH & Co. KGaA, Weinheim
ISBN: 3-527-40383-3

The main components and design characteristics of a rotating energy converter are shown in the diagrammatic sketch (Figure 12.1), where a ferromagnetic rotor is placed in the bore of an iron circuit which is excited by a superconducting permanent magnet (SPM). Conductors are fixed on the rotor, and comprise the armature winding. The source of the armature currents determines the operation mode (motor or generator) of the machine.

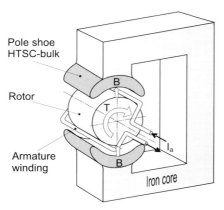

Figure 12.1 Sketch of a rotating energy converter with a rotor placed in an iron circuit. The conductors are situated in the air gaps of the slotless rotor where they interact with the magnetic field excited by an SPM. (T = torque; B = flux density; I_a = armature current)

This design illustrates the straightforward application of the HTSC in a *trapped-field motor*, which benefits from the field excitation capability of the SPM, which is superior to that of high-performance permanent magnets made from conventional materials. The use of bulk HTSCs in trapped-field motors was demonstrated in 1995 [12.20], but the problem of developing an effective technique for magnetizing HTSC permanent magnets *inside the machine* had not at that time been satisfactorily solved, which prevented fast progress in scaling up trapped-field motors.

Reluctance motors with iron poles in the rotor are improved by inserting bulk HTSC tiles into the space between the iron poles. This type of machine uses the field-shielding capability of bulk HTSC in order to focus the magnetic flux generated by the excitation unit into the iron poles. Reluctance machines are distinguished by high torque and low inertia [12.11–12.13]. A power increase of up to several 100 kW has been demonstrated in superconducting reluctance machines (Section 12.4).

The machine having the simplest design and operation of all the machines considered is the *hysteresis* or *induction* machine [12.9], [12.10]. This machine type is discussed first because it demonstrates different operation modes of machines with HTSC bulk materials.

Since the medium-sized machines considered here were typically tested in liquid nitrogen, it should be remembered that the viscosity of this liquid near its boiling temperature is comparable to that of air at room temperature!

12.2
Basic Demonstrator for Application in Electrical Machines – Hysteresis or Induction Machines

The simple rotor of this machine type (Figure 12.2a) consists of bulk HTSC cylinders or rings operating in a bath of liquid nitrogen. After cooling the machine in ZFC mode, the superconductor will be magnetized by the armature field, which has its main component perpendicular to the axis of the rotor. The aspect ratio of the rotor (length / diameter) is large enough to be considered in an infinite length approach. In addition, the large j_c in the ab plane limits flux penetration into the HTSC cylinder from the top and bottom circular faces. Therefore, flux has to penetrate through the sides of the HTSC cylinder.

A static, non-rotating field would induce a distribution of supercurrents consisting of two antisymmetric, crescent-shaped regions on opposite sides of the cross section of the HTSC cylinder. These two regions carry currents of magnitude j_c flowing parallel to the axis of the cylinder in opposite directions and denoted in the following as $(+j_c)$ and $(-j_c)$. A sinusoidal armature current changing with the frequency f_{feed} causes a field which rotates with the frequency $f = f_{feed}/p$ (where p is the number of pole pairs). The changing field modifies the distribution of the induced supercurrents. Various current and field distributions with a variety of curved boundaries between the $(+j_c)$ and $(-j_c)$ regions develop, depending on the revolution speed and the degree of flux penetration [12.15], [12.16]. An example is shown in Figure 12.2b, where dark and light shading represent currents flowing out of and into the plane, respectively. In a hysteresis machine, the instantaneous directions of magnetization within the HTSC cylinder and the applied field are displaced. Their interaction generates a torque that accelerates the rotor up to the synchronous speed. This is illustrated by the *snapshot* in Figure 12.2c showing the flux distribution at a certain moment inside the bulk superconducting rotor under the influence of a sinusoidal stator field [12.7]. Detailed finite-difference modeling [12.6] has shown that the superconductor is optimally used when the flux is a little higher than is necessary to penetrate the entire cross section of the full cylindrical superconductor, thus making possible a certain magnetic flux in the center. The current and field distribution for this optimum field penetration corresponding to a maximum torque per field energy is shown in Figure 12.2b. The situation where the penetration is below its optimum because of the restricted armature field is shown in the left inset of Figure 12.2d. This Figure further indicates that the torque can be increased by replacing the bulk with a hollow cylinder, which allows flux to enter the central part of the rotor, as shown in the inset right. This observation is positive proof that an adequate flux permeating the material is the most important factor in this type of machine.

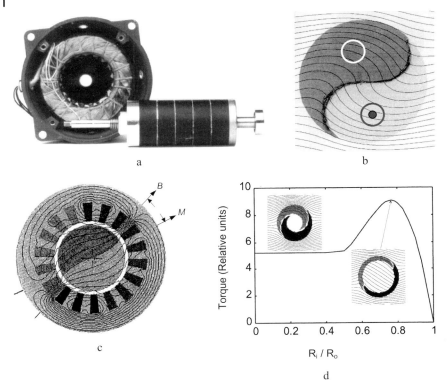

Figure 12.2 Hysteresis machine: **a** 500-W hysteresis motor with a rotor made up of stacked bulk YBCO rings (partially disassembled) [12.9]; **b** current and field distribution along the cross section of a bulk cylindrical rotor in a sinusoidal field applied perpendicularly to the cylinder axis. Dark and light shading show current flowing out of and into the plane, respectively. The case of optimal field penetration is depicted (taken from Ref. [12.15]); **c** snapshot of the operating hysteresis machine. The angle between B and M represents the instantaneous shift between armature field and rotor magnetization. The field shown in the figure is greater than the optimum in **b**. The analogous situation was analyzed in [12.7]; **d** development of the torque in a ring-shaped rotor with varying inside radius, normalized to the outer radius R_o under the same stator field B. Insets show the current distribution of the incompletely penetrated material – either full or ring (left) – and the optimal situation when a ring is completely penetrated (right) (taken from Ref. [12.15])

As long as the revolution speed differs from that of the stator field wave, each part of the HTSC cylinder exhibits individual magnetization loops. The tangential *Lorentz* force and the equivalent torque M_{hy} are generated by the hysteresis in the rotor [12.19], [12.38]. The machine type displays a high initial torque. This attribute is known from conventional hysteresis machines, which gave rise to the name for the machine design. (In contrast to a conventional ferromagnetic hysteresis motor, the diamagnetic magnetization causes the rotor to be pushed by the rotating field.) Since pole regions are not definitely distinguished by the design of the rotor, the machine characteristics can change considerably in the operating state. The torque increases with the applied field B_a according to a power law:

$$M_{hy} \propto B_a^n \tag{12.1}$$

A value of $n = 3$ was found for low penetration of the applied fields, changing to $n = 1$ for high penetration [12.15].

The considered machine is *running synchronously* with the armature field if the appearing Lorentz force F_L is compensated by the flux-pinning force F_p in the volume of the superconductor, i.e. $F_L \leq -F_p$. This can still be assumed in a *steady state mode* if fluxons move instantaneously to find a new pinned position. The application of this "magneto-static approximation" has been proven experimentally for slip speeds (i.e. the difference between the rotor and armature field frequencies) up to several tens of Hz [12.22]. The maximum value of the synchronous torque can be defined by the changeover from synchronous to slip mode due to external load.

The appearance of slip induces larger currents, which cause an increase in the local temperature, decreasing j_c and consequently giving deeper flux penetration, the latter resulting again in increasing torque, which is necessary for a self-adjusting mechanism [12.23]. Therefore, the term "induction machine" is sometimes proposed for machines running with a slip [12.22].

A high constant torque over a wide range of frequencies and self-starting and self-synchronizing capabilities are the advantages of this basic machine type with bulk HTSC, as well as its simple construction. The torque of a hysteresis HTSC rotor driven in a 2-kW stator was found to exceed that of a ferromagnetic hysteresis rotor by a factor of 6, in good agreement with theoretical considerations [12.19].

Unfortunately, the application of superconducting hysteresis machines is limited to low power demands because of the restricted field which can be generated in a conventional armature. Therefore, the high quality of recently developed bulk HTSC materials *cannot be optimally exploited in machines of this type*. Furthermore, their function involves an inherent thermal loss, which necessitates intensive cooling.

The lack of any fixed boundary between pole areas in the magnetized superconductor significantly distinguishes the hysteresis machine – even if it is operated in a synchronous mode – from the machines described in the later sections.

12.3
Trapped-Field Machine Designs

Machine types which are designed for synchronous operation using bulk SPMs will benefit optimally from the application of HTSCs. In this case, the bulk HTSC has to be magnetized, preferably up to magnetic flux densities above 2 T. Magnetizing and cooling the bulk superconductors assembled on the rotating part of a machine needs a very elaborate procedure, which has prevented large-scale application so far. Several approaches have been tested using both the armature winding and separate coils to magnetize the SPM.

The principle of the trapped-field motor was first demonstrated by *Itoh et al.* [12.20], who ran a mobile disk-type motor which was magnetized in the axial direction. Two rings of disk-shaped rotating YBCO SPMs were stacked on a common horizontal axis. The armature winding was located between these two rings. Details are shown in Figure 12.3. Each rotor ring carries 10 cylindrical bulk YBCO superconductors and is surrounded by a separate coil for pulse magnetization in an evacuated containment made from stainless steel. The test motor ran at 65 K after magnetization with applied pulses of 1.65 T at 65 K or 2.1 T at 30 K. The yielding torque was measured at 2.8 Nm and the power output was about 3.5 kW at 65 K [12.20], [12.21].

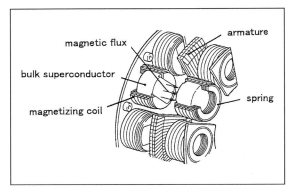

Figure 12.3 Detail from a trapped-field motor. Shown are two rotor rings with disk-shaped YBCO cylinders surrounded by the magnetizing coil and the armature winding between the rings [12.21]

A test motor [12.17] with a bulk cylindrical YBCO rotor (diameter 70 mm, length 45 mm) is shown in Figure 12.4a. In this case, the YBCO rotor was magnetized by currents flowing in the armature winding. Both DC and pulse fields were used for magnetizing the rotor. Figure 12.4b represents the stator current and the magnetic flux in the air gap during the magnetizing procedure due to a DC field. The field starts to penetrate the YBCO rings, following the sinusoidal development of the magnetomotive force (mmf) in the stator, which generates a sinusoidal development of the flux in the gap and consequently also in the bulk YBCO. Whereas the applied field in the gap reached 1.2 T, the remaining trapped field on the rotor surface was 0.36 T.

Alternatively, the trapped field was almost doubled by applying pulse magnetization using the same armature winding [12.17]. Pulses of 300 V and 20–30 ms duration were generated by an external power source. Thus, one of the most practical and promising magnetization techniques of a bulk rotor is pulsed-field magnetization using armature windings [12.31].

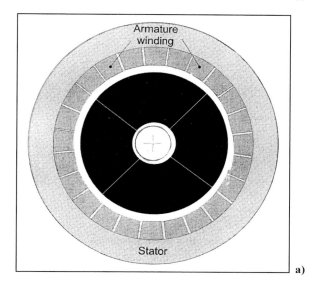

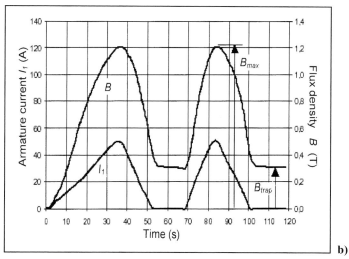

Figure 12.4 Synchronous trapped-field machine with *in-situ* magnetized SPM: **a** construction of the machine using the armature windings for DC or pulse magnetization; **b** armature current and air gap field during the magnetizing procedure by stationary and pulsed current. B_{trap} is the remanent magnetization on top of the HTSC [12.17]

Ring-shaped bulk YBCO rotors have also been used in laboratory test models of synchronous generators [12.28], [12.24], demonstrating the conversion from mechanical to *electromagnetic* energy.

The high critical current densities of 35–40 kA cm^{-2} achieved at 77 K in improved bulk YBCO material encouraged new machine designs for synchronous machines. A conventional synchronous machine with flux-collecting system and armature windings embedded in slots is shown in Figure 12.5a. From *Ampere's* law it is concluded that the distance between the rotor and the pole shoe of the stator (air gap g) can be increased when an increased magnetomotive force (mmf) becomes available. This allows one to design a slot-less stator [12.2], in which the armature windings can completely cover the increased teeth-less air gap space Figure 12.5b. Drives benefit twice from such design: firstly, the armature main field and hence the load-dependent saturation is drastically reduced, and secondly, the armature stray field of the conventional design (Figure 12.5a) is almost completely suppressed, as shown in Figure 12.5b.

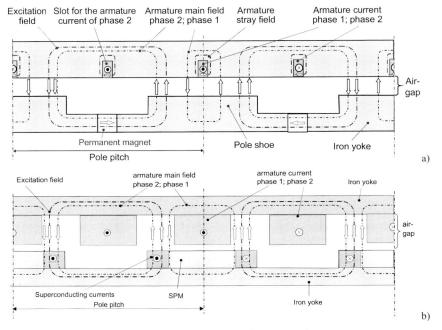

Figure 12.5 Linear representations of synchronous machines (two-phase design for simplicity): **a** conventional permanent-magnet-excited machine in a flux concentration scheme with armature windings embedded in slots; **b** slot-less SPM-excited machine according to Figure 12.6a

Slot-less machine designs with drastically reduced armature stray fields can be expected with the new high-current SPM even if they are air gap mounted (Figure 12.6a). As the flux is no longer concentrated on the teeth, the field strength of excitation in the air gap can be increased to 1.8–2.0 T. With a critical current density of $j_c = 40$ kA cm^{-2} in the SPM, a flux density distribution as shown in Figure

12.6b can be realized. This field profile has a wide plateau at 1.8 T corresponding to 25% of the maximum trapped field of the fully penetrated SPM. For comparison, a flux density of only 0.55 T can be achieved with the best high-performance permanent magnets (having a remanence 1.45 T) as shown in Figure 12.6c. Though this cuboid field shape is more favorable, the mean flux and the resulting torque amounts only one third of that which can be achieved with the SPM.

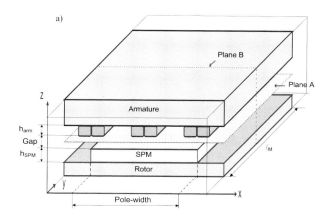

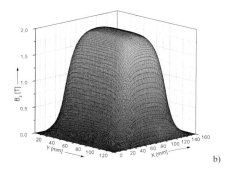

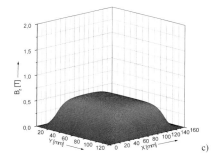

Figure 12.6 a Slot-less synchronous machine (dimensions: pole coverage: 2/3; h_{SPM} = 10 mm; air gap + h_{arm} = 15 mm). b Distribution of the normal component of the flux density (in Plane A) by use of SPM (j_c = 40 kA cm^{-2}). The plateau of the flux density at 1.8 T corresponds to 25% of the maximum trapped field of the fully penetrated SPM. c Flux density distribution by use of a conventional high-energy PM

Since restrictive design rules for conventional synchronous machines are no longer relevant, the complete air gap region of the teeth-less design can be used for the armature winding, thus increasing the effective magnetic air gap (heights of the winding and SPM in addition to the mechanical air gap). The armature reaction will be reduced drastically, allowing very high armature currents. The currents are restricted by Ohmic losses (IR^2) which causes temperature to rise. An extreme overload capability for short periods of time is characteristic for this machine type.

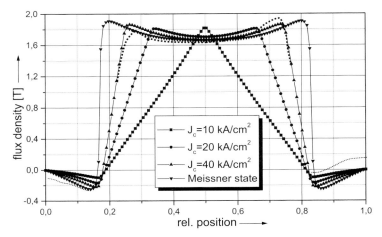

Figure 12.7 Distribution of the normal component of the flux density along the cross section line of planes A and B of Figure 12.6a for several critical current densities and the *Meissner* state in the SPM. The only minor influence of a load on the zero-load characteristics (symbols and solid lines) is demonstrated by the dotted line for j_c = 40 kA cm^{-2}. (The total current is kept constant for all curves)

Calculations of the field distribution along the line formed by the intersection of planes A and B in Figure 12.6a are shown in Figure 12.7 for several critical current densities. Full penetration of magnetic flux in the SPM and a peak field of 1.8 T are achieved for j_c = 10 kA cm^{-2}. For larger j_c values, a field plateau develops because of the reduced flux penetration and the restricted armature current. The width of this field plateau increases with j_c, whereas its field strength remains almost constant at 1.8 T. For comparison, the field profile for complete flux repulsion in the *Meissner* state is also shown. Numerical calculations of the driving forces have also shown that the obtained characteristics are almost independent of the load, thus indicating that armature reaction in this machine design can be neglected. It can be concluded that the large excitation field, up to 2 T, makes this type of trapped-field motor very attractive for further applications.

12.4
Stator-Excited Machine Designs with Superconducting Shields – The Reluctance Motor with Bulk HTSC

Machines which take advantage of the field-shielding capability of bulk HTSCs are simple to manufacture and do not need elaborate handling for the initial magnetization. They offer a reasonable alternative for the power range up to several

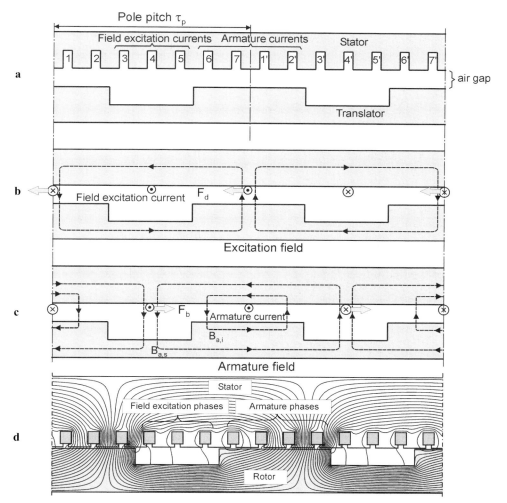

Figure 12.8 Principles of a conventional **St**ator **E**xcited salient pole **S**ynchronous **M**achine (SESM, reluctance machine): **a** magnetic configuration and armature winding of a polyphase SESM – combined current representation of the field excitation and armature phases; **b** excitation field and thrust forces; **c** armature reaction field and parasitic braking forces; **d** field distribution of a 7-phase SESM at full load (3 excitation and 4 armature phases). (F_d: driving force; F_b: braking force)

hundreds kilowatts. One of the most promising applications of HTSC shields [12.3] is the poly-phase, **S**tator **E**xcited salient pole **S**ynchronous **M**achine (SESM), the so-called *Reluctance Machine* [12.4].

The generation of magnetic fields and Lorentz forces in this unconventional type of machine is illustrated in Figure 12.8, which schematically indicates that the currents of phases 3, 4, and 5 operate to excite the main field in the position as presented, while phases 1, 2 and 6, 7 operate as armature currents.

When optimizing this machine with *conventional materials* (ferromagnetic rotor with salient poles), one meets a conflicting situation: highly efficient excitation requires a small air gap between rotor and armature because an increasing air gap causes parasitic braking forces to arise in the interpolar region (Figure 12.8c and d). Therefore, a reluctance machine (SESM) should be designed with a small air gap. On the other hand, a narrow gap results in an increasing armature feedback reaction and saturation. This is illustrated in Figure 12.8d by the high density of flux lines on the right hand side of the salient poles. The armature reaction contributes to significantly increased losses.

Bulk HTSC elements can now effectively suppress the armature reaction and protect the interpolar region from the parasitic braking forces if they are placed in the gap between the ferromagnetic poles of the rotor, as shown schematically in Figure 12.9.

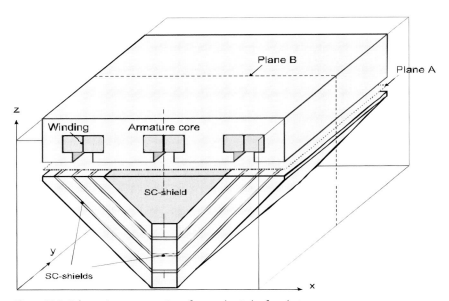

Figure 12.9 Schematic representation of one pole pitch of a *reluctance* stator excited salient pole synchronous machine (SESM) with HTSC shields in the iron core of the rotor

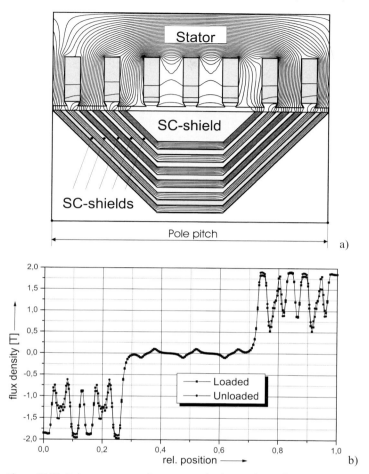

Figure 12.10 Reluctance motor (stator excited salient pole synchronous machine): **a** field distribution along plane B of the previous figure; **b** flux density distribution in the air gap (unloaded and at full load, position related to the pole pitch)

In Figure 12.10a, field distributions in this reluctance motor with HTSC elements are shown which have been calculated by a numerical FEM analysis. The flux is fully repelled from the interpolar space (no parasitic braking force). Since the armature reaction can be neglected in these machines, they can be operated with high air gap flux densities (peak values of about 1.9 T) and exhibit a high overload capability in contrast to conventional designs with homogeneous iron poles.

It should be noted that the energy conversion of a reluctance machine is generally less than that of the corresponding synchronous machine. This is a consequence of the interpolar space (in this case 3/7 of the pole pitch) which is necessary for proper operation.

On the other hand, this machine type is distinguished by its remarkably *robust passive rotor* with *low inertia* [12.4]. The use of the HTSC shields increases the torque by up to 100% compared with that of a conventional motor with salient poles [12.11], [12.12]. The machine is therefore designed to be used for *high speed and high-dynamic drives*, and under these conditions its special characteristics amply repay the effort required for its construction. Furthermore, the machine generates a reduced torque ripple compared with that associated with other high-speed machines [12.5]. The rotor with the "zebra"-striped design of Figure 12.11a incorporates a series of alternating iron and superconducting plates [12.11].

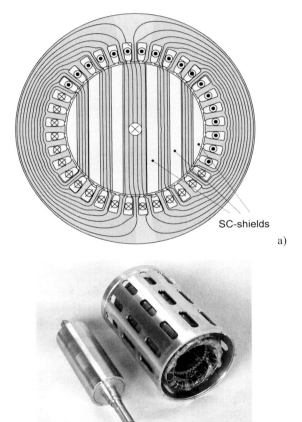

Figure 12.11 a Cross-sectional view and field distribution of a reluctance motor with "zebra"-striped arrangement of the superconducting shields [12.25]; b disassembled 200-kW reluctance motor. The stainless steel reinforced rotor is shown together with the stator. The diameter of the rotor is 147 mm and the length of the active part is 320 mm. Reprinted from [12.25] with permission from Elsevier

The complex rotor of this motor has been integrated into a hollow stainless-steel reinforcing cylinder to avoid desintegration due to high-speed rotation. The 200-kW (3000 rpm) machine shown in Figure 12.11b runs completely immersed in a bath of liquid nitrogen. This motor represents a prototype for reluctance machines which benefit from the applications of superconducting shields concentrating the exciting field into the iron core. The static and dynamic performance data have been determined using the test bench shown in Figure 12.12.

SESM-machine   Electronic survey and measurement equipment

Figure 12.12 Test bench for high-dynamic SESM equipped with load machine (DC brake), feeding and survey equipment [12.25]

12.5
Machines with Bulk HTSCs – Status and Perspectives

YBCO bulk materials can be used in energy conversion applications both as diamagnetic materials and as superconducting permanent magnets. The shielding capability of superconducting materials has been used to expel the stray field from the interpolar space in the iron rotor of high-dynamic reluctance machines. The dynamics of the machine is characterized by the angular acceleration, which is determined by the ratio $M:J$, where M is the torque and J is the inertia of the rotor (Table 12.1).

Table 12.1 Dynamics of a reluctance motor with bulk HTSC compared with conventional motors of the same torque (M = 500 Nm) [12.25]

Motor type	Reluctance motor SRE 150-320	High-dynamic asynchronous motor	Compact asynchronous motor	Standard asynchronous motor
Angular acceleration (s^{-2})	3100	1500	1000	300

Significantly increased power density has been achieved in reluctance machines by inserting YBCO plates between their iron poles. These machines represent the preferred type so far in which performances of several hundred kW have been realized (Table 12.2).

Table 12.2 Comparison of force densities in conventional and bulk HTSC motors

Machine type	Power range (kW)	Force density related to rotor surface (N cm^{-2})	Specific torque related to motor volume (Nm dm^{-3})
ac asynchronous	10 – 50	1 – 1.5	1 – 1.5
synchronous, PM	10 – 50	1.5 – 3	2.5 – 5
HTSC reluctance			
Ref. [12.13]	38	8.5	13.5
Ref. [12.25]	200	6	≈25

It is worth emphasizing that superconductivity in principle simultaneously affords bearing and drive. This has been demonstrated in laboratory scale models with disk-shaped rotors in self-levitating configuration [12.29] as well as with linear actuators [12.30].

In this context, a recent study should be mentioned which has demonstrated the generation of a strong force in a linear "launcher" on the laboratory scale [12.37]. The accelerating force results from primary pulse coils arranged in a central tube, while a hollow cylinder of bulk YBCO rings forms the secondary. The latter will be accelerated by the large Lorentz force in the axial direction which results from the high amplitude of the pulse field H and its fast decay dH/dt. Long-term plans for the design of an aircraft launcher using a sequential arrangement of pulsed primaries are under discussion.

Table 12.3 lists a number of machine designs which have been investigated, ranging from laboratory scale constructions to demonstration models ("demonstrators") for industrial application.

Outstanding potential for the improvement of machines with respect to efficiency, overload capability, and torque smoothness can be seen in designs exploiting the ability of bulk superconductors to trap strong magnetic fields. However,

Table 12.3 Characteristics of SC machines: demonstrators and test models

Exploitation of HTSC	Machine type	Power rating	Benefits	Remarks	Date	Reference
SPM	Trapped field		Slot-less, smooth torque, overload	Design	2003	[12.2]
Hysteresis	Hysteresis	500 W		Demonstrator	1995, 1997	[12.9]
Hysteresis	Hysteresis	1 kW	Self starting, simple control	Laboratory tests	1998	[12.10]
				Demonstrator	2002	[12.12]
Hysteresis	Hysteresis	2 kW		Demonstrator	2000	[12.19]
Hysteresis	Hysteresis, asynchronous		Linear actuator	Laboratory tests	2001	[12.18]
Hysteresis	Hysteresis		Bipolar bulk YBCO stator	Demonstrator	2003	[12.26]
Hysteresis	Hysteresis	4 kW	Drive for aircraft cryogenic pump	Demonstrator	2003	[12.14]
Shield	Reluctance	Up to 5 kW (20 K)	Drive for aircraft cryogenic pump	77 K: YBCO 20 K: laminate of BSCCO/Ag	2004	[12.8]
Shield	Reluctance		Axial flux induction	Laboratory tests	1997	[12.27]
Shield	Reluctance	150 kW 200 kW	High speed, low inertia mass	Demonstrator Prototype	2002 2002 2003	[12.12] [12.13] [12.25]
SPM	Trapped field		Axial flux induction	Laboratory tests	1997	[12.27]
SPM	DC machine Pulsed trapped field	3.5 kW, (65 K)	High efficiency Axial flux induction	Automotive demonstrator	1995 1997	[12.20] [12.21]
SPM	Stepping motor		Simple assembly, Self levitating	Laboratory	2001	[12.36]
SPM	Generator		High energy density	Laboratory test	2003 1998	[12.24] [12.28]
SPM	Generator	3 kW		Laboratory test	2003	[12.33]
SPM	Trapped-field motor	1.5 kW (30 K)	8 poles, outer rotor type	Refrigerator integrated	2003	[12.32]
SPM	Synchronous		Linear actuator	Laboratory	2004	[12.30]

one of the most important problems to be solved at the present time is that of devising practical methods of magnetizing the superconducting permanent magnets inside the completed machine.

Trapped-field machines can be designed for use as direct current machines with stationary SPMs (Figure 12.1) or as synchronous machines with rotating SPMs. They can be considered as potentially strong competitors of machines with superconducting windings in a medium power range [12.14]. The conceptual design of a hybrid linear motor with both HTSC windings and bulk HTSC permanent magnets marks the possibility of a linear machine type so far without equal [12.13]. This machine needs – in addition to high-quality bulk HTSCs – the new type of conductors with high current-carrying capability which are actually under development in the form of coated YBCO conductors [12.34], [12.35].

References

12.1 H. Weh, H. May, *Achievable Force Densities for Permanent Magnet Excited Machines in New Configurations*, in Proceedings of the International Conference on Electrical Machines, ICEM'86, Munich (Germany) p. 1107

12.2 H. May, R. Palka, W.-R. Canders, Przeglad Elektrotechniczny **2003** 667 (No. 10: Special Issue XII International Symposium on Theoretical Electrical Engineering ISTET '03)

12.3 H. Weh, H. May, *Field Conditioning by Superconducting Screens* in: Superconductivity in Energy Technologies, edited by VDI Technologiezentrum, VDI-Verlag GmbH, Düsseldorf 1990, p. 110

12.4 H. Weh, *Drive Concepts with New Machines* in: Proceedings of the Symposium on Power Electronics, Electrical Drives, Advanced Electrical Motors (Speedam '92) Positano, Italy 1992, p. 29

12.5 H. May, R. Palka, W.-R. Canders, M. Holub, *Optimized firing of inverter fed Switched Reluctance Machines for high speed and high power applications*, ISEF Conference 2001, Cracow, see http://www.iem.ing.tu-bs.de/paper/2001/mapa_01.htm

12.6 G. J. Barnes, M. D. McCulloch, D. Dew-Hughes, *Supercond. Sci. Technol.* **13**, 875 (2000)

12.7 A. Sfetsos, M. Pina, A. Gonçalves, V. Neves, M. McCulloch, L. Rodrigues, *Flux Plot Modelling of Superconducting Hysteresis Machines*, papers at http://www-seme.dee.fct.unl.pt

12.8 L. K. Kovalev, K. V. Iljushin, V. T. Penkin et al. Supercond. Sci. Technol. **17** (2004) S460

12.9 T. Habisreuther, T. Strasser, W. Gawalek, P. Görnert, K. V. Ilushin, L. K. Kovalev, IEEE Trans. Appl. Supercond. **7**, 900 (1997)

12.10 L. K. Kovalev et al. Mater. Sci. Eng. B **53** (1998) 216

12.11 L. K. Kovalev, K. V. Iljushin, V. T. Penkin, Supercond. Sci. Technol. **13** (2000) 498

12.12 L. K. Kovalev, K. V. Iljushin, V. T. Penkin, Supercond. Sci. Technol. **15** (2002) 817

12.13 B. Oswald, M. Krone, T. Straßer, K. J. Best, M. Söll, W. Gawalek, H. J. Gutt, L. Kovalev, L. Fisher, G. Fuchs, G. Krabbes, H. C. Freyhardt, Physica C **372-376** (2002) 1513

12.14 L. K. Kovalev, K. V. Ilushin, K. L. Kovalev, V. T. Penkin, V. N. Poltavets, S. M. Koneev, I. I. Akimov, W. Gawalek, B. Oswald, G. Krabbes, Physica C **386** 419 (2003)

12.15 G. J. Barnes, D. Dew-Hughes, M. D. McCulloch, Supercond. Sci. Technol. **13** (2000) 229

12.16 G. J. Barnes, M. D. McCulloch, D. Dew-Hughes, Physica C **331** (2000) 133

12.17 H. Gutt, K. Feser, A. Grüner in Supraleitung und Tieftemperaturtechnik, Tagungsband zum 8. Statusseminar, p. 62, full text: CD attachment, ed. VDI- Technologiezentrum Düsseldorf (ISBN 3 93 138444 6)

12.18 R. Muramatsu, S. Sadakata, M. Tsuda, A. Ishiyama, IEEE Trans. Appl. Supercond. **11**, 1976 (2001)

12.19 A. L. Rodrigues, *Hysteresis motor with conventional and superconductor rotors* in: Prcoceedings of the International Conference on Electrical Machines, ICEM 2000, Espoo (at http://www.seme.dee.fct.unl.pt)

12.20 Y. Itoh, Y. Yanagi, M. Yoshikawa, T. Oka, S. Harada, T. Sakakibara, Y. Yamada, U. Mizutani, Jpn. J. Appl. Phys. **34** 5574 (1995)

12.21 T. Oka, Y. Itoh, Y. Yanagi, M. Yoshikawa, T. Sakakibara, S. Harada, Y. Yamada, U. Mizutani, J. Jpn. Inst. Met. **61** 931 (1997)

12.22 H. Ohsaki, Y. Tsuboi, J. Mater. Proc. Technol. **108** 148 (2001)

12.23 Y. Tsuboi, H. Ohsaki, IEEE Trans. Appl. Supercond. **13** 2210 (2003)

12.24 Y. Tsuboi, H. Ohsaki, Physica C **392-396** 684 (2003)

12.25 B. Oswald, M. Krone, K. J. Best, M. Söll, M. Setzer, W. Gawalek, H. J. Gutt, L. Kovalev, L. Fisher, G. Fuchs, G. Krabbes, H. C. Freyhardt, in: Supraleitung und Tieftemperaturtechnik,Tagungsband zum 8. Statusseminar, p. 63, full text: CD attachment, ed. VDI-Technologiezentrum Düsseldorf (ISBN 3 93 138444 6)

12.26 A. Alvarez, P. Suarez, D. Caceras, X. Granados, B. Perez, J. M. Ceballos, Physica C **398** 157 (2003)

12.27 P. Tixador, A. Tempe, P. Gautier-Picard, X. Chaud, E. Beaugnon, IEEE Trans. Appl. Supercond. **7** 896 (1997)

12.28 J. R. Hull, S. Sengupta, J. R. Gaines, IEEE Trans. Appl. Supercond. **9** 1229 (1998)

12.29 X. Granados, J. Pallares, S. Sena, J. A. Blanco, J. Lopez, R. Bosch, X. Obradors, Physica C **372-376** 1520 (2002)

12.30 A. Sugawara, H. Ueda, A. Ishiyama, Supercond. Sci. Technol. **17** S176 (2004)

12.31 Y. Tsuboi, M. Ohsaki, in *Applied Superconductivity 2003*, ed. by A. Andreone, G. P. Pepe, R. Cristiano, G. Masullo, IOP Publishers, Bristol and Philadelphia 2004, p. 871

12.32 M. Hirakawa, S. Inadama, K. Kikukawa, E. Suzuki, H. Nakasima, Physica C **392** 773 (2003)

12.33 H. Tabuchi, M. Chiba, T. Nitta, in *Applied Superconductivity 2003*, ed. by A. Andreone, G. P. Pepe, R. Cristiano, G. Masullo, IOP Publishers, Bristol and Philadelphia 2004, p. 907

12.34 Y. Shiohara, Physica C **412-414**, 1 (2004)

12.35 M. Chen, L. Donzel,M. Lakner, W. Paul, J. Eur. Ceram. Soc. **24**, 1815 (2004)

12.36 M. Komori, S. Nomura, IEEE Trans. Appl. Supercond. **11** 1972 (2001)

12.37 P. T. Putman, Y. X. Zhou, H. Fang, A. Klawitter, K. Salama, Supercond. Sci. Technol. **18**, S6 (2005)

12.38 L. Kovalev, K. Ilushin, V. Penkin, K. Kovalev, Electrical Technology **2** 465 (1994)

13
Applications in Magnet Technologies and Power Supplies

Based on the different electro-magnetic properties of superconductors, a wide range of further applications (in addition to magnetic bearings and energy converters) are finding a technological or economic niche.

The extremely high magnetization and the field gradient of trapped magnetic fields can be used for magnetic separation. The capability of carrying large currents in combination with low thermal conductivity are fundamental to the application of HTSC rods as superconducting current leads. The resulting reduction in the heat load together with the availability of high-performance cryocoolers now allows the operation of low-T_c superconducting coils which do not need to be provided with liquid helium. Furthermore, the superconducting transition itself can be used to control the fault currents in electrical transmission and distribution networks. Finally, the shielding properties of zero-field cooled HTSC bodies make it possible to prevent the internal space of a cavity in a bulk HTSC element from being influenced by an external field.

13.1
Superconducting Permanent Magnets with Extremely High Magnetic Fields

The ability of HTSC bulk materials to trap magnetic fields of more than 10 T at temperatures significantly below 77 K (Chapters 5 and 9) can be used in different types and arrangements of apparatus. The useful application of HTSC bulk materials with a trapped field of 4 T at a temperature of 50 K in superconducting magnetic bearings is described in Chapter 11. Furthermore, HTSC materials with a trapped field of about 2 T drastically improve the performance of energy converters (see Chapter 12).

13.1.1
Laboratory Magnets

Magnetizing bulk superconductors cooled in refrigerator cryostats offers the possibility of constructing small laboratory magnets. Using a superconducting sole-

High Temperature Superconductor Bulk Materials.
Gernot Krabbes, Günter Fuchs, Wolf-Rüdiger Canders, Hardo May, and Ryszard Palka
Copyright © 2006 WILEY-VCH Verlag GmbH & Co. KGaA, Weinheim
ISBN: 3-527-40383-3

noid magnet to magnetize a bulk reinforced Sm123 disk (60 mm diameter), a trapped field of 7.5 T has been obtained on the surface of the bulk HTSC at 42 K. The corresponding magnetic field on the top surface of the vacuum chamber was 5.2 T [13.1]. By the use of pulsed fields to magnetize a bulk Sm123 disk of 36 mm diameter, trapped fields of 3.8 T and 2.0 T have been generated on the surface of the bulk HTSC at 30 K and outside the vacuum chamber, respectively [13.2]. A multi-pulse method (IMRA, see Section 5.5) has been used to efficiently magnetize the bulk HTSC.

Alternatively, a face-to-face type superconducting magnet system with two cryostats consisting of a pair of Sm123 bulks mounted on the cold stages of the *Gifford McMahon* refrigerators in each vacuum vessel has been developed. Trapped fields of 2.7 T (1.8 T) have been obtained in the open gap of 1 mm (12 mm) between the vacuum chambers [13.2]. Once magnetized, the equipment retains its magnetic induction over a long period.

Furthermore, a magnetic field source of 150 mm diameter has been demonstrated consisting of 7 pieces of Sm123, each of 50 mm diameter [13.3]. It should be noted that this design offers the possibility of tailoring the contour and degree of its saturation of the resulting field by changing the material and the size of individual cylinders. This option is essential for the performance of experiments in the gap of the field source.

13.1.2
Magnetic Separators

Magnetic separators, which have been known for over 100 years, have been widely used by the minerals industry. They utilize electromagnets, low-T_c superconducting coils, or permanent magnets in conjunction with iron magnetic circuits to separate magnetic particles from the host material. Low-T_c superconducting coils have penetrated into specific sectors of this market. They have been extensively used in high-gradient magnetic separators for kaolin treatment.

The criterion for the effectiveness of a magnetic separation set-up is the magnitudes of both the radial component of the flux density and its gradient at different distances from the surface of the excitation system. Technical advances in magnetic separator designs made possible by the use of HTSC **S**uperconducting **P**ermanent **M**agnets (SPM) are expected to lead to the commercial use of high-gradient magnetic separators for a wide spectrum of tasks. They are potentially applicable to the separation not only of strongly ferromagnetic but also of weakly dia- and paramagnetic particles.

A number of configurations of the SPM in relation to design and performance of magnetic separators have been considered [13.4, 13.5]. The proposed configuration of the SPM for a high-gradient "Drum-Equipped Superconductive Ore Separator System" [13.7] is shown in Figure 13.1.

An example of a High-Gradient Magnetic Separator System (HGMS) using bulk Ln123 magnets in a face-to-face configuration and the procedure to magnetize it by a pulsed-field method was demonstrated in Ref. [13.6]. The separation pipe contains stuffed filter matrices composed of ferromagnetic wires and is set between the poles of a superconducting permanent magnet system (see previous section), which causes magnetic saturation of the wires. The slurry containing colloidal fine magnetic particles is caused to flow through the pipe. Magnetic particles are collected under the influence of the steep magnetic gradient surrounding the saturated wires. Tests using finely powdered hematite (a-Fe_2O_3) revealed a separation rate of over 90%. The filters can readily be regenerated by passing a stream of water through them after the magnetic field is switched off.

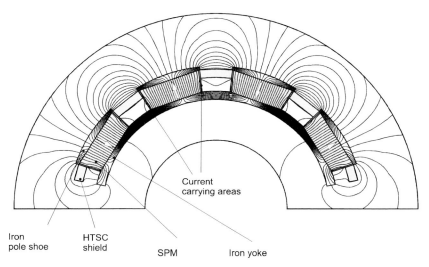

Figure 13.1 Magnetic field in the wet drum separator. Field excitation by high-field SPM with iron pole shoes and stray-field suppression by HTSC-shields

The advantage of magnetic separators with SPM is that they are small enough for mobile applications. A mobile magnetic separator system has been developed to remove water pollution and tested for water bloom removal [13.8]. The system is a contribution to the solution of an environmental problem in certain regions and a pilot test for related separation tasks in environmental and chemical technology. The conceptual design has two technological steps: (a) the chemical pre-process treatment of the goods to render them suitable for the magnetic separation, and (b) the magnetic separation using the magnetic separator system.

Major specifications of the conceptual design are listed in Table 13.1 together with some results of the test. The scheduled performance of the magnetic separator is 100 m³ per day. The first step depends on the actual task. In the case considered, water bloom removal, coagulants such as magnetite, other inorganic compounds of iron, and polymers were fed into the pre-process unit. Within 4 min, the colloidal biological load and suspended solids were flocked, together with the added Fe load.

Table 13.1 Specification of the mobile separator system

Capacity				100 m³ per day	
Mean residence time of water				5 min	
Pre-process cycle				4 min	
Bulk size and covered length				$33 \times 33 \times 20$ mm³ / 200 mm	
Superconductor temperature				60 K	
Impurity:	chlorophyll a	Load (µg/m³):	0.59	Removal efficiency (%):	94
	suspended solids		5		96
	N + P		10		90
	Fe		2.7		84
Biological oxygen demand			34		90

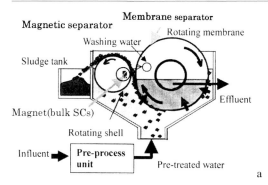

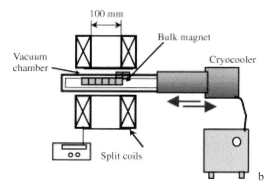

Figure 13.2 Mobile magnetic separator system for removal of water pollution [13.8]. a Combined separator system in which the superconducting magnet separator is the larger membrane separator drum. b Sketch of equipment for magnetization of the superconducting separator. Reprinted from [13.8] with permission from Elsevier

A superconducting separator system is responsible for the second step. The separator system is shown in Figure 13.2a. It combines two separation principles. The magnetic flocks which are deposited on the rotating drum of the membrane filter (large wheel, Figure 13.2) are removed under the influence of the washing water and the magnetic force exerted by the bulk HTSC inside the rotating shell of the smaller drum and are finally conveyed to the sludge tank. Seven YBCO blocks 33 mm square are cooled to a temperature less than 60K, thus enabling a field of 2–3 T to be generated at the surface of the cylinder. Cooling is provided by a Gifford–McMahon cryocooler.

Magnetization of the bulk superconductors in a mobile system is a further challenge [13.9]. The superconductor unit is shown in detail in Figure 13.2b together with the coils of a split-solenoid superconducting magnet which were used for magnetizing the HTSC bulks under a field of 5 T at a temperature of 35 K. Pre-processing and separator systems were installed on a 4 ton trailer.

13.1.3
Sputtering Device

The magnetic field generated by a bulk superconductor can be over than 10 times that of a conventional NdFeB permanent magnet in a standard magnetron sputtering facility. In the case of a ferromagnetic target, the magnet underneath the target should produce a field higher than the saturation magnetization of the target to avoid the suppression of the sputtering rates due to absorption of magnetic field in the target. The high trapped field of an SPM offers the possibility of constructing an alternative magnetron sputtering facility with high performance [13.10]. Figure 13.3 shows a sketch of the magnetron cathode equipped with an Sm123 bulk of 60 mm diameter magnetized at 40 K. A high magnetic field (4.2 T) at the surface of the superconductor coupled with a high target voltage (6 kV maximum) enabled discharge even at 10^{-8} bar pressure and targets of Cu, Ni, and Fe with a long target-to-substrate distance (300 mm). Uniform Cu films with high bottom coverage were deposited within a circle of 120 mm diameter on the substrate [13.33].

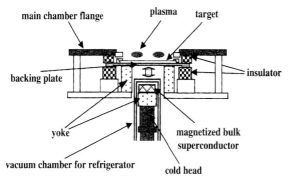

Figure 13.3 Schematic illustration of the magnetron cathode using Sm123 SPM. The minimum distance between the surface of the vacuum cylinder and the surface of the backing plate is 5 mm [13.32]

13.1.4
Superconducting Wigglers and Undulators

For the generation of intense X-ray synchrotron radiation, the beam of charged particles (e.g., electrons) produced by an accelerator system passes through a rectangular tube. The radiation is generated by (in the longitudinal direction) alternating magnetic poles on both sides of the tube, forming dipoles which accelerate the charged particles orthogonally to the direction of the field and movement.

High-performance excitation systems in electron accelerators consist of low-T_c superconducting coils operating at $T < 4.2$ K. In the case of *undulators* these have been replaced in certain facilities by blocks of high-performance NdFeB permanent magnets, which are used at cryogenic temperatures (the chosen term "cryogenic permanent magnet, CPM" should not be confused with the superconducting permanent magnet, here abbreviated as SPM) [13.11]. Their substitution by HTSC bulk materials offers the possibility of using permanent magnets also at the higher field level > 5 T necessary for *wigglers* [13.11, 13.12].

The desired shape of the magnetic field may be achieved either by an appropriate pole contour of the SPM or by the form of the magnetizing coil and the magnetizing procedure. In the design proposed in Ref. [13.11], a Gd123 bulk superconductor is an equivalent to a coil with one turn.

Wigglers and undulators are without doubt an important and extremely interesting field for applications of superconductors. However, the real challenge for bulk HTSC magnets is to achieve the precision rather than the strength of the field. An extreme precision is indispensable if conventional permanent magnets in the accelerator line have to be substituted by HTSC permanent magnets.

13.2
High-Temperature Superconducting Current Leads

The introduction of superconducting technology is making it possible to use very high currents, which, however, have to be transferred from a grid working at ambient temperature to the low temperature of the superconducting equipment, which include superconducting magnetic energy stores (SMES), NMR coils for tomography and high-field NMR units, high-field magnets in high energy accelerator facilities, and fusion magnets. Even for the excitation coils of electrical machines, currents have to be transferred from ambient to the low working temperature of superconductors [13.13].

A remarkable demand originates from the National and International High-Energy Laboratories [13.14]. Significant power loss is caused by both the resistance of conventional metals and heat transport to the low-temperature facility. According to the *Wiedemann–Franz* law, the high electrical conductivity of appropriate conventional metallic rods (e.g., Cu, Al) is accompanied by a high thermal conductivity. In contrast, HTSCs have low thermal conductivity. Thus, the first commercial impact of HTSCs on superconducting magnet systems has been the provision of current leads with a significantly reduced heat load. In a 20 kV / 2 kA DC experiment with a 0.5 kA class current lead (3 mm in diameter, 30 mm long), the specific heat load to the helium bath was determined to be 0.1 W kA^{-1} and the heat transfer from 80 K to 20 K was found to be 163 mW. The benefit corresponds to a reduction in cooling power of nearly 25% [13.15]. Current leads with reduced heat load together with the development of *Gifford–McMahon* type high-performance cryocoolers are the fundamental innovations realized in a generation of cryogen-free LTS magnets [13.16].

HTSC current leads are suitable to be used between the 77 (or 65) K and 4 K terminal of superconducting power magnets. Thus, homogeneity and mechanical behavior of the material become of crucial importance because of the thermomechanical stress that arises. So far, both Bi2212 (preferably in the form of tubes) and Y123 have been developed and applied for power magnets. Since the lead should also resist a quench, the current lead usually consists of a number of HTSC rods in parallel with a safety shunt and copper terminals. Thus, the temperature in the circuit remains below about 150 K. Especially Y123 current leads should be oriented with their c axis perpendicular to the flowing current. For 0.5 or 1.3 kA class current leads, highly oriented YBCO materials have usually been grown by directional solidification or a zone melt technique in strong temperature gradients. Typical lengths of superconducting current leads are in the region of 100 mm. An example of a data set for a Y123 current lead [13.18] is represented in Table 13.2. This 13 kA current lead has been tested by CERN and has fulfilled CERN's specifications concerning current capacity, heat leak, contact resistance, and quench protection of the superconductor. Also, resin-impregnated bulk YBCO current leads have been reported for use in MAGLEV under a 1 kA load [13.17].

Table 13.2 Representative parameters of a 13 kA current lead [13.18]

Heat leak (W)		Critical current (kA)		Contact resistance HTSC / terminal (nΩ at 12 kA)		Heat production (W at 12 kA)	Duration of quench (s)	
65 → 5 K	50 → 5 K	5 → 65	77 K	4 K	65 K	4 K terminal	at 6 kA	at 12 kA
1.84	1.20	>12.1	6	<1	19	0.17	60	45

13.3
Superconducting Fault Current Limiters

Any accessory part designed for use in an electric power system has to resist mechanical and thermal stresses under the influence of a potential short circuit current I_{sc}, which can exceed the nominal current I_n (current under normal operation) by up to 100 times. A reduction of the limiting current I_{sc} to the prospective current (I_p appearing in the worst case in an unprotected circuit) is an important contribution to reducing costs and increasing safety in the power grid.

Superconducting fault current limiters (SFCL) offer an ideal performance, since in normal operation their impedance can be neglected and, in the case of a fault, the transition to the normal state limits the current passively and reversibly [13.19]. Two basic concepts have been followed in order to realize an SFCL: the resistive and the inductive concept (Figure 13.4).

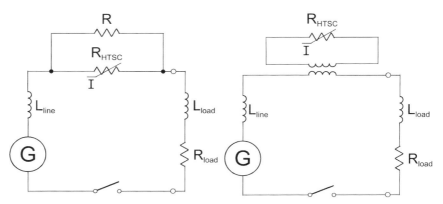

Figure 13.4 Scheme of the resistive (left) and inductive (right) types of SFCL

13.3.1
Inductive Fault Current Limiters

The inductive limiter concept applies the superconducting element to a transformer secondary, with the primary in the circuit (Figure 13.4b). With an appropriate layout, a fault current causes the superconductor in the secondary to quench, and the drastic change in the secondary impedance due to the quench results in a limitation of the fault current to a few tens of ms. Instead, an in-line choke is applied in the design of (Figure 13.5), where a bulk superconductor tube shields the iron core from the field of the sole winding during normal operation when the impedance is nearly zero [13.20]. The control ring in the design of Figure 13.5 has been added to shorten the recovery time.

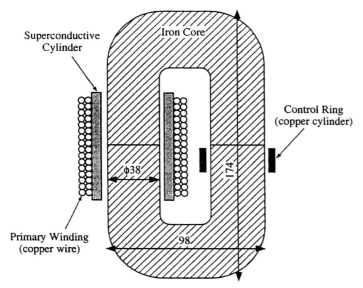

Figure 13.5 Design of an inductive SFCL using a choke with a BSCCO tube to shield the iron core [13.20]

13.3.2
Resistive Superconducting Fault Current Limiters

The resistive SFCL is directly in series to the line (Figure 13.4a). Cross-section and geometry have to be designed to minimize the interference with the network to a negligible amount under normal operation. In the event of a fault, the fault current pushes the superconductor into the normal state ("quench"), and resistance appears in the line. Different limitation characteristics have been distinguished in Figure 13.6 [13.19]:
 1. fast heating of a short-length superconductor
 2. "slow heating" of a long-length superconductor, which is implicitly realized in the case of materials exhibiting an expressed flux flow region (e.g., BSCCO)
 3. constant-temperature regime (infinite length approach).

In the first case, the superconductor heats up rapidly and the current density for quenching j_q is achieved after a few 100 µs, thus limiting I_{sc} to near to I_n.

The second case can be expected in materials like Bi2212, where the onset of flux flow (according to the superconducting phase diagram) limits I_{sc} for the first 10 ms. In contrast, Josephson barriers appear in multigrain YBCO material. The electric field over a length Δl amounts to $E = U/\Delta l$, and the power density jE has to be kept below a specific threshold value to avoid overheating which would result in injurious deterioration of the element. Any inhomogeneity in the material can especially be a source of "hot spots".

In practice, the superconductor is designed to have a critical current 2 or 3 times the value of the fault current I_{sc}. In a resistive SFCL, the superconducting element, rod, meander, or tube of the Ln123 or BSCCO type (see Chapter 6) is arranged parallel to a conventional resistive or inductive element. Then, during a fault, the resistance developed in the limiter shunts the current in part through the conventional element, which absorbs most of the fault energy. The SFCL recovers to its working state after recooling within hundreds of ms or seconds. The time span is such as to permit breaking the power line safely by conventional means (indicated by the switch in Figure 13.4).

In the third case, the power density is negligible, allowing a quasi-isothermal regime. However, the cost and size of long superconductor elements, the power losses (especially for AC use) in the operating state, and the long recovery time make this approach unfavorable.

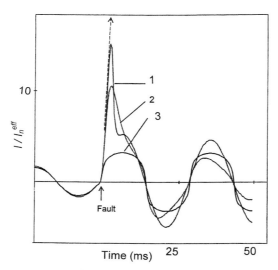

Figure 13.6 Different types of limiting behavior of an SFCL due to varying the conductor length and/or the choice of material (numerical simulation [13.19])

13.3.3
Status of High AC Power SFCL Concepts

Previous test and simulation results favor the resistive type of SFCL because of its lower power loss under normal working conditions, lower cost, and lower weight. Since most applications are to be expected in the AC field, the impedance is more important than the ohmic (DC) resistance, even in the case of an iron-free design.

Numerous engineering solutions and variants of a "hybrid" type have meanwhile been proposed and tested on the laboratory scale covering the boundaries between the two fundamental types. Promising examples of the two basic materials concepts are briefly considered in more detail here.

Resistive SFCL with Bifilar BSCCO tubes

This SFCL design applies bifilar BSCCO elements prepared by spin casting (Figure 6.15, Chapter 6), with a parallel metallic bypass. The resistance of the shunt must be low enough to protect the superconductor; on the other hand, extreme heating of the shunt results from too low resistivity. The electrical field limitation for deterioration of superconductivity can be assumed to be $E_{eff'} = 0.5$ V cm^{-1}. The maximum limitation time of the SFCL, t_l, should not exceed 100 ms. The metallic shunt in the considered SFCL [13.21] is in direct contact with the BSCCO tube, and therefore its temperature should never exceed a limit of about 300 K. Thus, the maximum heating rate T' is about 2300 K s^{-1}. From the equation $T' = E_{eff'}^2 (\rho\ C_v)^{-1}$, a CuNi alloy would be an appropriate choice (ρ = 40 µOhm cm, $C_{v,\ average}$ = 2.9 Ws cm^{-3}K^{-1}).

A series of 90 SFCL elements were designed to realize a 10 MVA 3-phase current limiter prototype, which has recently been set into operation in a German 10 kV power grid. Pre-prototype tests on the 10 kVA level (9 elements) have been reported in Ref. [13.22]. The quench characteristic was found to be similar to that in Figure 13.6 (curve 1), which in the considered case here is caused by the fast current transfer from the superconductor to the shunt coil. Tests have shown that the critical current and consequently the protected power increase by about 50% on reducing the working temperature from 77 to 66 K. A prototype for use in a 138 kV power grid is under development now [13.23].

Resistive SFCL with Melt-Grown YBCO Material

To realize an appropriate resistance parallel to *ab* planes in single-grain YBCO, a sufficient length of the current path is necessary, which preferably is obtained using a meander structure of planar superconducting elements (Figure 6.15, Chapter 6). Such element is coated with AgAu alloy or – more efficiently – joined by a NiCr by-pass reinforcement on both sides of the plate to reduce thermal stresses [13.24, 13.25]. A representative cross section of the current path is 2.5×0.8 mm^2, the length is tens of cm. The current under working conditions is rated at 1 kA with the voltage limit under fault conditions of 100 V [13.26].

Test results indicated that with an appropriate number of meander-shaped *single-grain* Y123 elements, similar limiting properties can be achieved to those for the Bi2212 tube variant. Loss was found to be independent of frequency and was therefore attributed to the pinning behavior [13.27]. Multigrain meander elements, however, are not acceptable because of the high losses caused by a high resistivity at grain boundaries [13.28]. The number of meanders (300) from single-grain YBCO to realize a prototype of a 30 MVA limiter or of bifilar tubes (90) for the 10 MVA prototype, respectively, give an impression of the size of the equipment [13.22, 13.26].

13.4
High-Temperature Superconducting Magnetic Shields

Engineers in electrical instrumentation and metrology are frequently confronted with problems of shielding sensitive circuitry and electronic or magnetic (e.g., SQUIDs) sensors from external electromagnetic fields. The use of materials with high conductivity enables easy shielding from high-frequency electromagnetic fields (Faraday's and Lenz's law) to be achieved. Generally, a wall thickness > 20 times the skin depth can be considered as sufficient for effective shielding. The use of superconductors is usually not required for shielding time-dependent fields.

In contrast, shielding of low-frequency or even d.c. magnetic fields would force the use of sometimes very thick magnetic materials. The best conventional material for such d.c. field shielding applications is a Ni77Fe15-based alloy with additives such as Cu or Mo (e.g., MuMetal). From Figure 13.7, one can conclude that for perfect shielding the opening of the cavity has to be directed orthogonally to the direction of the external field.

If the mass of such conventional shielding is not tolerable and the need for a cryogenic system can be accepted, HTSC shields offer a favorable alternative. If only very low d.c. fields are to be shielded, the superconductor operates in the Meissner state as an ideal diamagnetic material. Above the lower critical field H_{c1}, the inner space remains free of field until the penetrating applied field H reaches the inner wall. For a cylindrical cavity a wall thickness of $\Delta_{\lim} \triangleq H j_c^{-1}$ (H parallel to the axis) is required to shield this field. In multigrain materials of the 123 type, the shielding current is limited by the weak link behavior of grain boundaries whereas for Bi2212 type materials, the onset of flux creep has a limiting influence, making this material a better alternative. Large sized single grain Ln123 materials are now available for shielding even higher fields. The shielding factor of tubes with 5 mm single grains of YBCO exceeds the value for sintered YBCO by one order of magnitude in the low frequency range (1 – 200 Hz) [13.29].

The use of HTSC shields enables sensitive measurements to be performed outside a field-conditioned laboratory. Note, however, that first the HTSC cavity has to be cooled in a field-free room before it is then transported to the place of use. The internal shielding currents prevent the HTSC cavity from being influenced by the external field. The major part of the cavity remains shielded from the external fields even after opening the cavity to perform measurements provided that the open end faces in the same direction as the main component of the field to be shielded (Figure 13.7b).

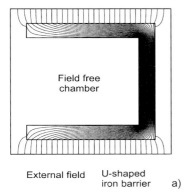 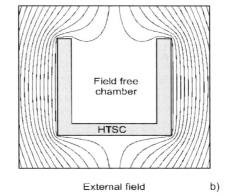

Figure 13.7 Field-free chambers made from (**a**) high permeability alloy and (**b**) HTSC bulk cylinders

Experience of the high-frequency magnetic shielding properties of HTSC cylinders and their use in combination with metallic cylinders is discussed in Refs. [13.30] and [13.31]. Magnetic shielding vessels with a closed refrigerator are meanwhile available commercially.

References

13.1 Y. Yanagi, M. Yoshikawa, Y. Itoh, T. Oka, H. Ikuta, U. Mizutani, Physica C **412**, 744 (2004).

13.2 T. Oka, Y. Itoh, Y. Yanagi, M. Yoshikawa, H. Ikuta, U. Mizutani, Physica C **335**, 101 (2000).

13.3 T. Oka, K. Yokoyama, K. Noto, IEEE Trans. Appl. Supercond. **14**, 1058 (2004).

13.4 J. H. P. Watson, Miner. Eng. **12**, 281 (1999).

13.5 J. H. P. Watson, I. Younas, Mater. Sci. Eng. B **53**, 220 (1998).

13.6 K. Yokoyama, T. Oka, H. Okada, Y. Fujine, A. Chiba, K. Noto, IEEE Trans. Appl. Supercond. **13**, 1592 (2003).

13.7 H. D. Wasmuth, K. H. Unkelbach, Miner. Eng. **4**, 825 (1991).

13.8 H. Hayashi, K. Tsutsumi, N. Saho, N. Nishizima, K. Asano, Physica C **392–396**, 745 (2003)

13.9 N. Nishijima, N. Saho, K. Asano, H. Hayashi, K. Tsutsumi, M. Murakami, IEEE Trans. Appl. Supercond. **13**, 1580 (2003).

13.10 T. Matsuda, S. Kashimoto, A. Imai, Y. Yanagi, Y. Itoh, H. Ikuta, U. Mizutani, K. Sakurai, H. Hazama, Physica C **392–396**, 696 (2003).

13.11 T. Hara, T. Tanaka, H. Kitamura, T. Bizen, X. Marechal, T. Seike, T. Kohda, Y. Matsura, Phys. Rev. Special Topics – Accelerators and Beams **7**, 050702 (2004).

13.12 H. Matsuzawa, J. Appl. Phys. **74**, R111 (1993).

13.13 K. Maehata, K. Ishibashi et al. Cryogenics **28**, 744 (1988).

13.14 P. Komarek, Supercond. Sci. Technol. **13**, 456 (2000).

13.15 R. Endoh, H. Kato, T. Izumi, Y. Shiohara, Physica C **392–396**, 1167 (2003).

13.16 A. Hobl, D. Krischel, M. Poier, R. Albrecht, R. Bussjaeger, U. Konopka, IEEE Trans. Appl. Supercond. **13**, 1569 (2003).

13.17 M. Tomita K. Nagashima, T. Herai, M. Murakami, Physica C **372–376**, 1216 (2002).

13.18 J. G. Larsen, B. Kristensen, J. G. Sommerschield, E. Frost, IEEE Trans. Appl. Supercond. **9**, 487 (1999).

13.19 W. Paul, M. Chen, M. Lakner, J. Rhyner, D. Braun, W. Lanz, Physica C **354**, 27 (2001).

13.20 M. Ichikawa, M. Okazaki, IEEE Trans. Appl. Supercond. **5**, 1067 (1995).

13.21 S. Elschner, F. Breuer, M. Noe, T. Rettelbach, H. Walter, J. Bock, IEEE Trans. Appl. Supercond. **13**, 1980 (2003).

13.22 J. Bock, F. Breuer, H. Walter, M. Noe, R. Kreutz, M. Kleimaier, K.-H. Weck, S. Elschner, Supercond. Sci. Technol. **17**, S122 (2004).

13.23 L. Kovalsky, X. Yuan, J. Bock, S. Schwenterly, *Superpower HTS Matrix Fault Current Limiter in Superconductivity for Electric Systems*, Annual DOE Peer Review 2004 (www.ornl.gov)

13.24 M. Morita, O. Miura, D. Ito, Physica C **357–360**, 870 (2001).

13.25 L. Porcar, D. Buzon, E. Floch, P. Tixador, D. Isfort, D. Bourgault, X. Chaud, R. Tournier, Physica C **372–376**, 1639 (2002).

13.26 M. Morita, H. Hirano, H. Hayashi, K. Terazano, K. Kajikawa, K. Funaki, T. Hamajima, Physica C **412–414**, 750 (2004).

13.27 E. S. Otabe, T. Endo, T. Matsushita, M. Morita Physica C **357–360**, 878 (2001).

13.28 M. Noe, K. P. Juengst, F. Werfel, S. Elschner, J. Bock, A. Wolf, F. Breuer, Physica C **372–376**, 1626 (2002).

13.29 H. Fang, J. R. Claycomb, X. Y. Zhou, P. T. Putman, S. Padmanhabhan, J. H. Miller, K. Ravi-Chandar, K. Salama, IEEE Trans. Appl. Supercond. **13**, 3103 (2003).

13.30 K. Hoshino, Magnetic Shield of High Tc Oxide Superconductor – Application for Biomagnetism, ISTEC – Journal Vol 5, No. 2 (1992).

13.31 Y. Horijowa, A. Omura, K. Mori, M. Itoh, IEEE Trans. Appl. Supercond. **11**, 2387 (2001).

13.32 U. Mizutani, H. Hazama, T. Matsuda, Y. Yanagi, Y. Itoh, H. Ikuta, K. Sakurai, A.Imai, Supercond. Sci. Technol. **18**, S30 (2005).

13.33 H. Hazama, T. Matsuda, U. Mizutani, H. Ikuta, Y. Yanagi, A. Sakura, A. Sekiguchi, A. Imai, Jap. J. Appl. Phys. **43**, 6026 (2004).

List of Abbreviations

CN	coordination number
FC, ZFC	field cooled, zero-field cooled
FEM	finite element method
L	liquid phase
MTG	melt texturing growth
MMTG	modified melt texturing growth
MMCP	modified melt crystallization process (alternative solidification path)
MFC	maximum field cooled
OFC	operational-field cooled
OCMG	oxygen partial pressure controlled melt growth
OCIG	oxygen controlled isothermal growth
CCOG	concentration controlled in oxygen growth
PM	permanent magnet (ferromagnetic)
SPM	superconducting permanent magnet
SFCL	superconducting fault current limiter
SMB	superconducting magnetic bearing
SEM	scanning electron microscope
STM	scanning tunneling microscope
TEM	transmission electron microscope

Further abbreviations which have only locally been used are explained in the text.

High Temperature Superconductor Bulk Materials.
Gernot Krabbes, Günter Fuchs, Wolf-Rüdiger Canders, Hardo May, and Ryszard Palka
Copyright © 2006 WILEY-VCH Verlag GmbH & Co. KGaA, Weinheim
ISBN: 3-527-40383-3

Frequently used Symbols

B	magnetic flux density
B_{irr}, $\mu_o H_{irr}$	irreversibility field
B_p	penetration field
C_p	specific heat
F	force
F_p	pinning force
G	*Gibbs* free energy
H	[1] magnetic field strength, [2] enthalpy
H_c	thermodynamic critical field
H_{c1}, H_{c2}	lower and upper critical field, respectively
j, j_c	current density, critical current density
K_{IC}	fracture toughness
k	*Boltzmann* constant
M	[1] magnetization, [2] torque
n	amount of substance
p	pressure
$p(O_2)$	partial pressure (of O_2)
R	gas constant
R	linear growth rate
r	radius, ionic radius
S	[1] flux creep rate, [2] entropy
T	temperature
U	activation energy (flux creep)
α	coefficient of expansion
γ	*Sommerfeld* parameter
γ_{Gr}	*Grüneisen* parameter
δ	deviation from (ideal) stoichiometry, especially for oxygen
ξ	coherence length
λ	[1] penetration depth, [2] electron-phonon coupling constant
κ	[1] heat conductivity, [2] mole fraction
Φ	magnetic flux
Φ_o	flux quantum
μ_o	permeability of vacuum
σ	stress, tension
σ_{max}	tensile strength

Some symbols which have been used only locally or with a different meaning are explained in the text.

See section 1.1.2 for the nomenclature of HTSC compounds which is used in this volume.

Index

a
Abrikosov lattice   2
adiabatic approach   114 ff
Ag/LnBaCuO composites   166, 168
Ag/YBaCuO   167
Ag/YBCO composite materials   138
anisotropy   1, 15
– anisotropy parameter γ 6, 80
– B_{c2} anisotropy   78
armature current   261

b
BCS theory   1, 98
Bean's model   8, 194
Bi-2212   15
Bi-2223   15, 18
$Bi_2Sr_2CaCu_2O_y$ 80
Bragg glass   2, 80, 174
bulk Bi2212   151
bulk YBaCuO/Ag composite materials, improved mechanical properties   141
bulk YBCO
– inhomogeneities   120
– visualization of inhomogeneities   125

c
calcination   46
chemical potential   31, 32, 39
chemistry   24
– defect   26
coherence length   1, 3, 6, 16, 132
contacting   150
cooling modes
– field cooling   105 ff, 192, 194
– zero-field cooling   105 ff, 193, 195 ff
cracking   110 ff, 180, 183, 186 ff
critical current density   10, 19, 71, 171, 176, 191
– critical currents from magnetization loops   9
– depairing critical current   3
– intergrain   146
– magnetization measurements   83 ff
– transport measurements   72, 82 ff
critical field
– lower critical field   3
– thermodynamic critical field   3
– upper critical field   3, 6
critical state   8, 116
crystallization   46
current lead   148, 284

d
Debye temperature   91
defects   39, 47, 56, 159
– interstitial   26
– point   26
– vacancies   26
dislocations   47, 56
doping   24, 83, 129, 132
– chemical   26
– chemistry   133
– extrinsic   24, 27
– intrinsic   24
– Li doped   109
– Li-doped bulk YBCO   176 ff
– Zn doped   109, 185 ff
– Zn-doped YBCO   187
dynamic approach   114 ff

e
Ehrenfest relationship   100
elastic moduli of bulk YBCO
– shear modulus   95
– Young's modulus   95
elastic moduli of flux line lattice
– bulk (or compression) modulus   92

– shear modulus   91
– Young modulus E   92
elastic properties of flux line lattice
– modulus of isotropic compression   5
– shear modulus   5
– tilt modulus   5
electric field-current relation   13 ff
electrical machines
– generator   259
– hysteresis or induction machine   260
– motor   259
– reluctance machine   260
– trapped field machine   260
– use of bulk HTSC   273, 275
extrinsic doping   27

f
fault current limiters   148
– resistive   149
field profiles   9, 20, 106 ff, 187, 194 ff
finishing   143
flux creep
– bulk YBCO   86 ff
– HTSCs   11
– in low-temperature superconductors   11
– reduction of flux creep   87 ff
– SMB   227 ff
– YBCO single crystals   90
flux creep models
– Anderson-Kim model   12 ff, 14, 86
– collective creep   12, 86 ff, 177
– melt-textured YBCO samples   13
– plastic creep   90, 177
– plastic vortex creep   174
– YBCO single crystals   13
flux jumps   105, 113 ff, 183, 184
– first flux-jump field   114 ff
flux line   3
flux line lattice   2
– experimentally confirmed   4
flux motion   3, 7, 11
functional elements   144, 145

g
Gibbs free energy   31, 50, 56
Gibbs's phase law   32
Ginzburg-Landau parameter   4
grain boundary   47, 125
– low angle   56, 58, 131, 144
growth sector   50, 59
Grüneisen parameter   99

h
heat conductivity   114, 116
high-dynamic drives   272
holes   26

i
incongruent   47
infiltration technique   148
interstitial   26
ionic radius   33, 43, 130
irradiation technique   83, 136
irreversibility field   15, 17, 75, 80, 84 ff, 135, 191

j
joining   146

l
levitation, electrodynamic levitation   200
liquidus   40, 41
liquidus surface   39, 42, 45
Ln123   157, 165
– advanced processing   160
Lorentz force   6, 7, 262
losses   116

m
magnetic bearing
– active magnetic bearing   240
– feedback-controlled   200
– hybrid superconducting magnetic bearing   240
magnetic phase diagram   191 ff
magnetic relaxation   11
magnetic separators with SPM   280
– magnetization   282
– mobile separator system   282
magnetic shields with bulk HTSC   290
magnetic tensile stress   111 ff, 179
magnetization
– irreversible   7, 9, 10
– reversible   7, 9, 10, 78
magnetomotive force   266
mass flow   50 ff
maximum trapped field   9, 20
meander   149
mechanical properties   91 ff
– bending strength   94
– bulk modulus   93
– data for bulk YBCO   93 ff
– flexural strength   94 ff
– fracture strength   92, 95
– fracture toughness   92, 94 ff, 113

- shear modulus   93
- tensile strength   94, 111 ff, 184
- Young's modulus   93
Meissner effect   192
Meissner state   191, 192, 201, 205
melt cast process   151
melt texturing techniques   19
melt-texturing techniques   47
- composition control in oxidizing atmosphere for growing (CCOG)   163
- isothermal processing at variable oxygen partial pressure (OCIP)   161
- Melt-Powder-Melt Growth (MPMG)   48
- modified crystallization process (MMCP)   48
- modified melt crystallization   60
- oxygen-controlled melt growth process (OCMG)   161
- $p(O_2)$ 161
- post-growth annealing of Ln123   161
- Quench-Melt-Growth (QMG)   48
microcracking   15
microcracks   92
microstructure   53, 56
mixed state   7, 191, 194, 201
morphology   53
mosaic structure   59, 146
multi-seeded melt growth   144

n
Nd-based HTSC   175
Nd123   157
nomenclature   3
numerical calculations
- inverse field problem   120 ff
- j_c distribution   121, 124 f
- local critical current density field profiles   119 ff
- stress distribution   180 ff
numerical calculations – forces
- perfectly trapped flux model   250 ff
- vector-controlled model   253 ff

o
ohmic loss   6
order parameter   96 ff
ordering   159
oxygen   35
- chemical potential   36, 38
- ordering   38
- stoichiometry   26, 36, 37
oxygen partial pressure   63

p
pairing mechanism
- d wave   97
- d-wave pairing   2
- s wave   97 ff
- s-wave pairing   2
pancake vortices   2, 15, 80
paraffin, imbedding in –   152
particle inclusions   54
- size of   56
peak effect   68, 171
- bulk HTSC   175 ff
- field-induced pins   175 ff
- peak field B_p 173 ff
- single crystals   172 ff
penetration depth   4
penetration field   116, 117, 195
peritectic temperature   50, 130, 148
perovskite   23, 33
phase diagram   31 ff
- EuBaCuO   34
- GdBaCuO   34
- LaBaCuO   34
- NdBaCuO   34, 156
- $p(O_2)$ 36, 37, 41, 42, 44, 156
- SmBaCuO   34, 40, 43, 44
- subsolidus   34, 36
- YBaCuO   34, 40, 41
phase separation   38, 158
pinning
- collective pinning   69, 173, 175
- intrinsic pinning   72
- single vortex pinning   69
pinning features
- dislocations   68
- oxygen vacancies   81, 172
- single vortex pinning   83
- stacking faults   67 ff
- twin boundaries   71 ff
- Y-211 precipitates   68 ff
pinning force   7
- collective pinning   17
- elementary pinning force   17
- models   17 ff
- scaling   16 ff, 84 ff, 134, 135
- volume   134
- volume pinning force   8, 16 ff
plastic flux motion   18
$p(O_2)$ 35
precipitation, sub-micro particles   135
primary crystallization   39, 40, 45, 46, 49
pulsed field
- irreversibility fields   75 ff

- magnetized 117 ff
- magnetizing 115 ff
- rise times 116 ff
- upper critical field 78

q

quantum of magnetic flux 3

r

rare-earth elements 2
reactions 47
- armature 271
- congruent 43
- invariant 32, 42
- univariant 32, 39, 42
refinement of the 211 particles 70 ff
refinement of the Y211 precipitations 83
reluctance motor 145
- armature reaction 270
- dynamics 274
- low inertia 272
- overload capability 271
- torque 274
resin 145
resin impregnation 151
rings 144, 145

s

seed 49, 144
- alternative seeding techniques 165
shaping 143
single crystals 46
Sm123 157
SMB
- air gap 219
- cross stiffness 209
- cryocontainer 234 ff
- cylindrical 236
- cylindrical excitation system 212
- cylindrically shaped flux-collecting system 215
- electromagnetic excitation systems 215 ff
- field cooling 203
- flat magnet arrangement 219
- flat magnet system 217, 218
- flux concentration 214, 220
- flux concentration excitation 220
- flux concentration excitation arrangement 222
- flux concentration system 217
- flux creep 227 ff
- flux-collecting arrangement 221, 241
- flux-collecting excitation system 236, 239, 255
- flux-collecting magnet systems 215
- flywheels 240
- force activation modes 207
- force characteristics 216
- iron collectors 223 ff
- levitation forces 201 ff, 225 ff, 230 ff
- levitational pressure 202, 204 ff
- lift 242 ff
- linear excitation system 211
- linear field excitation systems 214
- linear motion 241 ff
- magnet systems for field excitation 211
- magnetic clutch 237
- maximum field cooling activation 219 ff
- maximum field cooling (MFC) 207 ff, 211, 254
- operational field cooling 207 ff, 220 ff
- operational field cooling with offset 209 ff
- operational field cooling with vertical offset 207 ff
- planar 236
- pole pitch 217, 219, 222 ff
- restoring forces 208, 210
- rotary motion 236 ff
- specific operation conditions 245 ff
- stationary levitation 213, 234 ff
- stiffness 209, 239
- stray field compensation 221
- superconducting motors 238 ff
- transportation systems 242 ff
- turbo machine 236 ff
- zero-field cooling 202, 217
SMB – rotary motion
- dynamics 247 ff
- hysteretic losses 247 ff
- resonance frequency 247 ff
- rotational losses 249
solid solution 26, 34, 35, 42, 43, 157
- $Ln_{1+y}Ba_{2-y}Cu_3O_{7\pm\delta}$ 155
solidification 46
- peritectic 47
Sommerfeld parameter 98
specific heat 96 ff, 98 ff, 114, 116
spinodal 158, 159
sputtering device with SPM 283
stacking fault 58
steel tube 179 ff, 182
- coefficient of thermal expansion 179
- Young's modulus 180
stoichiometry 25, 38, 42, 159
structure

– crystal 22
– electronic 22
structure element
– chains 26
– planes 26
structure families 23
subsolidus 39
substitution 28, 34, 35, 39, 129, 130, 131, 136
– intrinsic 28
superconducting bearings 21
superconducting fault current limiters
    (SFCL) 285
– bifilar BSCCO tubes 289
– inductive SFCL 286
– melt-grown YBCO material 289
– operation characteristics 288
– quench 287
– resistive SFCL 287
– status of concepts 288
superconducting magnetic bearings 199 ff
superconducting permanent magnets (SPM)
    9, 20, 280
– applications 279
– high magnetic fields 279
– laboratory magnets 279
– magnetized 21
– magnetizing 115 ff
– reinforced YBCO 179 ff
– resin-impregnated YBCO 184 ff
superconducting transition temperature 91
supersaturation 52

t

tensile stress 181
– in YBCO and YBCO/Ag 142
tensile stresses 110 ff
thermal conductivity 96, 98, 101 ff
thermal expansion 96, 99 ff
thermomagnetic instabilities 105, 110, 113 ff, 183
time-temperature transition (TTT) 159
time-temperature-transition diagrams 44
torque 262
transition temperature 2
trapped field 145
– bulk HTSC 105 ff

– bulk HTSC at 77 K 109 ff
– bulk resin-impregnated YBCO 184 ff
– limitation of trapped fields 110
– low-temperature superconductors 105
– maximum trapped field 116
– neutron-irradiated YBCO 187
– proton-irradiated YBCO 188
– resin-impregnated YBCO 188
– standardization of measurements 108
– steel-reinforced 188
– steel-reinforced YBCO 185 ff
– temperature dependence 118, 181 ff, 188
– YBCO permanent magnets 179 ff
– YBCO sample 117
trapped field motor
– armature winding 264
– magnetization 264
– torque 267

u

undercooling 50, 52, 53, 63
undulators 284
upper critical fields 78 ff

v

vortex glass 2, 14, 75 ff, 79, 174
vortex liquid 2, 14, 75 ff, 79
vortex matter phase diagram
– bulk YBCO 75, 79
– melting line 173 ff
– YBCO single crystals 80 ff
vortex-creep phase diagram 174, 176, 177

w

weak links 19, 119, 125
welding 146
wigglers 284

y

YBaCuO, phase diagram 35
YBCO/Ag composite 95, 146, 186
YBCO/Ag composite materials
– improved mechanical properties 138, 140
– phase diagram 138
– processing 138